Engineers' Guide to

Technical Writing

Kenneth G. Budinski

ASM
INTERNATIONAL

The Materials
Information Society

Materials Park, OH 44073
www.asminternational.org

ASM International staff who worked on this project included Steven Lampman, Acquisitions Editor, Bonnie Sanders, Manager of Production, Nancy Hrivnak, Copy Editor, Kathy Dragolich, Production Editor, and Scott Henry, Assistant Director, Reference Publications.

Library of Congress Cataloging-in-Publication Data
Budinski, Kenneth G.
Engineers' guide to technical writing / by Kenneth G. Budinski
p. cm.
1. Technical writing.   I. Title
T11 .B83 2001   808'.0666—dc21   00-046476

ISBN: 0-87170-693-8
SAN: 204-7586

ASM International®
Materials Park, OH 44073-0002
www.asminternational.org

Printed in the United States of America

This book is dedicated to my technical writing professor
Robert E. Tuttle of General Motors Institute.

Contents

Preface

THIS BOOK contains material used to train new engineers and technicians in a large manufacturing plant of a Fortune 500 company. It was not a company-training course. To the contrary, it was one engineer's attempt to get coworkers to document their work in a reasonable manner. Coworkers who attended my tutorial sessions went on to become effective technical writers. In my opinion, our department is the model for the corporation. Everybody finishes his or her projects with a report. There is a format for each type of document and a system for archiving these documents. Formal reports go to the corporate library, and they are available to all on-line. It is quite an effective system. The weak link, even today, is that not all technical people in the company know about the company's technical document protocol, and up to now, there was no text to explain types of technical documents and how to write them. That is the purpose of this book. We do not focus on the "laws of the English language." This book discusses most types of documents that the average technical person will encounter in business, government, or industry. The overall objective of the book is frequent, effective, written documentation and the cost savings produced by effective communication.

The first iteration of this book was directed toward college students in the sciences and engineering. The eight or so reviewers, selected by the publisher, were all English professors, and they thought that the text material was somewhat overwhelming for twenty-year-olds who never had full-time jobs. One reviewer said, "My students do not even know what an abstract is." I had to agree with the reviewers; I wrote this book from my tutorial notes and "my students" were all working technical people. Some were engineers with twenty years of experience.

I started with a clean sheet of paper and rewrote this book for the practicing technical person, but in my heart, I feel that this material is not too much for young minds. Young minds can handle anything. In my opinion, this book can help students as well as working technical people.

The first four chapters are intended to bring the reader on board—to convince him or her that it is worth the effort to become a reasonably good technical writer. There is also a chapter on how to conduct technical studies. I have encountered many experienced technical people who did not know the basics of conducting a scientific investigation. There is a chapter on how to make effective illustrations and one on how to make oral presentations. The remaining "teaching" chapters cover specific types of technical documents: informal reports, formal reports, proposals, correspondence, etc. The book ends with another philosophical chapter. This one is on how to discipline yourself to get writing tasks done in a timely manner. Grammar, punctuation, and report mechanics is relegated to the appendix. Readers who need help can use them. The appendix also contains examples of just about every kind of technical document that one would encounter including a complete technical paper and a patent.

In summary, this book contains the writing suggestions of a typical engineer with five years experience in the auto industry and thirty-six years in the chemical process industry. I have written over forty papers for archival journals, a teaching textbook, and a reference text, and I review papers for four technical journals. I became heavily involved in writing as part of a career in research and development. This book reflects what is needed in industry, and I believe these needs are common in business and government as well. Effective communications is a prerequisite for a successful technical career, and this book presents a system that has proven to be successful in making effective communicators.

This book is the culmination of five years' work, including lots of reviews and rewrites. I thank all reviewers for their honest and helpful suggestions, in particular, Steve Helba and Nancy Kesterson for their work in getting knowledgeable reviewers and to J.M.J. for ideas. I acknowledge the talent of Judy Soprano for the chapter and cover art and the hard work of Angela Leisner at Home-Office Connection in converting my handwritten text and illustrations into an orderly electronic file. I could not have done this book without her.

Kenneth G. Budinski

CHAPTER 1

What Is Technical Writing?

CHAPTER GOALS

1. *Show where technical writing fits into the spectrum of interpersonal communications*
2. *Illustrate how technical writing differs from other forms of writing*

TECHNICAL WRITING is a broad term that encompasses a wide variety of documents in science, engineering, and the skilled trades. The major types of documents in technical writing can be grouped into four major categories (Fig. 1.1):

- Reports and communications in day-to-day business
- Technical papers, magazine articles, books, and theses for purposes of education, teaching, and the sharing of information and knowledge
- Patents
- Operational manuals, instructions, or procedures

Most technical writing in day-to-day business involves the preparation of various "reports" (Fig. 1.1). Writing reports is common for many technical people because reports are a major part of the development and application of technology. Very few companies pay technical professionals a salary without written words to implement and evaluate what has been worked on or developed. For example, if an engineer spends a year developing a new transmission for a car, several types of reports are needed for the design, evaluation, and implementation of the new component. Engineering must also report to management on the viability of design, costs,

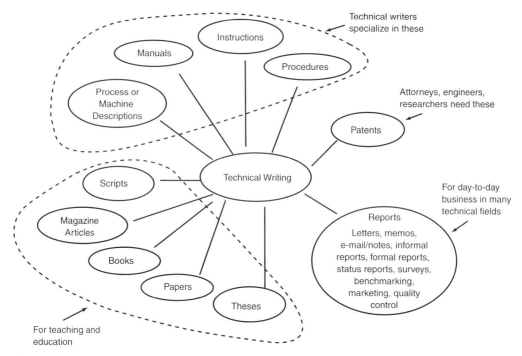

Technical writers specialize in these

Attorneys, engineers, researchers need these

For day-to-day business in many technical fields

For teaching and education

Fig. 1.1 Spectrum of technical writing

and work objectives. This usually requires a written document and related engineering drawings—a report.

A second category of technical writing includes documents for teaching and education (Fig. 1.1) in the form of scripts, magazine articles, books, papers, and degree theses. Scripts for videos, movies, magazine articles, or multimedia presentations are most often written and edited by professionals in these fields.

Books on technical topics are most often written by academicians, although technical professionals occasionally may write an entire book in their area of experience and knowledge. Writing a book obviously requires much more discipline than the writing of reports, but it still requires the clarity of presentation and purpose as in the reports and papers of day-to-day business. Chapter 4, "Writing Strategy," also has relevance for book authors. The key difference is that books are intended for a larger audience and should have unique and compelling features for the readers.

Papers and theses are more common forms of educational or informational documents written by technical professionals. Of course, many people in science and engineering write theses. However, they usually only do one per degree, and the formal writing style and related details are almost always rigorously dictated by the school involved. Papers are the other category in the grouping of types of technical writing that could be considered to be teaching or educational. This book includes information on writing a paper, because it is very possible that a technical person will write papers throughout his or her career.

Another category of technical writing is for manuals, instructions, and procedures (Fig. 1.1). This form of specialized writing is not addressed in this book because these kinds of documents often have legal/liability implications and are best left to trained technical writers. For example, if you invent a novel type of bicycle seat, a user who got hurt because he installed the seat pointing aft could sue you if you did not include in the installation and use manual a statement like the following:

> "The prow of the seat (point A in Fig. 6) should be positioned pointing at the handlebars (Fig. 7)."

Similar liability could be incurred by overlooking a safety or environmental concern in writing a heat treating procedure for a gear. If a particular career situation requires that you write these kinds of documents, appropriate references on technical writing are listed at the end of this Chapter.

Finally, patents require another key type of document in technical writing. Lawyers usually write patents, but not without lots of writing and searching on the part of the applicant. Thus, this book addresses the inventor's part of a patent application and the general criteria for patentability.

1.1 Purpose of This Book

With an understanding of what technical writing is and what aspects of technical writing are covered in this book, the reader can appreciate the purpose of this book. It is to give students and working technical people usable, easy to follow guidelines on how to write effective reports pertaining to all types of engineering, the skilled trades, and the sciences. The main emphasis is on engineering because of author bias [*that is what I know and do*].

There are many books and publications on technical writing; why is another needed? Forty years of personal experience in the engineering field have shown that in spite of availability of writing texts and courses, most engineers are poor and/or infrequent writers. In fact, some engineers never write any reports [*I used to monitor department reports and publish a listing in our newsletter. Some of our staff wrote 50 reports per year. Others had records as bad as zero for eight years*]. This aversion to written documentation undoubtedly happens in other fields. Chapter 2 cites some reasons why such behavior may not be in the interest of career progress or in your employer's interest. It is felt that a root cause of writing aversion is lack of writing skills. Some people were never required to take a writing course in college; others never practiced writing after college. Most writing texts are too detailed for self-study by working people in technical fields. This book provides a concise guide for self-study or classroom use that eliminates barriers to writing and addresses report writing in particular. The objective of the book is to promote the development of technical people with good writing skills and the benefits that this brings to the employer.

1.2 Attributes of Technical Writing

The remainder of this Chapter describes the specific attributes of technical writing and shows examples of how technical writing differs from other types of writing. In general, technical writing has a degree of formality, and it generally focuses on a specific subject with the purpose of making something happen or sharing useful information or knowledge.

Ten general attributes of technical writing are listed and described in the following sections:

- It pertains to a technical subject.
- It has a purpose.
- It has an objective.
- It conveys information/facts/data.
- It is impersonal.
- It is concise.
- It is directed.
- It is performed with a particular style and in a particular format.
- It is archival.
- It cites contributions of others.

There are probably more attributes, but the attributes in the above list define some key characteristics that distinguish technical writing from other types of writing.

Pertains to a Technical Subject

Technical writing must pertain to some aspect of engineering or the sciences in a given subject area such as the following:

- Philosophy, psychology, and religion
- History
- Geography and anthropology
- Social sciences
- Political science
- Law
- Education
- Fine arts
- Language and literature
- Science
- Agriculture

- Technology
- Health/medicine

Libraries usually categorize books into these subject categories, and technical writing may apply to any of these categories if the work contains engineering or science as the focus. For example, a paper on the acoustic/sound aspects of a piano could be very technical and end up in the music category. Similarly, a book on restoration techniques for antiques could be rife with chemistry and metallurgy, but it may end up in the fine arts category. The point is that technical writing can be on one of many different subjects if the subject is being described or evaluated in an objective fashion.

Has a Purpose

A technical document always is written for a reason, and the purpose of reports may be to explain what was done, why it was done, and/or the results of a study. The purpose of reports on investigations is usually to present the results of the study.

The purpose of reports and papers should also be clearly stated, as in the following example:

It is the purpose of this report to present the results of a statistical study on the failure rate of spring latches on a type D cardiology cassette. There have been a number of latch failures uncovered in the inspection cycle, and this work is the first step in reducing the latch failure rate to less than three ppm failure rate.

This excerpt identifies the purpose of the report as the presentation of results from a statistical study. Readers are also informed why the author(s) did the work. If the report is done correctly, it will also close with recommendations on what should happen next.

Has an Objective

The objective of a technical report is the overall reason for doing the work. In an industrial situation, the objective of any work is usually to make or increase profits. In the preceding example, the objective was to reduce failure rates to a level of less than three ppm. This will save money and increase profits. Discriminating between purpose and objective requires some practice, and this distinction is discussed in more detail again in the Chapters on strategies and introductions.

Conveys Information/Facts/Data

Technical writing should have substance in every statement. If a sentence does not convey information pertinent to a study, leave it out. Technical writing is focused on the technology under discussion.

A report without facts or scientific evidence to support an opinion also usually lacks credibility, and it is likely to be unsuccessful in achieving its purpose and objective. The following report excerpt illustrates reports with and without data. Which would persuade you?

No Data

A decision has been made to convert the machine shop grinding operations into a three-shift operation to increase efficiency and machine utilization.

Preferred—with Data

A study was conducted to improve the elapsed time required to grind a set of slitting knives. The average elapsed time for a regrind for the 1997 fiscal year was 11 days. A second study indicated that the largest time allotment in the 11 day regrind time was 3.4 days waiting for grinder availability. These studies were based on one shift (day). A three-week test with three-shift operation reduced the waiting for machine availability time to zero. The elapsed time for thirty knife sets that were ground in the three-week test time was less than one day. These test results suggest that three-shift operations should be implemented.

The use of data and factual information makes the work a technical report. The communication without the data is not much different than a water cooler discussion between coworkers. If the author is the leading expert of the world on grinding, his or her opinions may make the report persuasive, but most people are not infallible authorities on subjects.

Most reports need facts or data to support conclusions and recommendations, and the verbs listed here are probably associated with factual statements:

- Determined
- Solved
- Built
- Accepted
- Rejected
- Completed
- Passed
- Failed
- Broke
- Approved
- Cancelled
- Invented
- Designed
- Developed
- Discovered
- Uncovered
- Deduced
- Studied

Verbs that are often not associated with factual statements include words like the following:

* Think
* May be
* Suggest
* Appear
* Suppose

Impersonal (Third Person) Voice

The use of first person pronouns is usually discouraged in technical writing. The intrusion of "I" makes the work less authoritative. Similarly, it is inappropriate to use names of people and/or trade names unless there is no other way to describe the item.

> Discouraged
>
> I ran a series of hardness tests on the valve seals for Bob MacArther from the shops division, and I found that three of the seals were below normal. I also notified Harry Randall and Phylis Carter so that the two of them could do Rockwell measurements on future value seals.

The preceding excerpt from a report on metal hardness problems illustrates how not to write a technical report. Judicious use of personal pronouns is acceptable, but because a novice in technical writing may not know when it is acceptable, it is probably advisable to avoid the use of personal pronouns (I, you, me, we, mine) in formal reports and published papers. Writing in the third person is the style adopted in many journals and organizations. [*The text contains personal anecdotes that may use personal pronouns. I placed them within brackets so that I can follow the rule of no-personal pronouns in the remainder of the text. Consider these bracketed sections like the sidebars used in some texts to interject interesting facts, like biographical sketches, to keep the reader's interest. In my case, the first draft of this book was deemed "boring" by several reviewers. The second draft with personal anecdotes was not labeled boring by the second set of reviewers, just "rough." This third rewrite addresses the dislikes of all ten reviewers, and I left anecdotes like this in because, let us face it—English grammar and writing techniques are not the most titillating subjects.*]

With regard to using people's names in reports, it is not necessary and it reads "unprofessional." In addition, it adds length, and anything that adds unnecessary length to a document should not be done. If the intent of including names is to give credit, the correct placement of credits is not in the body of a report. Credits belong in end-of-document acknowledgments, which will be covered in a subsequent Chapter. Personal pronouns and names should be omitted because they are unnecessary. Trade names should be avoided because of liability considerations. The message can usually be conveyed fully without their use:

Preferred

A series of hardness tests were conducted on valve seals at the request of the Shops Division, and it was determined that three parts had abnormally low hardnesses. The appropriate individuals were notified so that they can request hardness testing on future valve-seal shipments.

Be Concise

Technical reports are usually written for business reasons. They are not intended to entertain; they communicate information to an identified person or group. Say what you want to say and get out! Wandering sentences and extra words reflect badly on the author and often have a negative effect on the readership that you are trying to reach.

Wordy

Polymer surfaces were studied to determine if physical surface changes occur with continued UV exposure. This program was necessitated to meet customer expectations for a longtime company with world-class name recognition. If surface degradation is in fact occurring, we need to ascertain and assess the severity of this degradation. Moreover, it is imperative that we address any product deficiencies so that the company image as a supplier of robust products is not denigrated.

Preferred

A study was conducted to quantify UV damage to polymer surfaces. This work was done to satisfy customer concerns about the weatherability of sun shields made from our outdoor grade of polypropylene.

Concision can become an acquired writing trait. There are text books on the subject, but a major source of extra words are phrases such as "it follows that," "in any case," and "nonetheless." It is often possible to replace these phrases with a punctuation mark.

Not Concise

The biopsy results were negative. Nonetheless, the nurse-practitioner sent a sample for retest to be sure.

Preferred

The biopsy results were negative, but the nurse-practitioner sent a sample for retest to be sure.

Concise writing is described further in subsequent Chapters, but every writer should strive to state his or her message with the fewest words. Invariably, the people who read technical documents are busy. Extra words mean extra work for them and that they like your document (plan, proposal, etc.) less.

Directed to Readers

Chapter 4 "Writing Strategy" discusses readership of reports, but at this point it is sufficient to say that technical reports must be directed to a particular readership. The author is responsible for determining the specific individuals or parties who will receive a technical document. Writing should be aimed at the readership. Directing a report determines the technical level of the writing. If you direct a report to your coworkers, you do not have to bring them up to speed on the organization of your department. They already know it.

Parochial Report

The attached procedure covers the operation of an infrared camera on the department's SEM. This equipment upgrade addresses the problem that exists in determining the exact location of beam impingement within the sample holder area.

The readers know what an infrared camera is, where it goes on the instrument, what an SEM (scanning electron microscope) is, and about the impingement problem, or they should know, if the document is correctly directed. If this report was to be circulated outside the department or to upper level management, it would be necessary to give background information and define terms.

Style and Format

The attributes of technical writing also include style and format. Style is the way that you write; format is the ordering and physical layout of a document.

The appropriate style for technical writing is objective. Technical documents present data, facts, calculations, test results, and theories, and these must be presented in an accurate manner that is not opinionated. Conclusions are inferred from test results; recommendations are the logical outcome of the conclusions.

Not Objective

The damaged gear train was removed in a bushel basket. Only a miracle worker could put this puppy back together. The operators must have fallen asleep at the controls.

> Preferred
>
> The damaged gear train was removed for inspection to determine the root cause of failure. At this point in the failure analysis, it appears that the unit cannot be returned to service. Testing will be completed by Wednesday.

The format (the basic elements and their placement) of technical papers and reports is a more structured one than that used for other forms of writing. Formal technical reports have basic elements and a structure as follows:

- Introduction (why you are doing the work)
- Procedure (what you did)
- Results (what happened)
- Discussion (what it means)
- Conclusions (what was learned)
- Recommendations (what is to be done with the new information or knowledge)

This style and format have been agreed to by international technical journals, most educational institutions that teach in English, and most industries or organizations that employ engineers and scientists. As shown in subsequent Chapters, all of these report elements may sometimes be put on one page.

[*I recently acquired a new supervisor who is not familiar with engineering or laboratory testing. He receives a copy of all my reports. He recently annotated one of my reports with "seems rather segregated." He is right; technical reports are segregated. The problem statement goes in the introduction; what you did goes in the investigation section. The results go in the results section, and so forth. Technical reports have a definite order.*]

In summary, technical reports have a standard style and format, and, as this book shows, this makes writing technical reports easy.

Archival

An intrinsic part of the value of technical writing is that it is written in such a manner that it can be archived and produce valuable and usable information in the future. Conversely, technical documents should not be generated on transient issues or subjects that will not be pertinent in the future.

> Not Archival
>
> The BCH perforators were shut down last Thursday because of a power interruption. The shutdown caused the loss of three master rolls of product. The root cause of the shutdown was determined to be a faulty relay in the control point of the perforating center. The specific details of the product loss are:
>
> _____ .

In summary, this production event was traced to an electrical problem. The BCH perforators will be permanently shut down and scrapped in two weeks, and production will be converted to Geneva mechanism machines.

Archival

It was determined that punch and die interference was the root cause of the tool breakage that has been occurring in the KCN blanking operation. Coordinate measuring-machine inspection measurements of 40 punches and 40 dies indicated that die holes were out of location by 5 to 20 μm. Measurements on the die machinery fixtures indicated that the C-dimension locating lug was 2° off axis. This caused the part to be skewed when the die hole was machined. The recommended procedure to remedy die inaccuracies is _____.

In the first example, the problem machines were slated to be scrapped, so there may not be a need to archive the report on the details of the shutdown. In the latter case, the problem was due to a fixture error. This kind of problem could reoccur. There is probably value in archiving this document. Most businesses and industries have guidelines on how long various documents need to be saved, and, if a technical paper is published in an archived journal, the document will be available for as long as the journal is kept in libraries. Thus, technical writing often results in documents that have value in the future and should be archived.

Attributions

Formal technical reports and papers must show sources of information and recognize contributions of others.

No References

The problem with the cracking of generator bellows on the gelatin mixers was determined to be stress-corrosion cracking from ammonia fumes that were generated by a nearby autoclaving operation.

References

Secondary ion mass spectrometry performed in the KRL Analytical Laboratory indicated a high nitrogen profile on the surface of the failed bellows. Fellows and her coworkers (Ref 7) have used a similar technique to verify absorbed nitrogen on surfaces of yellow brass. A number of investigators (Ref 8–10) have shown that ammonia concentrations as low as 30 ppm can cause stress corrosion of 70/30 yellow brass _____.

Formal reports also provide the opportunity to cite contributions or funding in an acknowledgment section at the end of a report. In summary, the proper use and citation of the work of others is another attribute that sets technical writing apart from other types of writing.

1.3 Other Types of Writing

In general, technical writing has a degree of formality, and it generally focuses on a specific subject. Reports and other technical documents are written to share useful information and knowledge or to make something happen. Technical writing should have substance in every statement. Technical writing also has a style and structure that sets it apart from other types of writing. Most important of these characteristics is that it be objective and supported by facts and data and that every attempt is made to ensure that the information is correct (as well as the presentation).

The most writing that some people will do in their lives is during their formal education. We learn the basics in grade school and practice this skill in the remainder of our schooling in countless themes, essays, term papers and [*shudder*] exams with essay questions. This type of writing may be termed "school writing." The teacher dictates the format. Each teacher wants certain elements, and, as we all have learned the hard way, your writing had better conform. In the working world, industry, business, and health, there are no Notre Dame nuns to tell you how to write. If your company does not mandate a format and style, you must make the decisions.

In this concluding section, other types of writing are briefly described as a further illustration of what technical writing is and is not. The writing examples in this last section do not conform to a particular format or style. Each is different. This is the situation in most writing. Each writer does something different as does each magazine, each newspaper, and each essayist. Technical writing, on the other hand, is done with a singular style and format that describes:

- Why you are doing the work (introduction)
- What you did (procedure)
- What happened (results)
- What it means (discussion)
- What was learned (conclusions)
- What is to be done with the results (recommendations)

Instructions

Instructions for use of a product or a process are often considered technical writing. They are often pertaining to a technology, and they are written in a style that has many of the attributes of technical writing.

> Your Excel inflater can be turned in two ways: by pushing on the red "Trigger Switch" (this is the most common way to operate the inflater) and by turning the silver "Lock/Off" button on the bottom of the handle (see diagram A).
>
> Most inflations can be done without the air hose. Simply push the nozzle onto the object to be inflated (see diagram B) and press the red button...

Advertising

> Rhinehard Tools are the best money can buy! Manufactured exclusively for us to our specifications for over two decades in Germany and exceeding the highest international standards, Rhinehard Tools are properly hand forged from the finest steel. Blades are individually tested (other brands batch tested) for proper Rockwell hardness (61–63 HRC)...

This writing is not considered technical writing because it is not factual. The claim of "best" could only be substantiated by testing every tool in the world using every possible test criterion. "Superior" statements pervade advertising on all types of products and services. Superlatives almost never can be substantiated, and thus they are not factual and should not be a part of technical writing.

Creative Writing

Creative writing requires saying things in words that create illusions or that establish a mood or other desired affect. It can be just humorous or entertaining like a murder mystery. A characteristic of this type of writing is the use of expressive and descriptive language.

> The night was as black as the bottom of the Marianas Trench. We huddled together and thought about happy times in Scranton. We were like teenagers at a drive-in movie, when we would...

Poetry (an acquired taste) also fits within this category of writing.

Opinion

Personal opinions are not part of technical writing. They belong on the editorial pages of newspapers.

> Opinion
>
> The most recent addition to the far-ranging list of idiomatic expressions in this country is "you're all set." I get this at the store checkout, at the library, at the car wash. I got it last week after giving blood. What does "all set" mean? Is my body homeostasis proper? I suggest replacement of "you're all set" with "thank you for your patronage, please come again." I know what this means.

Opinions and fads may also cross into technical subjects. For example, business and finance publications can be based on economic theories and mathematics, and thus they could be considered technical writing. However, many business books on management are mostly opinions rather than fact. Even many books in the hard sciences, like physics and biology, are based more on subjective interpretations. These forms of writing are not classified as technical writing.

Information

Documents that convey information can be technical writing if the information has the technical writing attributes. The following example of a meeting notification is certainly not part of technical writing. The program chairperson would probably archive the note for his or her annual report, but otherwise this meeting announcement would have no value as a technical document to others. Newspapers and news magazines also convey information. Most are not technical writing because they usually do not address issues in engineering or science. Newsletters, similarly, often do not match the attributes of technical writing.

> The Instrument Society of Western New Hampshire will meet at the Harding Hotel at 7:00 P.M. The speaker for the monthly meeting will be Harvey Ruft, and his subject will be programming of dedicated controllers for fractional horsepower servomotors...

Administrative

Most employees in large firms receive several memos or e-mails like the following each day. They are from bosses, secretaries, teammates, and colleagues; they deal with administrative matters such as meetings, deadlines, safety issues, and so forth. They seldom match the attributes of technical writing.

> There is an opportunity for those of you who have not attended the "presenter" training to familiarize yourselves with the audiovisual (A/V) equipment for the Engineering Conference on Tuesday, October 12 at the Riverside Convention Center. The A/V team will be available in Room 107 to show you how to work the projection equipment and laptop computer connections...

Entertainment

Much writing is intended simply for the entertainment of the reader. Countless books and magazines have reader entertainment as their primary objective. This is not technical writing. Stories and novels may not be factual. The information is often made up.

> John is a hypochondriac in a small town in Texas. He has been to every doctor in town and visits one or two regularly. When a new doctor came to town, he immediately made an appointment. When he arrived in the new doctor's office, he noticed a sign on the wall: $35 for initial visit, $15 for subsequent visits. Being frugal, John tried to behave like this was not his first visit. When the doctor came in, John said, "Hi, Doc! Good to see you again." The doctor examined John, and John said, "What do you recommend?" The doctor said, "Keep doing what I told you to do at your last visit."

Even books on real events may not pertain to technical writing when the purpose is to entertain rather than to serve a scientific purpose. For example, biographies and novels on real incidents usually do not pertain to engineering and science. Even those that do are often not technical writing, because their objective is to entertain, amuse, or engage readers. The same applies to business and finance publications. Some books on technical topics are written in a way to entertain or capture the imagination of readers. This form of writing is not technical writing.

Summary

This Chapter was intended to define technical writing and to start the discussion of the features that discriminate it from other types of writing. Some factors to keep in mind are the following:

- Technical writing communicates issues in engineering and the sciences.
- Technical writing has form and style requirements that are different from those of other types of writing. Reports need to include definite elements.
- Technical writing does not employ humor or slang.
- Technical writing is objective oriented.
- Technical writing does not blame people.
- Technical writing requires facts or data.
- Technical writing never hides facts.
- Technical writing deals with nonadministrative issues.
- Technical writing is never used as advertising copy.
- Technical writing is impersonal—it does not use personal pronouns or name people who performed parts of the work.

Important Terms

- Readership
- Technical Writing
- Objective
- Directed
- Technical Report
- Paper
- Concise
- Fiction
- Instructions
- Impersonal
- Nonfiction
- Concision
- Trade names
- Factual
- Directed

For Practice

1. Write a nonfactual paragraph and convert it to a factual account.
2. List six verbs commonly used in making nonfactual statements.

3. Take a paragraph from any work and make it more concise. Show the original work.
4. Define the readership for the following: (a) newspaper, (b) technical journal, (c) patent, (d) resume, (e) internet chat line.
5. Write a paragraph using personal pronouns; then rewrite it with the pronouns eliminated. Did the intended result change?
6. List the advantages and disadvantages of using trade names in a document.
7. List four differences between a fictional novel and technical writing.
8. Give five reasons for not using people's names in a technical report.
9. Describe the style of technical writing.
10. Write a note setting the policy of your company on vacation days. Show how it differs from a technical report.

To Dig Deeper

- G. Blake and R.W. Bly, *Elements of Technical Writing,* Macmillan Company, New York, 1993
- J.M. Lannon, *Technical Writing,* Scott Foresman and Company, Boston, 1988

Not Included in This Text

- Software
- Novels
- Plays
- Poems
- Maps
- Drawings
- Spreadsheets
- Medical Records/Forms
- Legal Forms
- Legislation

CHAPTER **2**

Reasons for Writing

CHAPTER GOALS

1. *Explain why report writing is a required part of a technical career*
2. *Understand the benefits of good writing skills*

WHEN YOU WORK or are in school, writing assignments are dictated. A typical humanities assignment may be a 1,000-word essay on World War II, trees, or famous people. In engineering and the sciences, writing is mandated in most courses with laboratory projects. Laboratory reports are probably the best examples of technical writing for young people in technical professions. The teachers who request essays and laboratory reports generally expect them to have a particular format, style, length, and so forth. Probably all laboratory reports conform to the report structure (introduction, procedure, results, discussion, conclusions, and recommendations) in the objective style that is advocated in this book.

Working scientists and engineers usually have the option of espousing a career filled with written documentation or not. [*In the laboratory where I work (about 25 people), we are required to give the department secretary a copy of all reports issued on jobs. At the end of the year, she makes a file for each individual, and the authors must decide which documents are to be retained and discarded. This is part of the corporate records management directive. The number of reports in each file varies from 0 to a few to more than 50. Some people write reports; some do not.*]

This Chapter presents the pros and cons of report writing with the purpose of convincing you, the reader, that writing technical documents is worthwhile and necessary. First, several reasons (or excuses) for not writing reports that are commonly given by working engineers and scientists are described and rebutted. These are followed by descriptions of the benefits of written documentation.

2.1 Excuses for Not Writing

Lack of Time. "I did not have time to write a report." This is probably the most common reason given by technical people with a dislike for technical writing. It is hard to conceive of having time to do a project but not having time to document the results, conclusions, and recommendations of the work. Government studies in the United States determined that the adult spends upwards of 20 hours per week watching television. An investigation report on a gear failure may take less than an hour to write. One would think that the required hour could come from the 20 television hours if there was not time at work. If people have the desire, they can find the time for anything. Lack of time is an excuse, not a valid reason for not writing.

Nobody Reads Reports. Sometimes people have the perception that reports are not read by the addressees, so why write them? If the reports have the right distribution, they should be read. Distribution lists for technical documents are discussed in more detail in Chapter 4, "Writing Strategy," and Chapter 9, "Formal Report—The Outline and Introduction." When technical people must get funding to do projects, certainly the sponsors should be on report distribution lists. The person approving funding wants a report and usually circulates it to others. [*It has been my experience that people always prefer written solutions to problems, and they often keep reports on department problems for years. People do read and want documentation.*]

E-mail has further invalidated the excuse of "Nobody wants to read them." It is now common practice to send e-mail messages to teammates when a person completes his or her phase of a project. You can attach a report to your note, and only interested parties will open it. An even simpler approach is just to send out an electronic note that the work is completed. Anyone who wants a copy of the report can reply in the affirmative to your note, and you can electronically attach the report in response to their reply. You will know what interest there is in your report. In this information age of electronic documents, you need to write a report to make this happen. Yes, you can send an e-mail message to your team asking if members want a copy of the report and not write one if there are no affirmative replies, but you must accept the risk of exposure if you receive an affirmative reply.

In summary, the perception that nobody reads reports is often wrong. In fact, you should always write as if your report will be published in a local newspaper. You never know who will read your work. Do not state anything that is bitter, nasty, libelous, unfounded, speculative, derogatory, exaggerated, sexist, or that in any way would offend others. People have lost jobs over memos "that nobody reads."

Reduces Job Security. Some people feel that they should keep their jobs because nobody else can do them. They have some skills or knowledge that others do not have. The company will always need these people because of their critical skills. They do not want to write reports because this may allow others to acquire their treasured skills.

Right or wrong, in the corporate world, nobody is indispensable. In a large company, only your immediate supervisor may know or appreciate your critical skills. In reorganizations, management and supervisors are often the first to experience cutbacks. If your immediate supervisor leaves because of downsizing or other reasons, knowledge of your critical skills may be lost, and there is no job protection. On the other hand, if you are diligent in documenting your work, even a new manager can read your reports and know what you offer to the organization. A record of active writing could preserve your job.

Trepidation. Some technical people lack confidence in their writing skills and do not opt to write technical documents for fear that they may contain writing errors that cause embarrassment. Some managers also do not know how to review a document. They make the authors mad or embarrassed or both with their annotated remarks on documents that they review. [*I received a "That's dumb" annotation from a reviewer on an end-of-chapter question that she did not like. If I was thin-skinned, this kind of reviewer response could discourage me. However, it really wasn't a very good question, so I wrote a new one and shrugged off the comment as helpful and valuable.*]

The best way to conquer a fear of making writing mistakes is by practice. Writing is a skill, and practice is required to build a skill. Poor writing or spelling will disappear with practice. It is also useful to get preliminary review comments and feedback through a colleague or support person that you trust. Document review is discussed in more detail in "Review Editing," Chapter 13. Getting a review of a document before it reaches one's manager prevents bad reviews while the author is honing writing skills.

E-mail Is Sufficient. Even though e-mail is a useful tool, e-mail does not replace the need for formal or even informal reports. With the ease of e-mail communication, there is a tendency to summarize work in brief e-mail messages. This is often an inadequate method for summarizing a body of work. E-mail is wonderful for conducting business and back-and-forth communications with team members, but these are not acceptable technical communications unless they meet the criteria presented in Chapter 1. Technical documents require facts, structure, and supporting data. If all of these are put into an e-mail message, the result is probably a technical document anyway.

In summary, e-mail messages without proper structure and supporting data do not constitute technical writing. Their use to replace technical documents should be discouraged.

2.2 Benefits of Technical Writing

Working technicians, engineers, and scientists may have the option of espousing a career that minimizes the need for written documentation, but there are several benefits and needs associated with technical writing. Effective writing skills are necessary for career success in most fields, and following are some basic points on the benefits of good writing skills.

The Boss Wants a Report. The most common reason for writing a report is that it is an expectation of the job. Mandatory reports are commonplace. Police and other law enforcement professionals may be the most prone to report-writing. They are ordered by their supervisors to write reports on crimes, calls for assistance, accidents, arrests, use of firearms, and so forth. Usually, there is a specific form for each incident. In these cases, reports are a critical need as they may provide documentation on litigation or criminal convictions.

In engineering and the sciences, reports may not be as critical, mandatory, or rigorous as in police work. However, many organizations use effective report writing as a performance measure. Some companies count the number of reports written and rate people on how frequently and well they write as compared to their peers. Often universities base tenure and status on a professor's record of publishing papers in archival journals.

In industry, reports are generally required for funding projects or purchasing capital equipment. These sorts of reports may include the financial benefits and other ramifications. In almost all cases, if there is no written proposal, there is no funding. Another common industrial situation is to use a technical report to document some type of work function. If one works in a testing department, the "product" may be the result of requested testing. Quality control functions often mandate reports on quality status or quality problems.

In summary, the preeminent reason for writing reports is that management wants reports written. Usually reports are written to keep management informed, but the reason may be broader, too.

Forces Analysis of Work. Even if you are not ordered by management to write a report, another important reason is that doing so helps you organize yourself. Report writing forces a systematic analysis of data, review of related facts, and organization of one's thoughts on a project. This applies to individual study in industry and school as well. The process of collecting and organizing data into reports is a very useful tool, even in learning a new topic or trying to grasp a complicated situation or process. This type of approach has been used effectively on a personal level by many authors to the real benefit of their readers.

Writing a report is the best way to find out if you are done with a technical study. It quickly becomes apparent if adequate data for making a decision is available. The path that you took is put into a procedure. Does it sound logical? Did the work meet the objective? Do you have clear conclusions and definite recommendations? Sometimes the initial writing of a technical report shows that only half or part of the work is done. Sometimes writing may point you in another direction than the one you initially identified. Thus, there may be many things that you may not see until you put pen to paper (fingers to computer keyboard).

Report writing as a way of checking technical work may be the most important reason for technical writing. Report writing is the process of organizing accumulated facts and distilling them into a coherent whole that adds value for you, your colleagues, business, or maybe even society. It can be an internal check on what you did, or it can help you check the work of others. Data supplied by others may be flawed. It appeared valid until you started to check it or plot it for a report. If the work is done right or wrong, the preparation of the report may help show this. The process of technical writing is also a useful tool in the evaluation of completeness. If the report appears complete and self-contained, finish the report and the job is done. If report writing uncovers some flaws or open questions, put the report aside. Fix the flaws and then finish the report with the flaws removed. The process of technical writing is thus a useful way to improve the quality of work.

Completes a Job. Industrial downsizing often means fewer people to do the same or increased volume of work. Many technical professionals find themselves juggling multiple projects, maybe ten or twenty. Reports can be used to end a project. A report can show your boss that you met the goal of the project; it tells others what to do about a problem. You have documented what you did. If a similar problem arises in the future, your report will show what has been done.

Future work may also be based on what you did. Coworkers will know where the problem or issue stands. In fact, it is almost unconceivable that a project can be ended without a concluding report. Every body of work needs a purpose and objective. Reports show that the objective has been achieved. If the project was to design and build a machine, building the machine does not end the project. Management will want written documentation to show that the machine met its objective. How does the actual cost compare with the estimate, and where do we go from here? A technical document will supply this information. Just as an approval signature from a boss starts a project, a written document ends it. You can go on to the next challenge.

Unreported Work Can Be Lost Forever. Many times in engineering and science, people spend considerable time and money investigating a new approach to a problem, designing a new mechanism, and formulating a new chemical, and for some reason, the work is stopped. If a report is not written on what was done, the work may be unnecessarily repeated by others, or, worse yet, it may be lost forever. Even if a project is completed, work not documented may be repeated by others who did not know the work was done. [*Just this week, I was consulted on a problem that occurred with chromium plated rolls. We used to plate them in our own plating shop, but the plating operation was outsourced about three years ago to save money. The plating experts went with the tanks. The problem that was described to me had occurred twenty years ago on rolls plated in England. We traced the problem to chromium hardness. We developed special techniques to measure chromium hardness. I told the new engineer to look for reports in* *our library written by J. Smith on the subject. He returned and found none. A key part of solving this problem was development of calibration curves to convert indentor penetration measurements to hardness. They could not be found in archived documents. The calibration work was tedious and expensive. It will have to be repeated—for the lack of written documentation.*]

Almost every engineer or scientist can cite examples where work had to be repeated because previous work was not properly documented and archived. With the frequent changes that occur in business and industry, adequate documentation of work is imperative. Enormous sums of money and significant lead time can be lost to reports that are not written.

Oral Statements Can and Will Be Altered. An important reason in favor of written documentation is that the alternative—an oral report—can be unreliable, risky, or even dangerous. Verbal communication can be altered each time it is repeated.

Television game shows have performed live experiments in which a statement was made to a person and that person was asked to pass it on to another person. The second person passes it to a third and so on. By the time the statement reaches the sixth or seventh person, it little resembles the original statement. Rumors are the best example of verbal modifications. [*Every time that we have a downsizing there are at least 12 official versions of the details. Some rumors have incentives, some have plant shut down, some have departments sold. After it is over, most times, the real version did not coincide with any of the 12 on the rumor mill.*]

In summary, verbal statements are prone to misinterpretation, alteration, or they may simply be ignored. A written document should always be used to close a study or project. It gives credit to the person(s) who did the work, and it shows responsibility for the work. They have completed their task.

Necessary in Global Businesses. The trend in business and industry in the 1990s and at least for the start of the new millennium is to do business on a worldwide basis. Those companies that do not tend to remain the same size or fail. English is a second language for much of the world. The largest countries in the world, China and India, both use English as a language for international business. Most Europeans speak at least some English. There are English language television stations in many foreign countries, which helps mastery in those countries. What this means is that companies in English speaking countries can do business in English with customers and business partners in many non-English speaking countries. This global scope of business places even greater needs on effective writing. Written documentation for all agreements, contracts, procedures, specifications, and so forth is more important when dealing with nonnative speakers of any language. Written comprehension is usually better than the spoken word because of pronunciation differences.

Being an effective writer, thus, is a must for doing business on an international basis. Writing in French when dealing with a French customer is, of course, preferred, but, chances are, English is an acceptable language for business communications. First, however, one must have reasonable writing skills. Poorly constructed documents with misspelling and poor grammar may confuse people who normally speak a language different from that of the writer.

[*For the past two years or so, I have had weekly video conferences with engineers in our French division on a joint project. I am amazed at the number of French people who are fluent in English and embarrassed that my French is so poor. We use English for communications; most people at the plant understand at least written English. I can e-mail and write reports in English and they will be understood by most. The same is not true of oral communications.*]

Writing Is Necessary for Standardization. Along with the trends of downsizing and globalization, there is a trend to standardize many aspects of science and engineering. Companies that make machines are trying to standardize components in the machine or subassemblies. For example, document copiers contain many rollers; in the past, these rollers were often each different from the others. Now, engineers have mandates to use only one size roller from one material with the same bearings. Similar standards are established for assembly procedures, installation procedures, maintenance procedures, and the like.

Carefully crafted "standard" documents provide information on how to make standardization possible. Use of standard designs greatly reduces engineering costs and elapsed time in new products. Using standard materials (bearings, motors, rollers, frames, etc.) allows economies of scale and use of preferred suppliers. This saves costs on purchased items. Standard manufacturing processes reduce training and capital equipment.

Good writing skills are needed to write standard methods, tests, and practices. And most businesses, governments, and industries want standards. They do not want to have ten different ways of purchasing, five different performance appraisal systems, and sixteen different ways of supporting rollers. "Let us reduce the number of ways that we do things" is the mantra of management. If you want to be part of the team that reduces costs, you need writing skills. You cannot tell everyone in your factory how to use only a particular bearing without the writing skills to communicate how to accomplish this. Minimalist designs and business practices require well-written communications.

Career Advantages. There are many basic benefits in having good writing skills, not the least of which is the beneficial effect on an entire career. We do not have a survey that has facts and figures to prove this contention, but "Help Wanted" ads in the newspaper and technical journals are irrefutable evidence that communications skills are a significant factor in obtaining employment in a technical field. Excerpts of engineering employment advertisements that follow illustrate expectations:

- *Senior scientist:* You must have strong PC, communications, and technical writing skills, with attention to detail …

- *R&D Metallurgist:* ... you must have a strong generalist background in physical metallurgy along with excellent communication, problem solving and computer skills
- *Metallurgist:* ... perform failure analysis using SEM-EDS, prepare reports on customer problems, write quality specifications and prepare audits of suppliers and special processes.
- *Thermal Spray Engineer:* Responsibilities include process qualification, tooling concepts, writing procedure manuals, inspection reports on customer problems and robotics programming.
- *Materials Application Engineer:* Strong communication skills and the ability to work in a team-oriented environment are essential.
- *Assistant Professor:* ... including demonstrated effective interpersonal, verbal and written communication skills

Without exception, technical job descriptions include statements like the following:

- Must have excellent communication skills
- Must be effective communicator
- Must have ability to interface with clients
- Good communicator
- Computer skills (word processing and spreadsheet software)

Effective communication is a job expectation for any technical (or business) position. It is not an option. Some job descriptions even ask candidates to supply a list of publications, the papers that they wrote for archived journals.

The other way that effective writing promotes career success is the perception on the part of managers that writing skills or lack thereof, reflect personality traits. Poor writing skills can be associated with undesirable personality traits and vice versa. Figure 2.1 lists some personality traits that may be associated with the practice of report writing. Needless to say, a good report writer may still be lazy, sloppy, unsure, and so forth, but the perception on the part of management types tends to be like that shown in Fig. 2.1. Lack of communication skills can have a negative effect on career because it is an expectation of most jobs and a lack of these skills can have bad connotations regarding personality traits. This may not be right, but it happens.

Summary

The purpose of this Chapter is to describe the many benefits of good technical writing in day-to-day business and some of the reasons (or excuses) for not writing. The overall goal is to engage the reader in the importance of good writing; that is, the purpose is to make you, the reader, committed

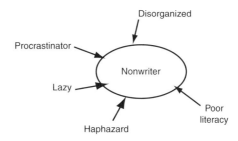

Fig. 2.1 Personality traits associated with writing and not writing

to the notion that your life and career will be better if you become proficient in technical writing.

[*The excuses for not writing were essentially those used by a coworker who was one of our brightest engineers. He was a mechanical genius and earned his salary many times over each year with his problem solving. Unfortunately, he never documented these accomplishments and he resisted formal laboratory studies that needed written proposals and results reports. He was a casualty of downsizing after 18 years with the company, probably for his "writing resistance." The irony of this situation is that he is a good writer. He writes a newsletter for an auto racing organization. He did not like to write and it cost him his job. As fate would have it, he now works for a consulting company where report writing is absolutely required.*]

Effective writing skills are necessary for career success in most fields. The following are some important points to remember from this Chapter:

- No time to write is never a valid excuse. A motivated person will find time for anything.
- People want written documentation on work of interest to them. They read these documents and often save them for future reference.
- Writing enhances job security by increasing your value to your employer. People with demonstrated writing skills are preferred in hiring, and they are perceived by many managers to be more valuable than nonwriters in times of staff reduction.
- Writing skills are acquired just like other skills—by practice. Peer review will help build skills.
- Computer mail is usually not the appropriate medium for a technical report or document. E-mail is usually not archival.

- Managers usually consider effective technical writing an expectation of most technical jobs.
- Writing technical reports and documents is the most effective means for concluding a project and transferring new information or knowledge.
- If technical work is not documented in writing in a timely manner, the work is often lost and may be repeated by others.
- Oral reports on technical work will be altered each time a person shares the results. Written reports do not change when transferred to others.
- International business transactions require written communications, and in most cases, English can be used as an acceptable language.
- Standardization of designs, business practices, materials, and so forth requires reasonable writing skills on the part of the author. Standardization is the wave of the future in business and industry.

As a final argument in favor of writing reports and other technical communications, writing the results of a project greatly increases job satisfaction. How so? When a job is concluded with a report or suitable document, the writer can clear details from memory knowing that these details can be recalled from the written document if needed. A completed project report can be a great stress reliever. The job is done and one can move on to the next challenge. It feels great. [*Finishing a project with a distributed report is the most satisfying part of my otherwise wretched existence. Yes, I have been told to get a life.*]

Important Terms

- Communications
- Analytical skills
- Standardization
- Career advancement
- Responsibility
- Oral reports
- Effective writing
- Credit
- Globalization
- Documentation
- Organization skills
- Accountability
- Job skills
- Results oriented
- Job expectation

For Practice

1. What writing would improve your workplace? Why?
2. Cite an example in business or industry where oral communication could create a problem. Show how written documentation would have prevented the problem.
3. List three irrefutable excuses for not writing reports. See if classmates can refute excuses.
4. List five technical jobs where good writing skills are mandatory.
5. List your most prominent writing weaknesses (two or three) and describe what you need to do to remedy them.

6. Write a short paragraph about your technical writing expectations in your selected career. What kinds of documents might you need to write?
7. Describe a personal situation where lack of written documentation produced some negative effect. (Share these in class.)
8. Describe how you use computers/e-mail for work communications. Are they adequate?
9. Write a paragraph on how written communications can reduce cycle time on a project.
10. You work for an engineering consulting company. What role would technical writing play in a proposed project to build a pedestrian bridge over a busy road?

To Dig Deeper

- T.E. Anastasi, *How to Manage Your Writing,* The Maqua Company, Schenectady, NY, 1971
- G. Blake and R.B. Bly, *The Elements of Technical Writing,* Macmillan Publishing Company, New York, 1993

CHAPTER **3**

Performing Technical Studies

CHAPTER GOALS

1. *Introduce newcomers to the methodology of performing technical studies and experiments in business and industry*
2. *Outline to all readers proven ways to acquire data for writing a report*

MOST SCIENTISTS AND ENGINEERS spend their careers on a variety of projects. For students, engineering and science careers usually contain some resemblance to the career steps illustrated in Fig. 3.1. You are hired as a junior scientist or engineer, and you can work your way up the career ladder by going the technical route or management route. These are the typical career paths in most of the larger companies. If you are an entrepreneur and start your own company, you may probably have to travel both the technical and management paths concurrently.

Of course, after first exposure to engineering or science, one may opt out. Life may be simpler. Some technical people also go part way through the steps in Fig. 3.1 and then return to academia to teach. Their lives will probably not be simpler, because many higher level teaching institutions require teaching staff to also bring in research dollars to support graduate students. They are like the entrepreneurs in that they run their research business and their teaching "business" concurrently. God bless them; they will have busy lives.

The purpose of this Chapter is to describe how engineering and science projects should be approached so that students who have not yet interned or acquired technical jobs get some exposure to what lies ahead. This Chapter discusses the types of studies that typically make up science and

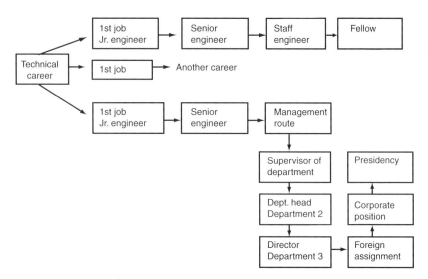

Fig. 3.1 Potential engineering career paths

engineering careers. Discussion of the general methodology of a technical study follows.

This text is aimed at a readership of practicing engineers, and it certainly applies to students in engineering, science, or other technical studies. Experienced engineers and scientists may choose to disregard this Chapter, as many practicing technical people develop their own systems for conducting experiments and investigations. However, this information might serve as an appropriate review for experienced engineers and scientists returning to technical work after an excursion into another field.

3.1 Types of Technical Studies

Technical studies are what most engineers and scientists do for a living. Working scientists and engineers are well aware of technical studies. They spend most of their time on studies. In fact, many engineering and science careers involve a project-to-project lifestyle.

A typical engineer's day may look like that shown in Fig. 3.2. [*Yes, it is my typical day. It is not glamorous, and there is little time for deep thought. A typical engineer may juggle anywhere from three to twenty projects/ studies at the same time. Forty years ago, there was the luxury in some companies of having one big project at a time. You could spend a day thinking about a step. This is not the case in the 1990s. Downsizing has pared engineering staffs to the bone. You are expected to work well over your official 40 h work week. We have been asked to give 48, minimum.*]

A research scientist may have fewer projects than an engineer, but reduced time to market edicts have gobbled up the lead time for inventions and new research and development projects. The saving virtue of a technical career is that it is dynamic, and you will not be bored. The rewards are adequate, and if that is where your talents lie, you will not be happy selling real estate.

Now that 30% of the student readers may be considering a business path instead of a technical career, the type of studies/experiments that technical people do will be discussed. What types of projects and technical studies are likely to be encountered during a technical career? Figure 3.3 shows a

Typical day in Engineering

7:00–7:05	Get coffee from department pot
7:05–7:30	Answer e-mail
7:30–8:00	Discuss work progress with technicians
8:00–9:00	Prepare for 9:00 meeting on project xyz; make overheads
9:00–10:30	International video conference with project team members
10:30–11:30	Write up test requests and address action items from video conference
11:30–12:00	Check on test in lab
12:00–12:20	Lunch
12:20–1:00	Answer phone messages
1:00–3:00	Perform calculations on test measurements; start results section on report
3:00–3:30	Answer e-mails again
3:30–5:00	Prepare for tomorrow's 8:00 AM project status meeting
5:00–6:00	Commute
6:00–7:30	Dinner with family
7:30–8:00	Goof off
8:00–10:00	Do planning for next day and work on report in progress

Fig. 3.2 Typical day in the life of an engineer

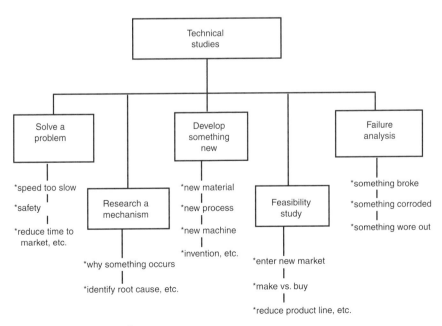

Fig. 3.3 Typical types of technical studies

spectrum of studies that technical professionals encounter. These basic types of studies are described in the following sections.

Solving Problems. The purpose of most technical professions is to solve problems. The problem could be in manufacturing; it could be related to a safety or health hazard that has been identified. It could be related to marketing, such as reducing the time to market for new XYZ products. Problem solving often requires testing, calculations, or both. Solutions should be data driven, not based on intuition, experience or opinion.

[*As an example of a problem, I was recently sent a polyurethane roller with which a customer in England reportedly experienced conveyor tracking problems with film strips in a processing machine. This was all the information that we were given with the roller. It came with two small reels of film. One film did not cause tracking problems on the roller, the other did. The complaint suggested a traction problem between one of the films and the rollers. The first step taken was a microscopic examination of the roller. It was a ground surface characterized by transverse grooves resembling tree bark. This is a typical surface on rubber that is ground to finished dimension. It was also noted that the surface texture valleys of the rollers were filled with some contaminant. The contaminant was locally removed and sent to chemical analysis. It was a photographic emulsion. Friction tests were then conducted with the two different films sliding on the contaminated rubber surface. One had high friction; the other had low. We cleaned the rubber surface to remove the contaminant and the friction of the two films was the same. We wrote a report recommending the use of cast polyurethane rollers with a smooth surface that will be less prone to the contamination that produced friction differences in the films.*]

This was a very specific problem, and the engineering analysis required only a few standard tests. The work required less than a day for testing, and the lapsed time for the analysis was only about a week. Some problems are much more complex. Some can take a year or more to solve.

Essentially, engineering problem solving involves performing various tasks that appear to have potential for solving the problem. If the problem is a broken power transmission shaft, think of all the factors that could have made a shaft fail. Then test for each until the most likely cause of failure is determined. Engineering projects almost always include:

1. State the problem (get all details)
2. Select one approach to solving the problem
3. Search literature and perform preliminary experiments
4. Perform refined experiments
5. Analyze data
6. Decide if problem is solved or more work is needed
7. Perform more studies
8. Analyze data
9. Write report

In summary, engineering investigations start with researching all details of the problem and completing preliminary experiments. The initial experiments are the investigator's best guess on tests that will lead to a solution. Simple problems may be solved at this point. Others provide many additional hours of challenge.

Research a Mechanism. Determining the mechanism or reason why things happen in the various sciences is often the daily fare of research scientists. However, in the lean, mean industries and institutions of the 1990s, nobody researches mechanisms without economic significance. Sure it would be nice to know what makes day lilies close up at night, but if the answer is not going to produce increased profits, it will not be researched. Mechanisms have been researched in the 1990s so that computer models could be developed to improve some operation or mechanism so that it would not fail or would work faster. Mechanism studies are supported when they can lead to increased profits or new products.

[*A mechanism study that I worked on concerned wear of tools that perforate the sprocket holes in film. For as long as anybody could remember, the perforation dies deteriorated in use by the formation of a shallow channel around the die hole. This channel would start about 10 μm away from the cutting edge and grow deeper with use. Eventually it would grow to the point where it deteriorated the edge. This wear phenomenon was called cratering.*

Many different tests were conducted to explain why cratering occurred. The tests suggested that craters started away from the cutting edge because the plastic film behaved like a fluid when cut, and the surface of the film would slide on the die when the punch contacted the film. The film bulged around the hole. Computer modeling of the perforating operation using finite element techniques confirmed that crater depth and location correlated with relative slip of the film on the die. The crater started where the slip velocity was greatest.]

The economic value of determining the mechanism in the example cited is that once it is known what causes the problem, design solutions can be developed to minimize or eliminate it.

Develop Something New. A study to develop a new material, process, machine, product, and so forth may be the toughest type of technical assignment. The probability of success is probably less than that for other kinds of studies. Often, management is essentially asking for an invention, and you may not have a sublime inspiration to produce an invention. Most inventions, in fact, are not the "Eureka, I have found it!" type. Most take hard work and the patience to try countless different approaches. Most people are aware that the light bulb, one of Thomas Edison's most useful and enduring inventions, took hundreds of attempts to arrive at a filament material that would last long enough to be useful.

In the 1990s in U.S. industry, most large assignments were given to teams. Needless to say, it is quite difficult to invent something as a team.

More often than not, incremental improvements by the team lead to a so-lution or product, but it is not a "microwave oven" type of innovation. Teams, however, tend to be more successful than individuals if the team leader summons the cumulative talents of the individual members. Team development of something new requires each individual to perform as if he or she is doing the project alone.

[*One of my toughest assignments was to develop something new—a light-weight cassette for x-ray film. A cassette is the folio-type container that holds a sheet of film tightly against an intensifying screen. If you think that you have a broken ankle, you go to a hospital for an x-ray. They put the cassette under your ankle and shoot x-rays through your leg to expose the film and check for fractures. The new product asked for by the hospital cus-tomers of x-ray cassettes was a lighter cassette. The larger cassettes weighed about two pounds and radiographers often carry a number of these at a time, and it becomes very fatiguing.*

A diverse team was organized to develop a light-weight cassette. After many brainstorming sessions, we set out to a change from aluminum to magnesium. The end product reduced weight by 35%, but it was a manu-facturing nightmare because of the poor formability of magnesium. We eventually had to stop making these cassettes. The team went into inactive status. A second team, a year later, took one of the ideas from the first team—an aluminum/plastic/aluminum composite sheet—and made it work. It saved 50% in weight without the formability problems of magnesium. We obtained a patent on this new way to make x-ray cassettes.]

Most projects in the development of something new require significant effort, time and talent. They often involve collaboration with others. A proj-ect to develop something new usually starts with an idea or concept. It could be yours or someone else's. As with all science and engineering projects, the starting phase involves information gathering.

If you have an idea for a new paperclip, you would probably start with a patent search. You would find that the first one was patented in the late 19th century, and scores of patents have been issued on different types. New paperclips are being patented as this book is written. Make sure that your concept was not previously invented. If you clear this hurdle, you may want to do a market survey to see if the idea can be made financially suc-cessful. If this study is positive, the real work starts. Build prototypes and make sure that your device works as intended. When all this is done, the inventor may want to start patent work and develop a manufacturing strat-egy. How will these be mass produced?

In summary, developing a new concept usually takes all of the hard work and patience that Thomas Edison used when he developed the electric light bulb. It was not a "Eureka" invention. It was many months of tedious work punctuated with an ample supply of failures. Do not be discouraged. If you have faith in your idea, go for it.

Feasibility Study. Many big engineering and science projects started with a feasibility study. The purpose of such studies is usually to gather facts that can be used to make business decisions. A heat treating company may want to expand into the field of high-volume induction hardening. These are the types of questions to be addressed in a feasibility study:

• What is the market?
• What is the long-term stability of the market?
• Does this venture match corporate strategy?
• What does the equipment cost?
• What are the manpower requirements?
• What are the space requirements?
• What are the safety/government/health/liability concerns?
• What is the expected return on investment?

This kind of project involves information gathering from a wide variety of sources coupled with mathematical analysis of the business economics. Usually the feasibility team makes the decision on the object or recommends a particular course of action. A feasibility study is performed by first listing decision factors and then accumulating enough information on factors to allow a decision to be made. It may be very hard to make the studies unbiased, because you may have a preconceived opinion on the subject. Do not let it shade your study. Be objective!

Failure Analysis. In the materials engineering business, all parts fail by one or a combination of three deterioration processes: wear, corrosion, or fracture. It wears out, corrodes, or breaks. Failure analysis in this field amounts to detective work to identify which of these maladies precipitated the failure and why. [*The last failure analysis that I did—this afternoon— involved a small punch and die. The die fractured into three pieces, and they brought me the pieces and the punch that was used with the die. Optical microscopy showed that the edge of the punch was chipped. The chips suggested that the punch hit the die; I had scanning electron micrographs (SEM) taken of the periphery of the die hole and the cutting edge of the punch. The SEM micrographs very clearly indicated that the punch hit one side of the die. This caused the die to explode into pieces. Since the punch and die were replaced and the machine was back running, it was not possible to see if there were some peculiar conditions in the machine to cause the punch to deflect 40 µm. A complete failure analysis would say that the die exploded because the punch hit it and how the punch moved 40 µm and hit the die. It may have deflected from a piece of tape, and so forth, that inadvertently went through the tooling. We had to end the analysis at simply saying that the die exploded because the punch hit it.*]

The generic steps in a failure analysis study are:

1. Acquiring background information—details on materials of manufacture, heat treatment, drawings, service history, operator comments
2. Planning the investigation—what is going to be tested? By whom? For what?

3. Testing—performing chemical analysis, hardness tests, metallography, mechanical property tests, x-ray inspection, ultrasonic inspection, and so forth
4. Analyzing acquired data—establishing a theory on the root cause
5. Testing the hypothesis—checking with parties involved to see if your hypothesis will "hold water"
6. Writing a final report—pinpointing the root cause and making recommendations to prevent recurrence

There are many other types of failures in science and engineering, but the same types of concepts apply. If you work for the state department of transportation, then a failure may be a constantly clogged off-ramp that feeds to another expressway. The failure analysis in this situation may involve the same basic steps in the above list. In fact, you may still have a testing phase. You may install temporary barricades to change upstream traffic flow to see if this reduces traffic congestion. You may gather data by interviewing commuters. Failure analysis tasks befall most technical people. The watchword for these studies is thoroughness.

3.2 General Methodology

Performing laboratory projects during courses in engineering and science are probably the closest a student gets to technical studies. Some colleges provide work-study programs, which give technical students more exposure to life after school. In most cases, however, the realities of a technical profession are not fully appreciated until a first job in a profession is underway.

How does one attack one of these projects? The answer to this question is to dig into them with a methodology that has evolved (over centuries) as the logical way to address technical tasks. The described methodology of handling a typical project may be new to newcomers to science and engineering, and old hat to experienced scientists and engineers.

The purpose of this section is to discuss the general methodology of performing a technical study. The objective is to gain an understanding of where and how to start a study and how to ultimately produce a successful study. A successful study does not always result in the achievement of an ultimate goal, but if a study is conducted properly, it can be an end product. For example, you work for a manufacturer of oil-well drilling tools. The working ends of the drills use cemented carbide with a cobalt binder for the cutting edge. In offshore wells, saltwater attacks the cobalt binder, and you are assigned the job of finding a more saltwater-resistant binder. You initiate a formal project to investigate nickel-chromium alloys as an improved binder. You test this binder in every imaginable way and learn that the corrosion problem is solved, but carbides made with this binder are so brittle that they are unusable. This is a successful study. Your company now knows that they must go down another road. If you publish this work, you have made a significant and permanent contribution to your technical field.

The methodology described in this section is not based upon any rules of science. It is simply a logical way to approach almost any project. The key elements are:

- Proposing a project (when it is part of your job to do so)
- Gathering background information
- Planning the project or test
- Conducting experiments
- Analyzing data, reaching conclusions, making recommendations, and reporting your work

Proposing a Project—The Idea

In the 1950s, engineers were the managers for most manufacturing and industrial operations. All of the top management, with the exception of a finance manager and general counsel, had a firm understanding of new product needs and how to manufacture these products. In this heady era of smokestack industries, new technical people were hired as project needs dictated. Technical managers assigned you to a project, and you did them under the manager's guidance. You did not have to propose a project; managers decided to do a project and hired people to do it.

Nowadays, it seems more common to have technical managers with a business background who do not have a background that allows them to assign specific projects to new engineers and scientists. The research director may know that the market is ripe for a cell phone that can be dialed by voice, but he or she may have no idea on how to develop new products. You may be hired to address a perceived need but not necessarily assigned a full plate of meaningful projects. After hire, you may also be required to start proposing projects that produce an annual return of three times your salary. Many industries and businesses in the United States are run this way, and this is where proposals come in.

Where do the ideas for a project come from? What does a project proposal look like? The idea for a project comes from inspiration, research, observation, somebody's comment—almost any place. [*In my early years in the automobile industry, my job was to design machines or processes that would reduce labor costs. I would get project ideas by observing production line processes. If there was one station on the assembly line that was troublesome (always slowing or stopping the line) I would make a rough sketch of a design to solve the problem and then gather cost data to be used in obtaining funding to make the improved machine. If the problem station added to the cost of the part at the rate of 0.02 h per part, I would have to design a device that did the operation in 0.01 h per part. I would also use downtime savings in the justification.*]

Technical talks, conferences, and trade shows can produce ideas for projects or studies. As you take notes, an idea may flash through your mind on how to use something that the speaker talked about. Quickly jot down this idea in your logbook and put an asterisk on it so that you can find it later.

Brainstorming sessions with team members are another way to come up with ideas. Sometimes it helps to gather a diverse group together with a facilitator who uses one of a number of idea-generating techniques. There is one [*I think that it was called synectics—designing by madness*] system that worked by having every person think about how a particular task or problem would be solved in nature. For example, if the problem is adhesion of a coating to a substrate, how would something stick on something else in nature? Trees exude sticky substances; burdocks have barbs that make fibrous things stick. Flies stick to the tongue of a frog, and so forth. Surprisingly, going away from the specific problem at hand and then coming back often brings out ideas that will potentially solve a problem.

[*My best ideas often come from thinking in the "alpha zone." They teach in psychology that people normally use only one side of their brain for most of their activities. The other side is seldom used, but that is the side that produces creative thought, ideas, inventions—it is your genius side. Unfortunately, we do not have a switch to turn on the creative side of the brain at will, but there is a natural "switch" that you can use. I have been told (probably by the National Enquirer) that you use the creative side of your brain when you are in the transition from sleep to awake. When you are ready to fall asleep at night and when you rise in the morning, you can invoke your alpha zone and produce some good ideas. I get my best ideas in the shower in the morning. I am not fully awake, and a hot shower helps my thinking. I keep a notepad nearby and jot down any good ideas that come from my morning shower. All this may sound weird, but I suggest trying it. There is no risk and if you do not get a good idea, you will at least come out of the shower clean.*]

Finally, ideas from projects and products can come from patent searches and thorough research of a subject. Read everything written on the subject; talk to users. Talk to manufacturers. Find out what is wrong and what is right about the product, machine, or process that is the potential object of the study. Sometimes a good project idea requires a combination of all of the above.

Writing a Project Proposal. After you have an idea for a project or study [*whether it was assigned, or the product of your genius*], you will need to get support and funding to proceed. Project proposals can be easy or almost more work than the study depending on the organization. The easy-funding project proposal may only require filling out a form like that shown in Fig. 3.4. In industry in the United States, funding for research

```
┌─────────────────────────────────────────────────────────────┐
│                   LAB DEVELOPMENT PROJECT                    │
│                                                             │
│   Date: _____                     │
│                                                             │
│   Title: _____        │
│                                                             │
│   Scope of Work: _____        │
│   _____         │
│   _____         │
│                                                             │
│   Anticipated Benefits: _____         │
│   _____         │
│   _____         │
│                                                             │
│   Support Staff Required: _____         │
│   _____         │
│   _____         │
│   _____         │
│                                                             │
│   Recommended Plan: _____         │
│   _____         │
│   _____         │
│                                                             │
│   Project Costs: _____         │
│           Labor:  Engineering _____         │
│                   Technician  _____         │
│           Material:      _____            │
│           TOTAL:         _____            │
│   Completion Date: _____         │
│                                                             │
└─────────────────────────────────────────────────────────────┘
```

Fig. 3.4 Simple project proposal

and development work usually comes from corporate profits. Research and development projects are deductible from corporate profits as long as the work is experimental and aimed at new products, processes, and materials.

In the "easy funding" proposal form, the project justification is simply a sentence on the value of the proposed project. Project justification can be far more complicated. It may be necessary to show detailed cost savings and to prove with calculations that your proposed project will produce more than the profit that could be made by simply investing the money in securities, real estate, or other investments. Most financial people claim that low risk investments could conceivably produce a 10% return. Thus, a project should show a return of about 20% per year to be attractive to corporate financial people.

On the other end of the scale are proposals for funding by the National Science Foundation (NSF), the source of many science projects at American universities. Figures 3.5 to 3.7 show the mandatory forms that must be filled out. These projects usually are for amounts of less than $100,000 per

NATIONAL SCIENCE FOUNDATION
Program Solicitation: NSF 98-153
Closing Date: December 14, 1998

STTR PHASE I -- PROPOSAL COVER PAGE

TOPIC	EMPLOYER IDENTIFICATION NUMBER (EIN) OR TAXPAYER IDENTIFICATION NUMBER (TIN)		
PROPOSAL TITLE			
NAME OF COMPANY	NAME OF RESEARCH INSTITUTION		
ADDRESS (including zip code)	RESEARCH INSTITUTION ADDRESS (including zip code)		
REQUESTED AMOUNT $	PROPOSED DURATION 12 months	PERIOD OF PERFORMANCE	

THE SMALL BUSINESS CERTIFIES THAT:	Y/N
1. It is a small business as defined in the STTR solicitation	
2. It qualifies as a socially and economically disadvantaged business as defined in the STTR solicitation. **FOR STATISTICAL PURPOSES ONLY.**	
3. It qualifies as a woman-owned business as defined in this solicitation. **FOR STATISTICAL PURPOSES ONLY.**	
4. It will exercise management direction and control of the performance of the STTR funding agreement.	
5. NSF is the only Federal agency that has received this proposal (or an overlapping or equivalent proposal) from the small business concern. If **No,** you must disclose overlapping or equivalent proposals and awards as required by the STTR solicitation.	
6. It will perform _____ percent of the work and the collaborating research institution will perform _____ percent of the work as described in the proposal.	
7. The primary employment of the principal investigator will be with this firm at the time of award and during the conduct of the research.	
8. It will permit the government to disclose the title and technical abstract page, plus the name, address and telephone number of a corporate official if the proposal does not result in an award to parties who may be interested in contacting you for further information or possible investment.	
9. It will comply with the provisions of the Civil Rights Act of 1964 (P. L. 88-352) and the regulations pursuant thereto.	

PRINCIPAL INVESIGATOR / PROJECT DIRECTOR

NAME	TITLE	
SOCIAL SECURITY NO.	HIGHEST DEGREE / YEAR	E-MAIL ADDRESS
TELEPHONE NO.	FAX NO.	WEB ADDRESS

RESEARCH INSTITUTION INVESTIGATOR)

NAME	SOCIAL SECURITY NO.	TELEPHONE NO.

COMPANY OFFICER (FOR BUSINESS AND FINANCIAL MATTERS)

PRESIDENT'S NAME	YEAR FIRM FOUNDED	NUMBER OF EMPLOYEES AVERAGE PREVIOUS 12 MO.: CURRENTLY:

PROPRIETARY NOTICE: See Section 7.5 for instructions concerning proprietary information.

Fig. 3.5 Formal project proposal form of the United States National Science Foundation (NSF Form 1304)

National Science Foundation
Small Business Technology Transfer (STTR) Program
Program Solicitation No: NSF 98-153

PROJECT SUMMARY

FOR NSF USE ONLY
NSF PROPOSAL NO.

NAME OF FIRM
ADDRESS
PRINICPAL INVESTIGATOR (NAME AND TITLE)
TITLE OF PROJECT
TOPIC TITLE
PROJECT SUMMARY (200 words or less)
Potential Commercial Applications of the Research
Key Words to Identify Research or Technology (8 maximum)

NSF Form 1304 (STTR 12/97)

Fig. 3.6 National Science Foundation project proposal summary form (NSF Form 1304)

SUMMARY PROPOSAL BUDGET

(SEE INSTRUCTIONS ON REVERSE BEFORE COMPLETING)	FOR NSF USE ONLY	
ORGANIZATION	PROPOSAL NO.	DURATION (MONTHS)
		Proposed / Granted
PRINCIPAL INVESTIGATOR/PROJECT DIRECTOR	AWARD NO.	

A. SENIOR PERSONNEL: PI/PD and Other Senior Associates (List each separately with title, A.6, show number in brackets)	NSF Funded Person-mos.	Funds Requested By Proposer	Funds Granted By NSF (If Different)
	CAL	$	$
1.		$	$
2.		$	$
3.		$	$
4.		$	$
5.		$	$
6. ( ) OTHERS (LIST INDIVIDUALLY ON BUDGET EXPLANATION PAGE)		$	$
7. ( ) TOTAL SENIOR PERSONNEL (1-5)		$	$
B. OTHER PERSONNEL (SHOW NUMBERS IN BRACKETS)			
1. ( ) POST DOCTORAL ASSOCIATES		$	$
2. ( ) OTHER PROFESSIONALS (TECHNICIAN, PROGRAMMER, ETC.)		$	$
3. ( ) GRADUATE STUDENTS		$	$
4. ( ) UNDERGRADUATE STUDENTS		$	$
5. ( ) SECRETARIAL - CLERICAL		$	$
6. ( ) OTHER		$	$
TOTAL SALARIES AND WAGES (A+B)		$	$
C. FRINGE BENEFITS (IF CHARGED AS DIRECT COSTS)		$	$
TOTAL SALARIES, WAGES AND FRINGE BENEFITS (A+B+C)		$	$
D. PERMANENT EQUIPMENT (LIST ITEM AND DOLLAR AMOUNT FOR EACH ITEM EXCEEDING $500.) (Do not use for Phase I)			
TOTAL PERMANENT EQUIPMENT		$	$
E. TRAVEL			
1. DOMESTIC (INCL. CANADA AND U.S. POSSESSIONS		$	$
2. FOREIGN (Do not use for Phase I)		$	$
F. PARTICIPANT SUPPORT COSTS 1. STIPENDS $_____ 2. TRAVEL _____ 3. SUBSISTENCE _____ 4. OTHER _____			
( ) TOTAL PARTICIPANT COSTS		$	$
G. OTHER DIRECT COSTS			
1. MATERIALS AND SUPPLIES		$	$
2. PUBLICATION COSTS/DOCUMENTATION/DISSEMINATION		$	$
3. CONSULTANT SERVICES		$	$
4. COMPUTER (ADPE) SERVICES		$	$
5. SUBAWARDS (PROIDE A SEPERATE NSF FORM 1030 FOR EACH SUBAWARD)		$	$
6. OTHER		$	$
TOTAL OTHER DIRECT COSTS		$	$
H. TOTAL DIRECT COSTS (A THROUGH G)		$	$
I. INDIRECT COSTS (SPECIFY) TOTAL INDIRECT COSTS			
J. TOTAL DIRECT AND INDIRECT COSTS (H+I)		$	$
K. FEE (If requested)		$	$
L. TOTAL COST AND FEE (J + K)		$	$

Fig. 3.7 National Science Foundation project budget form (NSF Form 1030A)

year. The average proposal contains about 30 typed pages, and most have the following format:

1. Identification and significance of problem
2. Background and technical approach
3. Technical objectives
4. Work plan
5. Commercial potential
6. Principal investigator and senior personnel
7. Consultants and subcontractors
8. Equipment, instrumentation, competitors, and facilities
9. Equivalent or overlapping proposal to other federal agencies
10. Current and pending support of principal investigation and senior personnel
11. Budget

The forms in the illustrations are for businesses, but university researchers must fill out similar forms. Obtaining funding for technical projects is a significant part of every research professor's life. These proposals are so detailed that you must have the project almost completed before asking for funding. In any case, the first step in performing a technical study is getting the funding. Most organizations require some sort of report for this step. These reports can be single page or the 30 to 50 page proposals required by literary agents and government organizations.

A big advantage of the formal government proposal forms is that they force the researcher to have the project planned to the nth degree. It almost seems like these projects cannot fail. The goal is to plan every step.

[*My projects are usually evolutionary. I know where I want to end up; I know where I will start. What happens between the early experiments and the end usually depends on the outcome of the initial experiments. However, the proper way to start a project is to develop detailed plans for the steps that are required in the government proposals.*

One of my projects this year was to develop a high-speed perforation test for polyester films. It meant building a piece of equipment. The planned steps were:

1. *Obtain basic punch press from the production department.*
2. *Strip the press down for conversion to robot operation.*
3. *Design the tool package.*
4. *Design the x-y-z transport system.*
5. *Purchase the x-y positioner and the z direction locator.*
6. *Submit the tooling package to the shops for fabrication.*
7. *Build the transport roll system.*
8. *Assemble all components.*
9. *Write software to produce operation.*
10. *Debug the unit.*

11. Complete speed trials.
12. Write report on attainment of goal.

Note that a formal report concludes the project. As can be seen from the procedure, there is considerable time between steps. To keep things moving, written milestones with dates are reviewed by the boss. Short projects have steps and timelines, and we keep them moving with a simple planning form (Appendix 13) for tracking the tasks and completion dates of a project.

Gathering Background Information

The engineering approach to projects involves gathering data throughout. There are many sources of technical information: libraries, the internet, government agencies, reference books, journals, patents, standards and specifications, and industry product literature. Another basic source of background information is contacting individuals who may have knowledge on the subject. If you are designing a machine or solving a manufacturing problem, interview individuals who can provide insights and knowledge. In manufacturing, machine operators and department managers often supply very valuable information. The same can be said for senior peers. Leave no stone unturned.

When gathering information on significant studies, always do a literature search. In fact, this must be done at the funding level for government proposals. The 30 page proposals that go to government agencies all have significant literature searches. In the case of funding for product development, a patent search is required.

Computers have made literature searching infinitely better than in the past. There are many electronic sources of information, and a person can read and transfer information from countless sites of universities, professional societies, and government agencies. Libraries can also help in researching by directing people to a wide variety of useful sites or information services. A few examples of reference information sources are listed in Table 3.1. Of course, Table 3.1 is just a snapshot sample in the fast-changing arena of information services.

Some commercial organizations supply databases on their products on CD-ROM. These are excellent sources of information. The ones that are given here, however, are usually aimed at selling the products in the database and the data is not objective. You also must be aware that some types of Internet information come with a cost. For example, all the thousands of ASTM standards are available on the Internet, but there is a charge.

You can get an idea of the type of organization that operates a Web site by the last part (or extension) of the Internet address:

- .org = a not-for-profit organization
- .com = a commercial enterprise
- .edu = an educational institution
- .gov = a government agency

Table 3.1   Examples of information sources for technical reference information

Reference source	Description
Patents	
APS, Automated Patent System	APS, produced by the Patent and Trademark Office, is a database containing the full text of patents, except diagrams, back to 1971. APS is an enhanced service that will be available soon on a cost-recovery fee basis.
CASSIS, Classification and Search Support Information System	CD-ROM Index produced by the U.S. Patent and Trademark Office. Using CASSIS, patent titles may be searched by key word back to 1969. Classifications and patent numbers can be searched back to 1790.
PTDL, Patent and Trademark Depository Libraries	A PTDL is designated by the U.S. Patent and Trademark Office (PTO) to receive and house copies of U.S. patents and patent and trademark materials, to make them freely available to the public, and to actively disseminate patent and trademark information.
United States Patent and Trademark Office, www.uspto.gov	Bibliographic searching. U.S. Patent Full-Text Database, Full text content of U.S. patents issued from 1 January 1976 to the most recent weekly issue date, including full-page images
Library catalogue services	
Online Computer Library Center Inc. (OCLC), www.oclc.org	OCLC is a nonprofit, library computer service and research organization founded in 1967 by university presidents to share library resources and reduce library costs. In addition to the online Cataloging service, OCLC offers a Web-based copy cataloging service (CatExpress), the Cooperative Online ResourceCatalog (CORC), and other online and offline services.
The Research Libraries Group (RLG), www.rlg.org	The Research Libraries Group is a not-for-profit membership corporation of over 160 universities, national libraries, archives, historical societies, and other institutions. The RLG develops and operates databases and software to serve the information needs of member and nonmember institutions and individuals around the world. The Web site at www.rlg.org describes its activities and services including the Research Libraries Information Network (RLIN).
Research Libraries Information Network (RLIN)	The RLIN service is a bibliographic information system of the RLG. The RLIN indexes let you search on subject phrases and words, title words in any order, personal names, conference titles, corporate names, and more than 40 specialized access points such as form and genre, and publication numbers (ISBN, ISSN). While the system is mostly used by librarians and archivists, individual researchers also value it.
Index and abstract searching	
Cambridge Scientific Abstracts (CSA), www.csa.com	Cambridge Scientific Abstracts (CSA) offers electronic searching of bibliographic abstracts. An Internet database service (IDS) provides Web access to over 30 major databases including METADEX (Metals Index) and others. CD-ROM, print, and document delivery is also available.
Engineering Information Inc., www.ei.org	Engineering Information Inc. (Ei) was established in 1884 by a community of engineers committed to sharing their own research results as well as learning from one another about new developments.
Engineering Index	One product is its Engineering Index, which is one of the most comprehensive interdisciplinary engineering information databases in the world with over 3 million summaries of journal articles, technical reports, and conference proceedings. The electronic form of this abstracts database is known as Ei Compendex and is available in print as *Engineering Index Monthly* and *Engineering Index Annual.* Engineering Direct allows immediate electronic access directly from CompendexWeb to the full text of 500 key Elsevier technology journals cited in Compendex.
The Institute for Scientific Information (ISI), www.isinet.com	ISI maintains a multidisciplinary bibliographic database for over 16,000 international journals, books, and proceedings in the sciences, social sciences, and arts and humanities. Products include searchable databases, alerting services, patent information products, chemical information products, research performance and evaluation tools, and full-text document delivery service.
Other general sources	
Sheehy's Guide to Reference Books	Available in most libraries, this book is updated regularly with listings of reference books in the sciences and humanities.
Linda Hall Library, www.lindahall.org	The Linda Hall Library is an endowed collection of publications on science, engineering, and technology. It contains more than 40,000 serial titles, 275,000 monographs, 1 million technical reports, and 1,110 standards and specifications and nearly the complete holdings of the United States Patents and Trademark Office.
National Technical Information System, www.ntis.gov	The National Technical Information Service is the U.S. government's central source for the distribution of scientific, technical, engineering, and related business information. Comprehensive subject searching of all NTIS collections is available at GOV.Research_Center Web site (http://grc.ntis.gov). In 1992, NTIS established the FedWorld (www.fedworld.gov) to serve as an online locator service for a comprehensive inventory of information disseminated by the federal government within the NTIS repository and information made accessible through an electronic gateway of more than 100 government bulletin boards.

With a continued proliferation of Internet sites, additional types of domain names may be considered in the future. For example, new Internet extensions could be ".bank" or ".library." New extension names like these are under consideration.

If you wish to cite references from the Internet, you need to list them similarly to periodicals and book references: Webster Online, Para. 315, Mar. 1999, Webster Dictionary, 14 Jan., 1998, <http://www.wd.com>

There are two obvious problems with using Internet references. The data may not be referenced to ensure correctness, and it may not be current. Some universities only put refereed material on their Web site, but in general, Internet information must be scrutinized. Compare data with other references and judge its accuracy based on knowledge of the subject and its agreement with other references on the subject.

Literature Searches. An essential ingredient of any significant study is a literature search to find relevant reference information from the multitude of journal articles, conference papers, patents, magazine articles, and books. This can be done by keyword searching of abstracts and/or titles in bibliographic databases. After getting search results (which may include full abstracts of the articles in search results), select the articles that you want to get from the library or a document delivery service.

Bibliographic databases are available from a variety of sources, and a few are listed in Table 3.1. These searches and document delivery cost money. This cost may not be trivial. Beware of translations; some can cost several hundred dollars. This is the down side of computer searches. Journals articles that are on-hand in most libraries usually cost less than $5 per article. Make literature searching a project planning item and budget monies for it.

If you are lucky enough to have a research librarian, it is possible to simply e-mail key words to your research librarian. Research librarians are not always available, and some technical libraries may not have access to some of the bibliographic databases. Professional societies such as ASM International will do searches on a contract basis if your company does not have a research library. Read the abstracts and check the ones where you would like the complete article. Send it back to the library for delivery of the full-text articles. This process may take just a few days for a targeted search and just a few documents, or it may take weeks if you search and require a large number of documents. It is also useful to have an appreciation of the different types of bibliographic databases available on the Internet or from the assisting library. Some databases are listed in Table 3.1, but more information can be found by digging deeper into the references listed at the end of this Chapter.

Needless to say, your choice of key words for the literature search is most important. Whenever possible, select words that do not need modifiers. Sometimes it is unavoidable to use modifying words, so choose them very carefully. [*Earlier this year, I started a project on knives for slitting photographic film. If I did a search on "knives," I could probably get three thousand hits. I used slitting plastic and slitting knives. These key words brought approximately 300 "hits," and I ended up ordering about 20.*]

Background articles from the library or Internet should be thoroughly reviewed, first of all, to make sure that what is proposed has not been done before, and, second, to see whether there are any hints in these articles that will help reveal the solution to your problem. You must now start to record data. List the articles that are pertinent and save them in a safe place. List these articles in the report references and, what is more important, compare your results with the results of previous workers in the field in the discussion of the report. This is a very important part of a paper to be published or a patent.

[*A trick used by veteran writers to obtain pertinent references on a subject is to get a few references on your subject and search their references for applicable references. Repeat this on each reference. You soon will have a very in-depth list of references to consult.*]

RULE

Always do a literature search on significant studies.

Designing Test Plans

If you have to use a simple proposal form like the one in Fig. 3.4 (or more formal ones like Fig. 3.7), you will already have a plan. If you were fortunate to get funding by providing less detail, you now must go into all the detail the formal proposal would have required.

A project plan starts with a thesis—what do you think will be the outcome of your study? Where are you going? For example, you made a proposal to use diamond coatings to improve the life of steel-rule dies used to blank out cardboard containers for rolls of film. You know that there are lots of commercially available coatings. Part of the project will be to measure the effect of process temperatures on the hardness of the steel-rule material. You need a project step to compare process details. Next you will need a step to compare the adhesion of candidate coatings. Some may not adhere. You also need to compare the effect of the coating on edge sharpness. Finally, you need to compare the abrasion resistance of candidate coatings. The final plan may resemble the following proposed project steps:

1. Literature search to establish a list of candidate coatings (assigned to KGB, due 1/3/00). Estimated hours: 20 engr, 2 tech
2. Compare available coating processes for applications (assigned to KGB, due 2/28/00). Estimated hours: 40 engr, 4 tech
3. Order test samples from six vendors (assign to KGB, due 1/30/00). Estimated hours: 2 engr, 4 tech
4. SEM cutting edges of coated blades (assign to ATC, due 3/15/00). Estimated hours: 4 engr, 20 tech

5. Conduct wear tests (ASTM G 146) on candidate coatings (assign to KGB, due 5/2/00). Estimated hours: 0 engr, 40 tech
6. Review laboratory data (assign to KGB, due 6/30/00). Estimated hours: 40 engr
7. Conduct production trials (assign to KGB, due 8/31/00). Estimated hours: 48 engr, 20 tech
8. Write project report/recommendations (assign to KBG, due 10/15/00). Estimated hours: 50 engr, 0 tech

This plan is based on the thesis that you will uncover a diamond coating that will stay on and provide a significant improvement in service life. Current dies last 15,000 to 25,000 cuts. You want those numbers increased to 150,000 to 250,000 cuts.

Statistical Design of Experiments. Sometimes you are faced with a very complicated process. There are ten functional chemicals in a coating. You want to improve the durability of the coating. Where do you start? The traditional approach to this kind of project challenge is to hold nine chemicals fixed and vary only the tenth with durability changes measured with a standard scratch test. There are now computer programs available that will guide you through statistical experiments and allow you to test multiple factors like the ten chemicals at the same time.

Statistical design of experiments is a technique that became popular in research and engineering in the 1980s and 1990s. It is a way of using statistics to solve a problem or understand a process. It starts by identifying independent and dependent variables in an experiment. The purpose of the experiment must also be established, because this affects the selection of independent and dependent variables. Independent variables are those that the experimenter controls; dependent variables are those measured in the experiment. In an industrial experiment, the process variables may be such things as time and temperature; these are the independent variables. The dependent variables may be percent retained austenite and hardness.

A designed experiment has several steps. The first step is a screening process to identify important independent variables and to show the relative importance of each. This step could have from 2 to 30 independent variables. The next step is to perform high/low experiments on the variables identified as important in the screening experiment. For example, if time and temperature are identified as important variables in the screening experiment, then a high value and a low value are assigned to temperature (sometimes a midpoint is also used). This type of experiment is called a two-level factorial experiment and in this case is performed to measure the effect of high and low temperatures and high and low time. The "response surface" from this experiment should provide information to develop some models for further testing. The final step is to conduct special experiments to refine a model. If the model is correct, you should be able to adjust an independent variable (or turn a defect on or off) to produce some desired effect in a process.

Factorial design experiments are identified by a notation such as "2^3." This means that the experiment will have three independent variables at two levels (usually the high and low that can be anticipated in the process). This experiment determines the importance of each variable. This type of experimentation is contrasted with experimentation where one factor (variable) is changed at a time and all other variables are held constant.

Mathematical analysis of the experimental data indicates which variables are most important and if there are interactions of one variable with another. There are all sorts of rules on sample size, sampling errors, and the like, but by the late 1990s, most of the designing and analyzing had been performed by software for the design of experiments. The software prompts the user on identification of independent and dependent variables as well as actually conducting the experiment. Needless to say, the computer "crunches" the data and produces the system model.

It is outside the scope of this text to present usable details on how to use this technology, but the readers should be aware of these methods in the study of complex processes with as many as 20 independent variables. If the project involves life testing of products or machines, another statistical process called Wiebull statistics makes sense of the highly variable data that comes from life testing.

Summary. Project planning may involve an intuitive plan like the diamond coating project, or a statistical planning technique may be required for more complicated projects. In all cases, you should estimate costs and develop a timeline for completion of steps.

RULE

Technical projects should be divided into specific tasks with a start, execution, and end.

Performing Experiments

Unlike a physics lab, real-life experiments, tests, designs, and so forth take much more time than a four-hour laboratory session. Probably the handiest part of a technical study involving testing or laboratory work is tracking the logistics of getting samples fabricated and scheduled into various tests and ensuring that all aspects of the project keep moving. It may take two weeks to get test samples machined. Then, in the case of the prior example for evaluating diamond coatings, you need to arrange to send them to six coating suppliers. This involves purchasing contracts and writing specifications for the coating. This coating step may involve another three weeks. Testing may take three more weeks. It may take months to conduct the necessary preparations and experiments, which can easily be delayed in any step. Keeping experiments progressing over months of elapsed time can be a daunting task.

Use of a notebook with permanent pages is recommended for tracking experiments and experimental data. It is your journal, your diary. In some organizations, these books are numbered, and there is a place for signing and witnessing each page (Fig. 3.8). This is important for product development work. What do you record? In some instances your notebook should

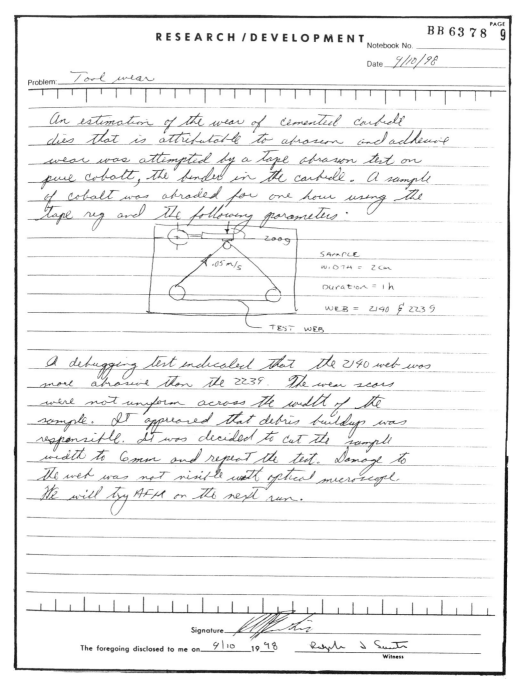

Fig. 3.8 Sheet from a project log book used to document experiment details

be used as a diary. Whatever you do each day on a project is recorded. Computer-generated data is taped or glued into the project notebook. If you conduct an experiment with new chemicals or with some novel conditions, record everything in the notebook. Bring the notebook to project meetings and record decisions and action items.

You will also need to establish a file, maybe even a file drawer, for supporting information such as the literature survey and supplier catalogs. Record the information critical to your project in the project notebook. If you have the assistance of technicians or students, require them to also maintain project notebooks.

Most development projects take months, sometimes years. It is nearly impossible to remember all tests and experiments without good record keeping. If you are working on something that is potentially patentable, you need to be proactive about signing and witnessing notebook pages. The first step in filing for a patent often requires submission of copies of notebook pages. Occasionally coworkers may file for a patent on something you were working on. Detailed and signed pages in a notebook can show who was first to discover what is claimed in the patent. I have had coworkers patent ideas that I gave to them. If you want to give away ideas, do it in writing. You then have claim to any patents that others write using your idea. There are coworkers in every organization who are essentially without ethics. Notebooks and published reports will prevent stealing of your ideas and findings.

RULE

Keep a log book of work and results on significant studies.

[*Since writing the above rule, my employer has started testing electronic notebooks. I have also learned that the U.S. government is developing software so that engineers in various physical locations can keep a collaborative notebook on a server. All collaborators could enter their data in this electronic notebook. So, electronic notebooks may eventually become an effective alternative to paper notebooks.*]

Reporting Results

If you have been working on a project for six months or a year, you will probably have more information than anyone would be willing to read. Digest the information and put it into a report. Most organizations mandate a formal report on project completion. If funding was for a one-year work increment, you must say what you did even if the problem is not solved. You need a proposal to get additional time and money. How do you condense a full notebook and a file drawer full of supporting records into a few pages for management (customers) to read? Publishing results of large projects in an appropriate technical journal automatically limits the size of the document to 10 to 15 pages including illustrations. This means that you can only use about five pages of graphs, photos, and supporting tables. You may have a 12 in. pile of data; extract maybe four or five graphs, two photos, a test schematic, and one or two tables that best summarize the work. Write a formal technical report using these data and the references used early in the study.

[*I usually write an internal report before even considering publishing the work in the open literature. If there is nothing proprietary in the report, it can be easily altered for publication.*] An internal report usually contains references to departments and buildings. These cannot be used in an article. Also, papers should be written with the metric International System of Units (SI) units on measured properties. In-house reports usually contain the units used in the operations discussed. Often some units are not in the SI metric system.

A significant reason for publishing work in the open literature is that it gives your company insurance that you can continue to use the processes and materials cited in the paper. For example, if you describe the use of silicon nitride for the construction of emulsion extrusion dies and somebody patents this idea, you can continue to use this material without patent infringement. Your published paper establishes that your use predates the patent. In fact, this very thing has happened. [*One of our competitors patented something that we had been doing for years, but kept it a trade secret. When the patent came out, we had to stop using our trade secret. Articles can be part of a patent strategy.*]

Small projects need reports to bring them to a close. All of the data gathering that we discussed applies to one-month projects, only during these, there is less data to be concerned with. A one-page informal report may be the appropriate closure document. Often, you need to make proposals and submit a plan for every $2,000 project. The steps are the same.

RULE

Close significant projects with a report.

Summary

This chapter discusses the types and general methodology of technical studies, so that students and others develop an understanding of how technical projects are conducted. The project sequence is as follows:

1. Articulate idea/concept
2. Research idea/concept/plan
3. Seek funding (report or presentation media)
4. Acquire funding
5. Implement plan

6. Revise plan (original was naive)
7. Continue work; record data
8. Analyze data
9. Present findings/results (report/presentation)
10. Close project (closing report needed)

As you can see, report writing is an integral part of this activity. That is the reason for this book. Technical writing is necessary, but it must be done in a certain way to be effective. Technical projects/studies need to be done correctly if they are to be successful. This book cannot give you project ideas or directions on solving a problem, but it can explain how recording data and documenting experiments well are a significant help in making projects successful.

Finally, this chapter concludes with some key items pertaining to the experimental/engineering process:

- Project funding is often a new engineer's (scientist's) first professional task.
- Funding requires convincing others that your idea will work and generate profits.
- Proposals must be supported by a search of what was done in the past.
- The minimum proposal usually includes scope of work, the benefits, the plan, the costs, and the time required.
- Literature searches must be done on a project.
- Be conscious of costs in literature searches.
- Experimental data should be recorded in a project journal.
- Record everything, including failures and blind alleys.
- Publish ideas in reports if there is risk of coworkers stealing them.
- Sign, witness, and date notebooks where patents may be involved.
- Publishing results brings closure to a project and protects intellectual property.

Important Terms

- Patent
- Funding
- Management
- Journal
- Experiment
- Peer review
- Literature search
- Project

- Milestones
- Proposal
- Deliverable
- Closing report
- Project plan
- Data condensation
- Project budget

For Practice

1. Write a proposal to get funding to develop a new bicycle seat.
2. What journals would you search for background information on bicycle seats?

3. Write a detailed project plan to develop a bicycle seat.
4. Describe the benefits (including savings) of a project to develop a passive rear seat belt system for automobiles.
5. Write a project plan for the development of a lightweight magnesium bicycle helmet.
6. Establish milestones for the helmet project.
7. State five benefits of keeping a development notebook.
8. What is peer review, and why is it important?
9. Describe what information should go into a project log.
10. Describe in a short paragraph the type of technical projects that you do or anticipate doing and the role of technical writing in those projects.

To Dig Deeper

- D. Beer and D. McMurrey, *A Guide to Writing as an Engineer,* John Wiley & Sons, 1997
- R.E. Burnett, *Technical Communication,* 4th ed., Wadsworth Publishing Company, Albany, New York, 1997
- R.A. Day, *How to Write and Publish a Scientific Paper,* 3rd ed., Oryx Press, New York, 1988
- V. Gibaldi, *MLA Handbook for Writers of Research Papers,* 5th ed., The Modern Language Association of America, New York, 1999

CHAPTER 4

Writing Strategy

CHAPTER GOALS

1. *Understand how to develop a writing strategy*
2. *Understand the importance of directing your technical writing to specific individuals or groups*
3. *Understand how to select appropriate readers*
4. *Understand how to write to a particular readership*

LIKE MOST PROJECTS, writing a technical document should start with the development of a writing strategy as shown in Fig. 4.1. This requires definition of three basic elements:

- The readership (the who) of a document
- The scope (the what) of a document
- The purpose and objectives (the why) of a document

Defining the readership simply means identifying the person, persons, or organization to be reached with the technical writing. The scope of a document is defined by boundaries of what needs to be discussed. Are you going to write about all punch presses or only the one that you spent the last four months redesigning? The purpose and objective of the report includes intention and a description of why the work was performed. In most cases, concurrent decisions must be made when defining the three elements of a writing task.

The purpose of this Chapter is to describe the factors that should be considered in thinking about a writing strategy. These factors include an analysis of readers, understanding the scope, purpose, and objective of a document, and how to write to a particular type of reader. These issues must be resolved before one sets pen to paper. It is your plan. Thoughtful development of a writing strategy will make the job easier and the product better. Strategy is

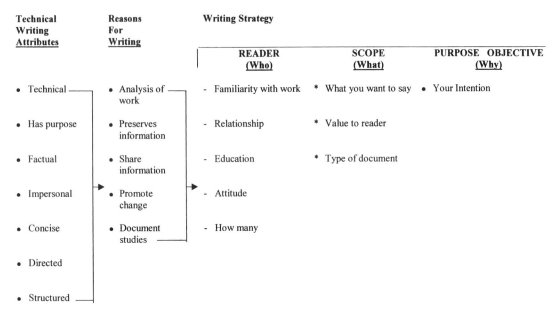

Fig. 4.1 Elements of writing strategy

important for any technical document, but it is even more critical for large documents. Even experienced authors of technical books can benefit from careful planning of writing strategy.

4.1 Analysis of Readers

Analysis of readership is as important to report writing as a game plan is to a football game. You must decide on your intentions in writing. Who do you want to communicate with? What level of writing is appropriate? Readership analysis comprises all of these factors in any form of writing, but it is more important to give readership special treatment in technical writing than in other forms of writing. Writing above (or well below) the reader's comprehension may produce a negative result, one counter to the writer's intentions. Technical reports often request action on the part of others. If the report is written at the wrong level or if it is sent to the wrong people, it may not have the intended effect. Give this phase of report planning adequate attention. Here are some points to remember:

- Readership is your intended audience.
- The intention of your report determines readership.
- The readership selected determines writing level.
- The level of writing must be such that it is understood and is useful to all readers.
- Selection of a circulation list (your readership) should respect organization hierarchy; include contributors, sponsors, and potential users of your work/recommendations.

Importance of the Reader

The previous Chapter listed several reasons why technical people should include written documentation as an essential part of their job. Some of the more prominent reasons for writing were discussed in the previous Chapter.

- Analysis of a process/system, and so forth
- Preservation of information
- Sharing of information
- Promoting change
- Documentation of work

All of these reasons for writing are focused on a particular readership. The following examples illustrate the importance of readership in getting results from technical writing. Only specific readers can implement your recommendations. Only certain readers are interested in your work. Sending work to an uninterested party may have a negative effect on your intention. On the other hand, interested readers can promote whatever you wrote about. Thus, readers are very important, and they should be an essential part of your technical document strategy.

Analysis of a Process. An example of an analysis document might be a report regarding analysis of the efficiency for an assembly line of single-use cameras. The line was not producing as many good cameras as claimed by its designers. You studied the line and determined that there were three troublesome workstations producing most of the downtime. If you are recommending a redesign of these stations, it is very important to reach the appropriate readers if recommendations are to be implemented. Just sending the report to the boss will probably not do it. Part of your writing strategy needs to be identification of readers who can approve and fund recommendations.

Preservation of Information. You performed a complicated study on the mechanism of silver plateout in emulsion pumps. This problem has been plaguing the emulsion handling department for decades. You learned the root cause and the cure. If you want this significant discovery preserved and shared, identify readers who will disseminate it to people who buy, use, and repair pumps as well as to emulsion formulators so that they can design out the chelating agent that you discovered catalyzed the plateout. Finally, see that the document is archived so that future generations will know how to deal with this problem. The readers you select will determine if silver plateout problems will be eliminated because of your work.

Sharing of Information. In some consulting work, you learned that seizures of the spooling mechanisms in film spoolers were caused by a change in suppliers for porous bronze plain bearings. The purchasing department changed suppliers [*for lower costs*] and did not bother to notify design engineering. In your study, you determined that the tolerances on bearings from the new supplier are such that a significant fraction of these bearings [*there are hundreds used*] will end up with one-fifth the recommended operating clearance. These spoolers are used worldwide. You need to identify readers worldwide to remove the Brand B bearings, replace them with Brand A, and ensure that the purchasing department buys only Brand A for these units in the future. This will be a daunting reader selection task. However, if you do not reach all the right readers, more failures [*at $30,000 each*] will occur.

Promoting Change. You visited an outside laboratory and learned that its employees measured corrosion rates of metals immersed in solutions by analyzing the solution for dissolved elements with inductively coupled plasma atomic emission spectroscopy. The laboratory can analyze for sixty elements in the parts per million range in less than five minutes. Three laboratories in your company do this kind of testing but measure corrosion rates by mass change after long immersions. This new system could reduce elapsed test time from 30 to 3 days with greater accuracy and lower cost. You need to write a document to readers who can change the way corrosion tests are run. You also need to convince them.

Documentation of Work. You completed a six-month study of plastics for movie cans. The present cans wear when heavy rolls of film jostle in them. The wear particles can contaminate the film. You performed extensive laboratory tests screening 15 different plastics and then had prototype cans molded from five candidates. Further testing was done in a full size vibration test. You found a plastic that will solve the wear problem and have reams of data accumulated from the various tests. Now you want to order your raw data and distill it into a document that will recommend implementation of the winning plastic. The right readers can make your recommendations happen, and you will have documented your results in a way that can be archived and used by others.

RULE

Readers are important. Select appropriate ones and write to them.

Selecting Readers

How do you select a readership for a document? Commercial publications, like new magazines, spend large sums of money and significant research defining the target audience for a new magazine. [*I recently became a reader for a new magazine that arose out of a TV home repair show. The advertising in the magazine clearly indicates that the intended readers are tool nuts and wanna-be carpenters. They have centerfolds on types of nails. It is quite an expensive magazine, and the products advertised are mostly expensive.*] It is clear that the intended readership is upscale; this magazine is for people with better than average incomes. They probably designed the magazine for the reader who:

- Is male
- Has an income above U.S. average

- Loves tools
- Loves to work with hands
- Loves carpentry
- Likes to learn new things
- Owns a home
- Pursues building as a hobby (not a profession)

The people who compile the magazine have features written to this reader profile. They have established their intention (to make money), and the readership is those people who fit the readership profile. The articles are written to the level of a novice carpenter.

Because readership determines level of writing, decide on the readership as soon as you decide on your intentions. The readership for a technical document should include all people who could benefit from the document content, the people who helped with the work, and the management people who want to keep abreast of what you are doing. A trip report should be sent to all who need to know what you observed and concluded from the trip. A laboratory report needs to be directed to the department requesting work and appropriate management. A report on a problem should have a readership of the party who submitted the problem, management, and the people responsible for implementing the problem solution.

It is advisable not to send reports to people for political reasons or to send reports to people who will not be interested or are scantily involved. In addition, in business and industry, there is a hierarchy for reports. It is inappropriate in some organizations to send reports to people above your immediate supervisor without his or her approval on the report.

The recommended approach is to establish a readership list for every type of report that you are expected to write in your position. If you are a machine designer, the list may include:

- Your immediate supervisor
- Your department secretary (for filing)
- The three other designers who work on similar projects
- Your customer contact
- The machine shop that builds your machines

Make readership lists for your reports and clear these lists with your supervisor. You will appear well-organized; you will be acting in conformance with your supervisor. Your writing job will be easier because you do not have to bother with this step each time. Figure 4.2 shows an organization chart for a product engineering department in a large corporation. For an engineer in the 40 person machine design group, it would be inappropriate to send copies of reports to the group or division managers or to any of the other managers unless they had a specific involvement in the work. This is part of corporate culture, and this type of hierarchical culture exists in most organizations. Essentially, there is an unwritten rule

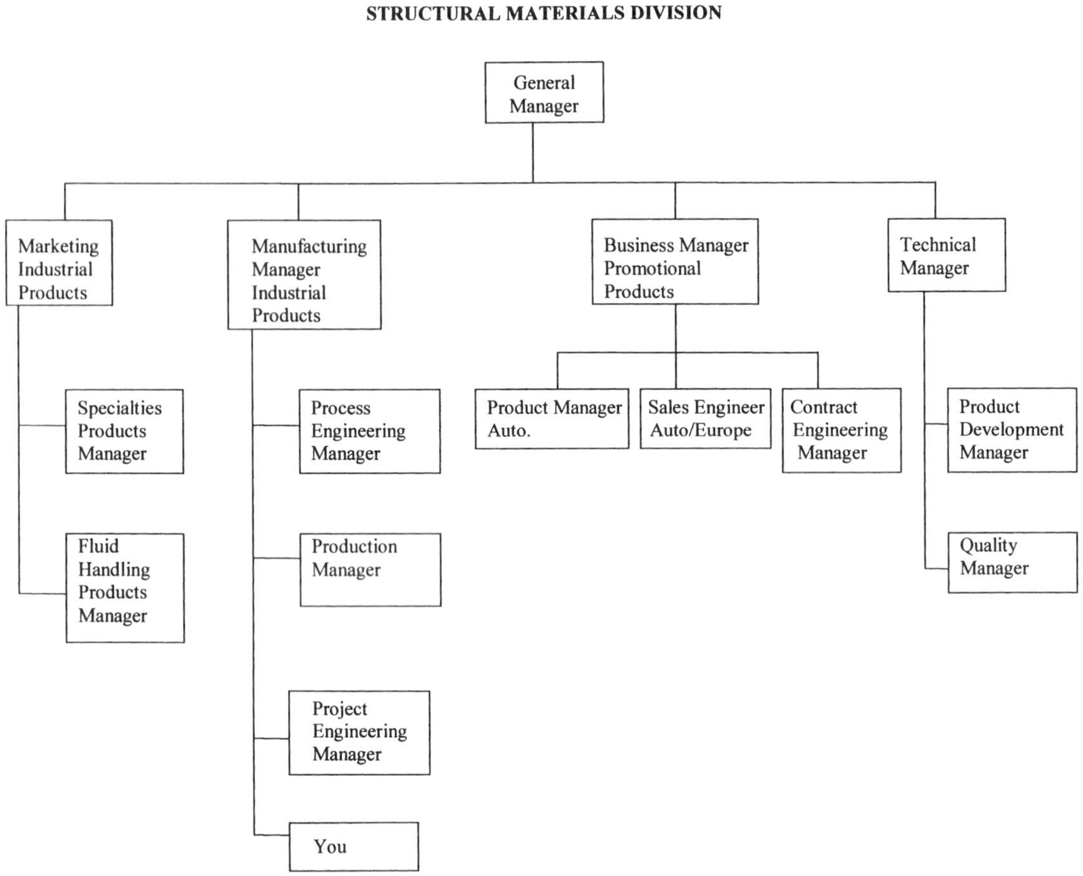

STRUCTURAL MATERIALS DIVISION

Fig. 4.2 Typical organization chart

not to go over one's immediate supervisor in any matter without his or her knowledge [*and approval*].

4.2 Scope of Writing

Every written document has boundaries on the depth and extent of coverage, and part of a writing strategy should be the definition of those boundaries. If you are writing about an investigation that involved many different studies, a report that included all of this work may be longer than anybody would want to read. Putting limits on what is included in a technical document involves consideration of the following factors:

- Number of ideas/experiments/studies/subjects
- Depth of writing
- Level of detail

Depending on the writing situation, there may be other factors to consider in determining the scope of a document. Most technical writing, however, requires a definition of scope in terms of the number of subjects, technical depth, and the level of detail. A suitable scope statement in an introduction might be like the following:

> This report describes the application of the loop abrasion test method to rank the abrasion resistance of six different test steels. They are candidates for P35 first-form tools.

This kind of statement may be placed in the scope or format sections of the introduction. It is not mandatory to discuss technical details in the introduction of a report if there is no benefit in telling the reader in the introduction; the report itself may adequately show this to the readers. However, you must make decisions on the level of technical depth and detail, and write accordingly.

Number of Subjects

Most jobs that require written documentation involve multiple tasks. The tasks that need to be addressed in reports must be identified. For example, a study of a tool failure may involve chemical analysis, surface texture measurements, hardness measurements, and optical microscopy of a metallographic cross section taken from the failed part. If the chemical analysis of the failed part indicates that the tool was made from the wrong steel, a decision needs to be made on whether the failure-analysis report should include the tests that were performed (even if you did not identify the root cause of the failure).

A similar situation occurs with ideas. You are writing a proposal to get funding for a project to use coatings on shafts to improve the life of mechanical seals. You really want to investigate physical vapor deposition (PVD) coatings, but it may increase your chances of getting funding if you include electrodeposition coatings and case hardening heat treatments. If you include them, you can show how your proposed PVD solution compares with coatings that are more familiar to readers. You could also bring up the negative environmental aspects of electroplating and salt bath heat treatments. Thus, the number of ideas included becomes part of a writing strategy. The same situation exists with number of experiments, or subjects, designs, and even thoughts.

Depth of Writing

When a child asks his or her mother, "Mommy, where do babies come from?" the reply can range from a biological description to "from the hospital." In the strategy phase of technical writing, you must also decide on the desired depth of technical discussions. How technical should you get?

Are you going to do a literature survey on your subject, discuss previous work, and compare theories in your technical document? The need for technical depth generally depends on the complexity or difficulty of a subject. Simple problems, if there are any, do not require the technical depth of an intricate or subtle problem.

If you have limited knowledge on the subject in your planned document, it is usually advisable not to attempt an in-depth treatment of the subject. State that you have limited knowledge on a subject, but explain why the subject is important. You may not be an expert on microgear pumps [*or whatever*], but it appears that they are not meeting the requirements of your process because they do not have the pumping capacity to meet your process needs. Most of the document can address the facts and figures of your process problem. You are the expert on how the process is supposed to work.

An expert in a particular subject can go into great depth on it. However, the depth of coverage should be directly related to the intended purpose and objective of the document. Too much depth and unnecessary technicalities should be avoided, if at all possible. If it is not necessary, it may bore readers and end up eliciting a negative reader reaction. You need to decide on the appropriate level and depth of the technical presentation for the stated purpose and objectives.

Level of Details

Like the issue of depth, a writing strategy must address the level of detail that needs to be included. If you are writing about an evaluation of five different molding machines that are under consideration for purchase, you can include the five pages of specifications on each machine. Then you can write a comparison on each item in the specifications, or you can list significant differences and limit discussion to these differences. As described in subsequent Chapters, concision is a desirable attribute in technical writing. Detailed descriptions of processes, designs, experiments, and so forth can make a document too long. On the other hand, archival journals require test procedures that are repeatable by others. For these instances, you need to include enough detail so that a reader can duplicate and check your test results. In making the decision of level of detail, it is usually better to err on the minus side.

RULE

Do not bore your reader.

4.3 Purpose and Objective

Most documents are written for a number of reasons, and a definition of these reasons is an essential part of writing strategy. A report on an investigation is written to present the results of the investigation. This is the purpose of the report. However, the body of work was sponsored by the National Science Foundation (NSF) [*or whatever*] to find a cure for drooping eyelid. This is the objective of the work. All documents have a purpose and an objective, and these are different.

[*A number of reviewers of this book had a problem with my insistence on a purpose and objective as elements in introductions. They said that purpose and objective mean the same thing. I looked up the dictionary definitions for purpose and objective in four dictionaries of different size and age, and all four listed goal, intention, purpose, and objective in the list of synonyms. So, the reviewers were right. Dictionaries and many people use these words interchangeably, therefore, we need our own definitions in technical writing that discriminates purpose, (n) in technical writing, the stated intention of the document, and objective, (n) in technical writing, the stated outcome of the body of work reported in a document.*]

Now that purpose and objective are defined in the context of technical writing, let us discuss how they fit into writing strategy. The objective of most engineering projects is increased company profit—money. The objective of most scientific studies is a finding that adds value to the sponsor or benefits the world. Objectives are the long-term reasons for writing. Almost always, the objective is determined by the sponsor or funding organization.

Purpose usually determines strategy. Purpose is more immediate. You want something to happen as the result of your written document. You may want more funding. You may want to present the status of a project, to present final results, or to introduce a problem or new design.

Whatever you want to achieve by writing a document, the definition of purpose is needed to establish the necessary scope. As shown in Chapter 6, "Criteria for Good Technical Writing," purpose and objective can be a part of a document introduction. In addition to the stated purpose or purposes, you may also have a subliminal purpose such as a raise for your good work or to get your name better known. These subliminal purposes should stay that way. Do not let them show through in your writing.

4.4 Writing to Various Readers

The following list indicates the types of readers likely encountered in a technical writing situation:

Active participants

- Customers
- Teammates

- Peers
- Immediate supervisor

General interest

- Customer management
- Your management
- Potential customers
- Library

Public

- Technical journal
- Handbook
- Presentation
- Book
- Trade magazine

This categorization is based on anticipated interest level. The people involved with the subject of a document will probably have a high degree of interest. They are likely to be familiar with the technical terms you would like to use.

Chemical Engineering magazine is written in chemical engineering jargon. The publishers do not define what a reactor or distillation column is. They use verbs like catalyze, smelt, and hydrogenate without defining them. This is accepted practice. This is what the readers want. If all the articles were written for non-chemical-engineers, they would be twice as long because of definitions and explanations, and the chemical engineers would find them boring. They would stop subscribing to the magazine.

Writing a technical paper is similar. If the paper is to be published in a mechanical engineering journal, freely using mechanical engineering terms (stress field, resonant frequency, Hertzian contact, etc.) without defining them is allowable. The caveat to this rule is if the work is to be published in a journal that does not specialize in your field. If a mechanical engineer publishes a paper in a general journal like *Scientific American,* writing at a different level will be necessary. Others than mechanical engineers read this journal.

Interested readers are likely to want more details than general interest readers. The technical depth level of the writing should match the background of the readers. Table 4.1 shows the readership and writing level for a variety of writing tasks. This illustration suggests that technical writing for publication should be written at the graduate engineer or scientist level, but a technical report may be written to production management or even a safety department. In these instances, it is necessary to mentally examine each person on a distribution list and rate their knowledge of the subject.

Table 4.1   Readership intention and writing levels for various documents

Document	Intended readership	Intention	Writing level
Daily newspaper	Everyone who reads	Profit for shareholders	High school graduate
Time magazine	Adults	Profit for corporation	College graduate
Chemical Engineering magazine	Chemical engineers	Profit for corporation	Chemical engineers
Ph.D. thesis	College examinations board	Get degree	Technical journal
Research proposal	Reviewers and fund administrators	Research funding	Technical journal
Technical report	Suppliers of funding, department requesting work	Solve problem	Lowest reader level
Resumé	Companies with job openings	Get a job	High school graduate
Paper for publication	Researchers in the field	Share technology	College graduate in field
Test report	Customer supervisor	Answer question	Lowest reader level
Bicycle assembly instructions	Worldwide buyers	Get the bike assembled	No manual skills

After doing this, write to the lowest level of the person/persons who really need to understand and take action as the result of reading the document.

The following illustrates various technical depths in writing:

Management Level

I will be on vacation most of next week. If any situations arise that require action from me, please see Barb. She will take care of things in my absence.

Correspondence Level

I am writing this letter to inform your company of a defective component on your AO 102 gas-fired hot water heater. After only ten months service, the plastic drain valve spontaneously broke in half, flooding my laundry room and office. I had the failure...

Technical Report Level

One of the most important applications of plastics is plain bearings. A suitable plastic for this type of application will have low wear rate. It should not abrade the shaft that runs in the bushing. Two plastic tests are commonly used for ranking materials for this type of application, the thrust washer and the bushing test [3, 4]. Both tests require significant contact area of the plastic to a metal counterface, usually a low-carbon steel in the annealed condition, ...

The following illustrate differences in details in writing:

Many Details

The paperclip friction test starts by placing the sample on the inclined plane. Wear rubber gloves and handle film samples only on slit edges. Avoid breathing on the sample since it may have a hydrophilic surface. Fasten the sample with the spring clips provided on the inclined plane. Place the paperclip rider on the test surface about 5 cm from the hinged end of the inclined plane. Raise the inclined plane with a slow continuous motion until the rider starts to move. Stop vertical motion and lock the inclined plane at the breakaway angle. Read the inclined plane angle to the nearest ...

Few Details

The paperclip friction test imposes the end of a weighted paperclip rider on a film sample affixed to an inclined plane. The inclined plane is raised to produce sliding of the rider and the tangent of the angle of the inclined plane at breakaway is the coefficient of friction.

Neither style is wrong. The level of details, number of subjects, and the technical depth need to match the anticipated readership. If a readership is predominately general interest, the detail level should be low. If a paper is written for a journal, the detail level should follow that required by the journal. As discussed in Chapter 10, "Formal Report—Writing the Body," the work performed during a scientific or technical study needs to be separated in sufficient detail to allow others to report the work. The rationale of this approach is to allow confirmation of new theories or models.

Fog level can be varied with the readership. However, it is really not necessary to write this way at any level. The following example from a company report shows this.

Original Form

Increasing exposure to ultra violet (UV) created a surface anisotropy that is manifested in reduced SFM pull-off forces and lenticular surface tendrils. The lenticular surface is enhanced by selective removal of crystalline and amorphous domains during oxidative plasma etching. The lenticular pattern is probably due to alignment of molecular chains in the drafting operation. Spherulites in polyethylene terephthlate are normally in the range of 3 to 7 μm while the surface tendrils were orders of magnitude larger than the typical spherulite size. Thus the conclusion that they are not the result of oriented crystallization...

Translated

Exposure of polyester sheet to ultraviolet radiation produces surface degradation characterized by microscopic ridges aligned with the long axis of the sheet...

The person who wrote the original example was a research chemist, and the words and style used are appropriate for a readership of peer chemists. However, the document from which this was extracted was sent to a large general interest distribution list. The writer may have wanted to impress these general interest readers, but their reaction may be more like, "What is this jerk talking about?" The message here is to pay attention to fog level. Avoid words like "stereospecificity" and "autogenous" if the document will be circulated to in-house readers who do not understand these terms. There is no fog limit for technical journals, but keep in mind a reviewer for a journal or other public domain medium may not work in your technical specialty. He or she could reject your writing on fog level.

RULE

Write so that all intended readers can understand.

Some readers may not be receptive to your written document because of attitude problems. A reader may dislike you, your department, the subject, or even reading. If you are aware of any of these attitude problems, deal with them in your writing or in your distribution list. Do not send your document to a reader who will just throw it in the waste bin. If faced with a bias from a particular individual or department, you can sometimes address it in your writing. If a particular department is known to have a bias against

beryllium copper because breathing particles of this alloy can be a potential health hazard, show that you did a thorough study to address this issue.

Addressing a Bias

After completing extensive laboratory testing, we concluded that the optimum material for the 102 Cassette latch is beryllium copper, CDA alloy 190. We are well
aware of the health concerns with this material, so we performed life tests on 10
latches. We removed all particles from the ambient air above the latch with a particle suction device during the test, and we used the same device in the cassette
to remove any particles that may have been generated in opening and closing the
latches. We sent the vacuum device filter for chemical analysis (ICP) and it was
determined that particulate levels were less than 0.1 parts per billion compared
to the 2 parts per million allowed by EPA standards.

These particulate tests demonstrated that the risk of generating beryllium copper
particles in this application is nil. We tested the cassettes to 10^7 cycles and the design life is only 10^5 cycles. Based upon the laboratory screening tests and life tests,
we recommend immediate conversion of 102 cassette spring to CDA alloy 190.

Without these extra tests and words in your document, your recommendations may never have been given serious notice.

This is an extreme example of a problem with reader attitude, but one is
likely to find attitude problems that could obliterate a message. Try to anticipate attitude problems and deal with them in the document strategy phase.

Summary

This Chapter is intended to promote the definition of a writing strategy for
every technical writing task. The basic elements of your strategy should be:

- Thoughtful consideration of who you want to read your document
- Establishment of a document scope and content boundaries (what you
  want to cover)
- Establishment of the purpose (immediate intention) and objective (ultimate outcome)

In most instances, you must make decisions on all these at the same time.
This is really "concurrent engineering." Defining the readership is the key-
stone of writing strategy. If you just want to inform your boss of the status
of projects, he or she is the reader; write accordingly. In writing to get funding for a project, there is a greater challenge. Some management readers
may not know anything about your technical specialty, and writing may need
to include some tutorial information to show the importance of the specialty.
Before writing to establish a discovery as an intellectual property, one must
decide how to do this. If a patent is decided on, write a patent document in

legal style. If public disclosure is chosen, write to the journal readership. The readers you write to must be consistent with your intentions.

Figure 4.3 is a checklist of questions that should be answered in planning a writing strategy. This checklist may serve as a guide in planning a technical writing task.

Checklist for Writing Strategy

1. **Why are you writing this document?**

   a) Personal reason _____

   b) Purpose of document _____

   c) Objective (end result of work) _____

   d) Other reasons _____

2. **Who are the intended readers?**

   a) Public _____

   b) Immediate supervision _____

   c) Management _____

   d) General interest _____

   e) Other _____

3. **Who is the primary reader?**

   _____

4. **What is the appropriate technical level?**

   a) peer _____

   b) management _____

   c) journal _____

   d) other _____

5. **Are there any reader attitude issues (bias, trust, credibility, etc.)?**

   a) with a reader _____

   b) with the company _____

   c) with the subject _____

6. **If the answer to Question 5 is "yes," how will these be dealt with?**

   _____

7. **What subjects are to be included?**

   _____

8. **What is the appropriate level of detail?**

   a) repeatable by others _____

   b) peer level _____

   c) other _____

9. **What is the appropriate document type?**

   a) letter _____

   b) informal report _____

   c) formal report _____

   d) proposal _____

   e) paper _____

   f) other _____

10. **Is the document to be published?**

   _____

11. **If "yes" to 10, do you have the author's style guide from the targeted journal?**

   _____

12. **Is the document to be archived?**

   _____

13. **If "yes" to 12, how will this happen (copy to library, etc.)?**

   _____

14. **What is the anticipated value to the reader?**

   _____

15. **Does the document address sponsor issues?**

   a) meets project objectives _____

   b) tells sponsor current status _____

   c) part of job _____

   d) other _____

Fig. 4.3 Checklist for writing strategy

Important Terms

- Readership
- Strategy
- Fog level
- Writing level
- Scope
- Attitude
- Intention
- Purpose

- Detail level
- Distribution list
- Objective
- Technical depth
- Subliminal purpose
- Organization
- Peers

For Practice

1. You are going to have a garage sale. Make a readership list for the notice.
2. You are a computer scientist, and you just wrote a program to calculate budget implications of production scrap rates. Who would you send the report to?
3. What is the intention of a report showing a budget overrun?
4. You are creating an announcement for a technical program for an engineering organization. What writing level do you use?
5. What is the correct way to deal with subliminal messages in a technical report?
6. You are a production manager in the organization of Fig. 4.2. You want to write a budget status report. Who do you send it to (readership)?
7. Write a reader profile for *Sports Illustrated* magazine. Share with other class members and develop a composite profile of the five most important reader characteristics.

To Dig Deeper

- J. Allen, *Writing in the Workplace,* Allyn & Bacon, Boston, 1998
- P. Anderson, *Technical Writing: A Reader-Centered Approach,* 3rd ed., Harcourt, Orlando, 1995
- C.T. Barosow, G.J. Alred, and W.E. Oliu, *Handbook of Technical Writing,* St. Martin's Press, New York, 1993
- C. Kostelnick and D.D. Roberts, *Designing Visual Language: Strategies for Professional Communicators,* Allyn & Bacon, Boston, 1997

CHAPTER 5

Document Options

CHAPTER GOALS

1. *Describe the types of reports and documents that are part of technical writing*
2. *Understand how to select a document type for a particular task*

ONE MAJOR GOAL of this book is to simplify technical writing and to make it easier and less onerous for writers. There are numerous ways to communicate ideas, facts, and other business information, and a key part of writing strategy is to decide on the type of document to be written. Therefore, the purpose of this Chapter is to define and describe some document options that are available. This Chapter will try to make you familiar with differences between different documents. Our objective at this point is to present enough information to guide the writer in selecting a type of document. Each serves a different purpose. The details on how to write these documents are discussed in subsequent Chapters.

A fundamental problem in lean business and industrial organizations of the 1990s was that few companies had editors to monitor the consistency and appropriateness of writing in an organization. [*Before downsizing, my company had people whose full-time job was reviewing and editing technical reports and material for presentation outside the company. The last of these people left in 1998. There are no technical editors reviewing our writing now. There is no one enforcing writing consistency or correctness. New hires write documents like they did wherever they came from. One of our new engineers came from an aircraft company. He writes documents that resemble government specifications. Our new engineer fresh from graduate school writes a combination laboratory report and thesis style, all electronic. He composes on the computer, imports digital photos, and in general writes in a way that resembles a thesis.*]

The problem with writing in any style and format is that such writing might appear disorganized to customers and clients (readers). Some documents may be of poor quality because important sections like test procedures are omitted; others may burden the reader with unnecessary detail. Every organization should have agreed-to document options. These save incredible amounts of time and money by eliminating ineffective communications.

Most technical writing in industry can be categorized as either formal or informal writing, as shown in Fig. 5.1. Formal technical documents usually have the following attributes:

- Refer to work of others
- Are written with a specific format
- Contain information with long-term value

Informal documents are essentially documents of lesser importance, significance, or long-term value. Table 5.1 lists typical life examples. In summary, formal documents report information of long-term value, while informal documents are for most other writing tasks.

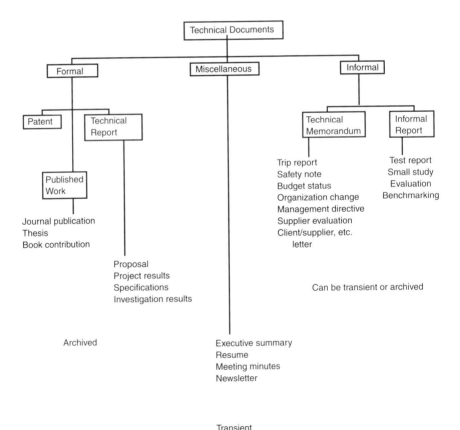

Fig. 5.1 Spectrum of technical documents in industry

Table 5.1 Probable useful life of various documents

Document	Probable life, years
Formal documents	
Results of an experiment/investigation	5
Results of a failure analysis	5–20
Machine design outline	5–20
Article for an archival journal	Indefinite
Patent	Indefinite
Market study	1
Product proposal	1–5
Business proposal	1–5
Health/safety issue	1–5
Informal	
Organization change	1
A strategy	1
Product evaluation	1–5
Test report	1–5
Benchmarking	1–2
Design/product specification	1–3
Yearly budget	1
Business letter	1

This Chapter describes various types of informal and formal technical documents and their hierarchical importance. Each document type serves a different purpose. Writers should be familiar with these differences. The objective at this point is to present enough information to guide the writer in the selection of a document type. The focus is on formal and informal technical documents. Newsletters, resumes, and other documents are classified here as miscellaneous documents and are discussed in Appendix 8. Technical writing of manuals, handbooks, and books are outside the scope of this book, as stated in Chapter 1, "What is Technical Writing?" The major types of technical documents listed in Fig. 5.1 are described. Subsequent Chapters go into more detail on the elements of formal and informal reports.

5.1 Document Hierarchy

A document hierarchy exists in most technical professions. Facetiously, one could say that a resume is the most important document of any career, because it is used to obtain a job. However, this book does not describe resume writing, because this subject probably requires a book by itself. Many standard guidelines for resume writing are also available. [*The subliminal reason for not dedicating much space to the resume writing in this text is that I guided my three sons in writing their resumes early in their careers, and they all blamed their unsatisfying first engineering jobs on my resumes. I no longer advise them in this area. Since I have only worked for four employers in my 42 years in industry, I am hardly expert at writing resumes.*]

Assuming that your resume worked and you have a technical job, whatever the field, your employer will have a hierarchy for written documents; some documents are deemed to be more important than others. In academia, a department report on a body of work is probably the least important. Publishing this work in a trade journal is next in value, followed by publication in an archival technical journal.

In an industrial organization, the report hierarchy probably correlates with the importance of the work and the value of passing the information to others. A letter to a vendor on a $500 item probably has neither, but if the letter is on a worldwide contract for supplying link pins for each four cylinder engine made, the importance is obvious. However, there probably is no value in sharing this information with many others, so it would not become a technical document for sharing.

Figure 5.2 is an illustration of typical document hierarchy in U.S. industry. The hierarchy may be different in other businesses, but a hierarchy will be present. Awareness of this hierarchy is necessary in selecting a document type. One may look foolish trying to write a formal report on a routine material selection problem. Similarly, sending an e-mail to teammates on an important development is not adequate documentation. [*At one time, our laboratory came under the control of a research and development organization. They wrote formal technical reports to conclude year-long studies. I started to write formal technical reports on some of our normal "part failure" types of studies, and I was duly informed by the division manager that*

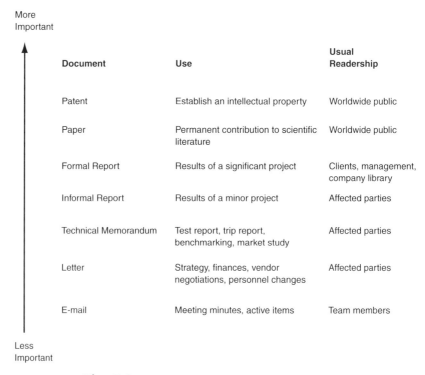

	Document	Use	Usual Readership
	Patent	Establish an intellectual property	Worldwide public
	Paper	Permanent contribution to scientific literature	Worldwide public
	Formal Report	Results of a significant project	Clients, management, company library
	Informal Report	Results of a minor project	Affected parties
	Technical Memorandum	Test report, trip report, benchmarking, market study	Affected parties
	Letter	Strategy, finances, vendor negotiations, personnel changes	Affected parties
	E-mail	Meeting minutes, active items	Team members

Fig. 5.2 Typical document report hierarchy in industry

I should use a "lesser" report form. An informal document (lab report) was more appropriate. Management wants the appropriate vehicle to be used.]

Think about your message when you start to write a document and decide where this document belongs in the hierarchy of technical documents. Selecting the wrong medium (hierarchy) could have a negative impact on writing purpose.

RULE

Establish the appropriate document hierarchy in any writing task.

Technical Memoranda. In engineering and the basic sciences, most projects and studies are concluded either by a formal or informal report. However, some writing tasks do not fit into these categories of reports. This type of document is termed a "technical memorandum" in Fig. 5.1. These documents generally pertain to business or ancillary aspects of a project or investigation.

A trip report is a good example of a technical memorandum:

Example of Trip Report

TO: John Smith, Shops Div.

FROM: K. Budinski

SUBJECT: Trip to PVD Coating Company

On October 22, I made a trip with Mary Cordow to the Belfast Coating Center in Buffalo, NY. We toured the tool coating center, but the main purpose for our trip was to assess the coating capabilities of their new continuous components coating facilities. We are considering a lubricious PVD Coating for 29 mm P20 shutter blades to address a sticking problem.

The new components coater costs about $6 million. It reduces processing time from 4 h to 1 h, and the racking system can accommodate 200 kg of parts. Belfast Coating estimated that they could apply a 1μm thick coating of MoS_2 on 6000 shutters per hour. This would mean a per piece cost of 11 cents.

This tour convinced us that Belfast Coating is capable of depositing a thin film lubricant coating on all production units, and we recommend a trial coating of two production lots for the June production run.

The readership and retention of this kind of document depends on the situation. A trip report may or may not be important, depending on what was learned, but this kind of document can be useful to share. Not everybody could go on the trip. Sharing your experience gives others the benefit of the trip. These documents are usually distributed to peers, and they are usually deemed low in the hierarchy of technical documents.

Safety notes are a form of technical memorandum used in many manufacturing organizations. They may be necessary to stave off litigation when accidents happen. In the United States, employers must inform employees of health and safety hazards, and products must be labeled to warn consumers of risks.

Safety Incident

TO: J. Obrien, Laboratory Center

FROM: S. F. Tee

SUBJECT: ATD incident

The ATD incident involved making up a metal etching reagent using picric and other acids. The chemist went to measure the prescribed amount of picric acid when she noticed that the liquid in the bottle had turned to a solid. It crystallized. Fortunately, she knew that crystallized picric acid becomes explosive and it can be set off by shock such as jarring the bottle. The city bomb squad was called, they removed the crystallized picric acid, and exploded it in a remote area.

Please survey your chemical storage areas for picric acid. If you have any, do not touch it. Call *XXXX* and a safety department representative will come to you to examine the material for signs of crystallization. Thank you for your cooperation.

This kind of document may be archived to show that chemists were duly warned of the potential risks of a particular chemical.

Budget, organization, and management directives are another part of technical writing, like it or not. They are necessary to make the business run.

Management Directive

TO: CP&R Department

FROM: F. Howell

SUBJECT: Personnel Change

Effective 11/10/99 T8/T9 technicians will be allowed to apply for advancement to Code 39-41 engineer/technologist. Interested technicians should fill out on-line form PMB 133 by 9/20/99. This change is being made to make company technicians' salaries competitive with their counterparts in peer industries.

The purpose of these kinds of documents is almost always to share job-related information with the workforce.

Letters to suppliers, clients, peers, and so forth are almost daily tasks. These are usually executed on e-mail. The retention time for these documents in servers is often dictated by company rules on electronic data storage. Some are definitely transient in nature and can be deleted from electronic files on a regular basis.

Supplier Letter

To: Roberts; Rose www@Roseel.com

From: R. Schwartz, Inspection Center

Subject: Motor/Controller Selection

Confirming our phone conversation of 11/10/99, we would appreciate a quotation on price and delivery on a direct drive DC motor and speed controller with a torque minimum of 900 inch-pounds in the speed range of 200 to 600 rpm. The motor needs to have flange mounting around the drive spindle and clockwise shaft rotation. The controller should have a minimum of an on-off switch and a rheostat for speed control. Your quotation is required by 11/20/99.

Thank you.

The kind of letter in the previous example probably does not need to be archived or shared with other readers. It is simple and to the point, and it is part of technical writing. These kinds of work letters are part of every engineer's and scientist's job. They are part of the informal writing aspects of the profession.

Informal Reports. Laboratory reports are generally a good example of an informal report. A report on bolt failure may have involved some laboratory work. A laboratory testing report is written. This document is probably more widely circulated than the trip report, and it may be more significant. If the failure occurred on a key piece of equipment, report distribution should include the managers involved, and the document may be archived for a number of years. This is still an informal report (Fig. 5.2), but it is a step above a memorandum (like a trip report) in corporate importance. Appendix 1 is an example of a laboratory report.

Formal Technical Reports. A formal technical report is the gold standard of technical writing in industry. In the United States in the 1990s, the typical cost of supporting an engineer for one year was between $150,000 and $200,000. If an engineer has been working on a major project for a year, managers surely will ask what benefits the company received from the work. What are the results? What value are the results? A formal technical report is appropriate. This report should be archived in the company library, and it should be complete enough so that the work can be repeated by others.

Published Works. In some organizations, the next step in document hierarchy is a technical paper for publication in an archival journal. Some companies encourage publication; some discourage journal publication, because it may open a channel to leak proprietary information. All organizations should have a formal review policy on anything published at any level by their employees. This includes nontechnical writing.

The review of manuscripts for journal publication should include the writer's immediate supervisor (or higher management at the discretion of the supervisor), corporate editors for writing correctness, and the legal department to ensure freedom from libel and maintenance of intellectual property. However, it is still possible that a "loose-lipped" researcher could divulge company secrets in the oral presentation of a paper.

[One of my colleagues was prone to be loose lipped about company secrets. I would have to nudge him under the table when he would say too

much at vendor meetings. His papers usually gave away too much information. The company started to offer a bonus for patents. He stopped writing papers and concentrated on patents. This is the appropriate medium for divulging secrets. It solved our intellectual property leak.]

The argument in favor of publishing papers is that this protects intellectual property (the 1990s euphemism for company secrets). As mentioned previously, publication in a public media like a journal allows the use of the disclosed technology, even if it is subsequently patented by others.

Patents. The most important document that you can write in many industries is a patent. One does not really write it alone. A lawyer participates, but it is your document. Patents are recognized here as documents with the highest priority and importance, because a patent can be the lifeblood of an organization. In these times, there are hundreds of companies with the capability to dissect and copy (steal) a product. The only hope to capitalize on an idea and product is to have it protected by appropriate patents.

5.2 Report Types and Selection

You are a machine designer, and you have just completed work on a significant project. You want to publicize results so they can be implemented by others in other business units of the company. What kind of report should be written? There are at least five options:

- Patent
- Technical paper
- Formal technical report
- Informal report
- Technical memorandum (letters or e-mail)

Ask yourself which of these options best meets your needs. The following is an example of a situation where a formal report (see Appendix 2) was selected for proposing a new concept for a face seal that was causing significant maintenance problems. An informal report of the same project is given in Appendix 3.

Example of Selection of Report Type for an Investigation of Face-Seal Material

The project that you just completed was intended to solve a manufacturing problem. There was a chemical process in the laboratory that used a face seal to prevent leakage from a vessel at the agitator feed-through (see Fig. 1 in Appendix 2). One of the causes of leakage was sticking of the spring. A department engineer had a clever idea to eliminate the spring. He made the web on the seal very thin and used flexure of this web to provide the spring load on the seal. This spring load is necessary to accommodate the waviness and wobble that exist in the system due to machinery tolerances and tolerance buildup. This system was put into production, but the hardened contacting surfaces did not provide the desired service life of 500 h. They wore significantly and leakage started at about 200 h.

You were given the task of finding a new, more wear-resistant material of construction for the seal members. You conducted laboratory wear tests and determined that a type C2 cemented carbide seal ring should provide the required 5× improvement in seal life. You had a prototype made up by an outside supplier. The rubbing surface was covered with carbide to a depth of 1 mm by a proprietary brazing process. The carbide deposit cracked and developed a rough edge in grinding.

The hardfacing process did not work. Do you want to let others know this? You need to decide which type of report to write so the appropriate people know that your proposed fix to the problem did not work. You did not get positive feedback from management on a patent for the coated flex seal, but you may want to write another type of document (such as a technical paper for publication, a formal technical report, an informal report, or a technical memorandum by e-mail). Because the carbide coating did not work as planned, it was felt that there was not sufficient reason not to write a paper on the test results. In addition, your company was not in the best of financial straits, and it was discouraging travel to attend technical conferences to present papers. You decide on a formal technical report.

Each type of report option described in this Chapter could be used to present the results of the seal project and to ask for more money. None is wrong. They present different levels of detail, and they will have different effectiveness in meeting the purpose of reporting on the status of the project, the conclusions, and recommendations.

RULE

Decide on the type of report before you start writing.

Patents

Patents, a very special type of technical document, are the ultimate examples of archiving; they have unique patent numbers and are permanently on file in the country in which the patents were issued. Patents can be good for 17 years. They are available to all, and they are kept available for searches indefinitely.

Every organization should have a strategy on when to apply for a patent. One company, Bud Labs, has adopted the strategy to only patent inventions that are in some way associated with the corporate mission statement (which is to be the worldwide leader in the development and manufacture of tribology equipment). Writing a patent for a device or process is an important option to consider, but make sure that it is in keeping with company strategy and meets patent criteria. A patent lawyer should also be contacted if a patent appears feasible and in the best interest of an organization.

In the example of the hardfacing seal, the seal coating cracked, and so it is not appropriate to submit the work for a patent. In addition, in large companies with patent attorneys, the cost of patenting this type of work would be about $30,000 in the United States. Patent costs and time are also a consideration with this report option.

Patentability is another basic factor. An invention can be patented only if it meets three requirements:

- It must be new.
- It must be useful.
- It is not obvious.

You can check to see if something is new by conducting a search of previous patents. A search is a government requirement to get a patent, and patent searches can be done via the web site of the United States Patent and Trademark Office (www.uspto.gov).

You also must demonstrate that your invention is useful. It must work and produce some desirable effect. Obviously, if a mechanism or process does not work, then a patent is unnecessary. The leaky face seal did not work, so the option of writing a patent is not relevant.

Finally, a patentable invention must be nonobvious. This means that you propose a unique outcome that is not predicted from common experience. For example, energy efficient homes often have light-switch sensors in the wall switches that turn the lights on when somebody enters the room. You design a light bulb that does the same thing. It is not obvious that a light bulb will sense a room occupant and turn on. It is useful, it reduces the cost of energy by a switch sensor, and it saves energy. It is also new, because this product does not currently exist.

Technical Paper

A technical paper is a document written for publication in an archival journal. A key word in the definition is "archival." If published, the work becomes a part of permanent literature, which is kept in some libraries indefinitely. For work that will be of value to others years from now and that adds to cumulative knowledge, select the technical paper report option.

Technical papers for archival journals generally have a standardized style and format. Most journals have essentially the same style and format requirements, but all have their own unique instructions for submission of a manuscript. When you read a technical journal, you know that the article starts with an abstract, then an introduction, and so on. All papers follow this format, so that readers receive a coherent presentation. The reader may not agree with what is written, but it will be clear and understandable. This is because papers in archival journals have a common format and are reviewed by three or more technical peers and the publication editors.

Selection of a specific journal may also require choosing between several journals. In tribology, for example, the journal choices could be:

- *Lubrication Engineering*
- *Journal of Tribology*
- *Transactions of STLE* (Society of Tribology and Lubrication Engineers)
- *Tribology*
- *Tribology Letters*
- *Wear*
- *Transactions of ASME* (American Society of Mechanical Engineers)

Archival journals also have a hierarchy. Workers in the field often have an opinion on the rankings in prestige to these journals. If you are a professor of metallurgy, the department chair probably has an opinion on which journal is the most prestigious.

A paper for publication was probably not appropriate in the study of the face seal. The work was not finished, and the author is recommending additional funding. This is not appropriate in a published paper. In addition, papers are intended to share a new discovery or to add to the body of knowledge in an area. This work was not fundamental enough for most journals. Papers usually propose models and explain in great depth what transpired in a particular test. Why did carbide self-mated perform better than the other couples? What is the mechanism for improved life? Most organizations have guidelines on publishing technical work. These need to be reviewed to guide writing decisions.

Formal Technical Report

A formal report is an archival document that contains technical information of potential long-term value. The essential elements of formal reports include an abstract, introduction, middle (procedures, results, discussion), and end (conclusions and recommendations). Formal reports should also include citation of publications with supporting technical information.

Appendix 2 is an example of a formal technical report on the seal project. It was distributed to the directors and several members of a customer division as well as to potential users in other divisions, the supervisor, two peers, and to the report library (for a five-year retention period). This job is now documented. The results were distributed to all parties concerned. If other people work on the seal-wear problem in the future, they will know what was done already and why a brazed carbide overlay did not work.

The formal report presents significant detail and is probably the more appropriate report option in most organizations where these kinds of laboratory studies are conducted. It is not too long for busy people to read, and the work is in the form for archiving. Others will be able to review this work five years from now and know what was tried, what worked, what did not. The informal report options may work in some organizations.

The report also recommends additional funding. Some managers may give the author another $40,000 based on a few e-mail sentences, but many will not. The author must make the decision. He or she knows the "mood" in their organization. An informal report is not recommended for what appears to be a significant project with significant cost.

Informal Reports

All reports that are not technical papers or formal reports fit into the category of informal reports (Fig. 5.1). Most of these documents concern temporal matters, and they are considered to be transient documents. Documents with an anticipated life of a year or less are termed transient; they are deemed to have no long-term value and do not need to be saved.

Most informal reports are not archived. However, some informal reports are of such a nature that they may need to be archived for five years or more. For example, a product evaluation report may be valuable for a number of years (Table 5.1). It may be on a product that is competitive with one of yours, and such information may have long-term value even though the document was very brief.

Product Evaluation

TO: R. Crisp, Poly-Baryta Div.

FROM: B. Gull, Materials Engr. Lab

SUBJECT: Evaluation of Onbracko Polyloc Cap Screws

The poly-baryta folio line was experiencing product width problems due to continual loosening of #2M × 6 cap screws that hold the cutoff knives. The conventional screws were replaced with #2M × 6 cap screws with an axial plastic strip embedded in the threads (Onbracko Polyloc). The thread interference produced by compression of the plastic insert is supposed to prevent loosening of the screws.

After three months of testing, it was determined that the use of locking screws did not solve the problem. It has been decided to test 12 point head screws made from a special high-strength alloy. These screws will have a higher seating torque and 25% higher bolt tension.

Things that have been tried in the past have a way of coming around again. There could be value, thus, in saving this informal document for as long as the subject machine is still used.

Informal reports have the same basic structure of a formal report with a start, middle, and end, but are briefer with less detail. Many times in informal reports, the required elements may be adequately covered in only one sentence. Purpose and objectives may be implied in an informal report, and the methods of testing or analysis are not described in detail for replication of the supporting tests or analyses. The following is an example of an informal report. In fact, it is really an abstract, an abbreviated account of the work.

Informal Report

A study was conducted to determine the course of stray arc defects in cutting dies with electrical discharge machining (EDM). Arc spots on the polished die surface render the tools unacceptable for use.

The laboratory study consisted of measuring the electrical conductivity of candidate dielectrics using ASTM P-143 procedure. Test cuts were made with high- and low-conductivity dielectrics, and it was determined that high purity water significantly reduced the amount of defects compared to the current dielectric, column-deionized water.

The recommendation resulting from this study is to eliminate the ion exchange columns and use triple distilled water as the cutting dielectric.

The preceding informal report has an introduction, body, and end, but they are only a sentence or two. A formal report would include all details. In a formal report, the introduction would contain a better description of the die defects, what causes them, what they look like, and so forth. Informal reports on problems and investigations have a definite format, and these documents are discussed in more detail in Chapter 12, "Informal Reports."

Another example of an informal report is in Appendix 3. This report is for the same seal project described in the formal report of Appendix 2. It contains fewer details and has a different purpose, but it still presents the basic conclusions, recommendations, and overall objectives of the study. In the investigation of the hardfacing seal, an informal department report could have been done. This type of report may not need supervisor approval. It is not archived in the company library, and the distribution is usually only to a few people.

E-Mail Messages

The lowest document in the hierarchy of reports is an e-mail note. It is transient. It can be saved, but in the computer world of 2000, its longevity depends on the cycle of the next computer upgrade. [*I was given my first corporate computer in 1985. By decree, my system has been completely changed five times: new computers, new systems. Data from previous systems are gone as are many pieces of software that I liked and needed.*]

The incompatibility of computer systems is probably the major disadvantage of recording work strictly in electronic format. Another disadvantage over a hard copy report is that graphics, photographs, and other supporting material may not be available in electronic format. This is becoming less and less a problem, but complete electronic distribution may take more time for preparation than hard-copy reports. Depending on the available capabilities, scanning of photos may be very time consuming; in addition, some intended recipients of a report may not have the appropriate software to receive attached files with photographs and other data.

However, sending reports as e-mail attachments is common, and it is a viable alternative to sending hard copies of reports to a distribution list. Thus one can write a formal or informal report and send it as an e-mail attachment or just send an e-mail note asking for the same funding as the longer reports:

To: R. Swartch, K. Lee, J. Fargo, R. Primus

From: K. Budinski, Materials Engineering Laboratory

Subject: Leaky gel reactor seals

The gel reactor seals in B4L3 are leaking and producing production losses (over 200K for 1997). The Materials Engineering Lab was asked to test other seal materials in the laboratory and to arrive at a seal couple that produced longer life.

Laboratory metal-to-metal tests identified six material couples that produced better wear resistance than the current seal couple. A prototype seal was made with a new material couple, self-mated cemented carbide, but the carbide on the flexible seal member cracked during fabrication.

With this memo, we are requesting an additional $40,000 and four months project time to fabricate and test another new seal configuration.

Summary

This Chapter describes the thought process that authors should use to make a decision on report type. There are pros and cons to each type. Subsequent Chapters explain how each type of report contains the same essential report ingredients but to different degrees. For example, the purpose and objective in an informal report may be implied, and testing details may be eliminated.

Overall, significant projects warrant a formal technical report. The following are some points to keep in mind in selecting a report option:

- Patents are for protecting significant ideas for new inventions of useful and unique devices or processes.
- Patents cost significant amounts of money and time.
- Papers in archival journals require in-depth treatment of a subject.
- Papers provide worthwhile peer review, but require significant effort on the part of the author.
- There is a hierarchy in technical documents in most organizations. Writers need to be conscious of this and select the appropriate document level.
- Formal reports include the details necessary for others to continue or reproduce your work.
- The abstract on formal reports, if well done, will serve as the executive summary for managers that are too busy to read reports.

- Informal laboratory reports are appropriate for minor projects but usually not for significant projects.
- E-mail messages are transient documents and usually are not adequate for documentation of a significant body of work.

Important Terms

- Patent
- E-mail
- Not obvious
- Informal report
- Hierarchy
- Useful
- Archive
- Abstract

- Technical memorandum
- Archival paper
- Executive summary
- Laboratory report
- Formal report
- Embodiment
- Editor

For Practice

1. Dissect one of your better ideas on paper and determine its patentability. Does your idea seem useful and nonobvious?
2. Write an abstract for a paper that you would like to write. Does it have the fundamentals required of a paper?
3. Cite an occasion when an informal report is appropriate—include details.
4. Cite an occasion from your work where a formal report was required.
5. Take an e-mail report that you have received and critique it for appropriateness for the situation.
6. Write an e-mail report on a recent project/lab assignment.
7. State five pros and five cons for writing a formal report on a major project.
8. What is the difference between a formal report and archival paper?
9. State five reasons why a formal report is more important than an informal report.
10. You are a design engineer who just completed building and installing a dial assembly machine. What type of report would be in order? Who would you send it to?

To Dig Deeper

- M. Market, *Technical Communication,* 5th ed., St. Martin's Press, New York, 1998
- S.E. Pauley, *Technical Report Writing Today,* Houghton Mifflin Co., Boston, 1979
- J.E. Sincler and N.H. Vincler, *Engineering Your Writing Success,* Professional Publication Inc., Belmont, CA, 1996

CHAPTER 6

Criteria for Good Technical Writing

CHAPTER GOALS

1. *Introduce factors that are important in writing a good report*
2. *Understand how your technical document will be reviewed and read by others*

GOOD TECHNICAL WRITING is the overall objective of this book, and the preceding Chapters explain why writing is an essential part of a technical career and why it is in one's career interest to continually document work and be proactive in writing. The remaining Chapters of this book address the specifics of writing documents, but this Chapter summarizes the expectations for good technical writing by readers of this book.

What does good technical writing look like? Figure 6.1 illustrates three major aspects of all technical documents:

- Content
- Presentation technique
- Use of the language

These attributes apply to most types of technical documents. You should master the necessary skills to meet these expectations in good technical writing.

The purpose of this Chapter is to introduce the basic terms and concepts of good technical writing; the working details of good technical writing (for example, how to achieve logical document sections) are described in subsequent Chapters. These basic expectations represent the document attributes

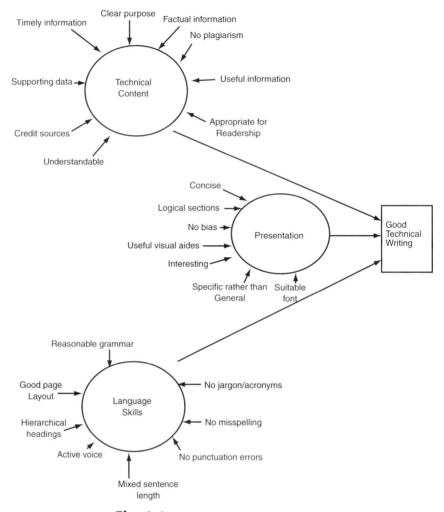

Fig. 6.1 Attributes of good technical writing

that one should master in progressing through this text and practicing technical writing skills.

6.1 Technical Content

Word Usage Appropriate for Readership. The technical content of a document depends on the type of document and the intended readership. Good technical writing dictates that a document is written for a particular readership. This requires an analysis of readers ("Writing Strategy," Chapter 4) and proper choice of words and subject matter for the intended readership. In writing a patent, proposal, or technical report, an author must choose the content and wording properly. A procedure for a local heat treating company can be written with typical words and content.

Procedure for a local heat treater

1. Preheat at 1000 and 1500 °F, 15 to 20 min soak each
2. Austenitize at 1925 °F for 15 to 30 min after part is at temperature
3. Quench in molten salt at 1025 °F; hold in quench until 1025 °F is reached throughout the work
4. Air quench parts to hand warm
5. Triple temper, 2 h each temper
6. Obtain minimum hardness 60 to 62 HRC

However, to convey the same information to a teammate in another country, a writer may need to change the wording to make it understandable to that person:

For a French teammate

Jacques:

The baseline design that was agreed to by the team specifies hardening of the XDBD ways on the crimper to make them like our machine. Please locate a heat treater in your area who can do salt bath hardening. The recommended heat treat procedure requires four salt baths.

The parts should be hung vertically during the heating operations. One salt bath is needed to preheat the parts to 380 °C, a second to preheat to 350 °C, and the third to 1250 °C. These baths raise the parts to the recommended hardening temperature. The parts only need to be immersed in the preheat baths until they reach the bath temperature (about 10 min). The time in the hardening bath should be 15 to 20 min after the parts reach bath temperature (10 min).

The fourth bath should be maintained at 520 °C and the parts are immediately quenched in this bath after the soak in the 1250 °C bath. The parts are held in the quench bath for 10 min and then air cooled until they are almost cool enough to touch. They are then returned to the 520 °C bath for 2 h. After a 2 h soak, parts are air cooled to room temperature. This step is repeated twice more.

The procedure, which is summarized on the attached, will produce a hardness of 60 to 62 HRC.

Both documents contain the same information, but the wording is changed to accommodate different readerships. Also note the use of different temperature units.

Useful Information for Selected Readers. Technical documents must provide useful information that the readers want. The author must determine what the readers want. Authors also have the opportunity to decide whom to send information to. Only send documents to readers who will find value in the information in the document. If a reader asks, "Why was this paper sent to me?" you have sent it to the wrong person. This can often have a negative impact on what you were trying to achieve with your document.

No Plagiarism. Plagiarism is the use of another's work without permission. What constitutes fair use is discussed in a later section, but basically you cannot use any copyrighted material in a document. In 1999, software was developed that can scan college essays for material plagiarized from encyclopedias and similar reference books. Most technical writing is done on computers, and many documents are distributed via e-mail. This means that computer policing of documents for plagiarism may make plagiarism easy to detect. Needless to say, the effect of being caught in an act of plagiarism can be devastating. One stolen paragraph has ended otherwise distinguished careers.

Factual Information/Supporting Data. Putting a statement in a report constituting what a reader knows to be an error in fact can cast the credibility of the entire document in question. Opinions from Alan Greenspan, the Director of the U.S. Federal Reserve Board, can cause significant changes in the U.S. economy, but few young engineers can claim that readers will respond to their opinions.

Most technical writers need facts supported by data to convince readers to take action or pay attention to the message in a document. For example, stating "This is a costly problem" will get less reader reaction than saying "This problem cost the company $147,000 in the first quarter of 1999."

Clear Purpose. A common problem with papers published in scientific journals is the absence of a statement of why the work is being done. Some papers are very scholarly and written elegantly. There is not a fault in the document. An introduction that states that, for example, the purpose of the work is to examine the dislocation density in the deformed surface layers when pure copper is slid on itself under high speeds and high loads. The author previews previous literature on the subject and goes into detail of the tests, deriving mathematical relationships that equate dislocation density with lost friction energy and concluding that the friction energy is the source of the activation energy needed to form dislocations.

So why was this work done? Nobody slides pure copper on pure copper in a machine unless they are trying to weld the two pieces together. It is not a logical engineering practice. The reader ends up with a puzzlement. This is all very interesting, but all science should lead to some benefit. What is it?

A clear, concise statement of the purpose of any document is mandatory. State what you are trying to achieve with a document. Make the purpose and objective clear. Provide information on the importance of the work described in the document. Readers' time is important; they need to know why they should read your document. What is the purpose?

Timely Documentation. As pointed out in Chapter 3, "Performing Technical Studies," many situations should be documented in the average work week of a technical person; you should write something. It is important that documents be completed in a timely fashion. Reports should be completed when the work is done or when action is needed; proposals for funding must obviously meet any submission deadline.

Timeliness is very important for a report on a problem or investigation. Problems are a big part of any industry or business, and most engineers spend a significant portion of their career solving problems. The solution to most problems should be stated in a written document. As in any report on work done, verbal statements can be unreliable. After completing a project or solving a problem, people (like your boss) will ask what the employee learned. You can tell them verbally, but someone else may not get the complete story or spoken results may be misinterpreted. State the outcome in your own words and provide them in documentation in a timely way, as verbal statements change with every telling.

RULE

Make all documents timely.

If you are doing fundamental research that has no timeline, what is a timely document? Even fundamental research needs to be documented as soon as significant results are achieved. Procrastination may cause work or learning to be published by someone else. Scores of examples in the great inventions of the last two centuries show where two people came up with the same invention (like the telephone) at the same time, but the person who was really second to make the discovery got his/her patent or was published first and got all the glory and rewards.

In addition, a more practical reason for timeliness is that you may lose data or forget important details if you do not write your report in a timely manner. [*Unfortunately, this has happened to me more times than I care to admit. When I waited a month or so to write up a project, many times, I uncovered a missing piece of data. We would have to go back and duplicate the setup to get this piece of data. If I started the report immediately, I would have quickly discovered the missing data while the equipment and testing staff were in place.*]

In summary, timeliness is one of the most critical criteria for a good technical document.

Sources of Information Credited. A good technical document gives proper attribution when the work of others is cited to make a point or provide background information. To use another's words without attribution is plagiarism, as mentioned previously, but to disregard citation of pertinent literature is an omission. Few technical journals will publish an article without references to similar or related works, and these references must be properly listed. Details on proper citing of references are described in a subsequent Chapter. The main point is that sources of information should be cited in every technical document.

Attribution of sources does not always have to be in the form of a published journal article.

Informal Attribution

As shown in Fig. 12, the unseating torque for the head bolts varied considerably around the periphery. Measurements were made in the Product Development Lab, and they were carried out with a data-logging torque measuring device.

This attribution recognizes that another group sent some unpublished work that the author incorporated into a document. If these data were used without attribution, readers would assume that the writer generated the data when he did not. Needless to say, people who generated the data would not be inclined to help in future projects. Another way to recognize help from others is to use a contributor line on the document header or in acknowledgments. Both techniques are addressed later.

> Materials Engineering Laboratory
>
> Report No. 203
> TO: David Doe, Product Engineer, Div. INC26312
> FROM: Ken Budinski, Materials Engineering Lab, INC21212
> SUBJECT: Die head bolt tensions
> Date: 11/12/20
> Contributors: M. Kohler

Understandable Statistics. Statistics are becoming a more prominent part of corporate life. This is fine, but this particular branch of science has a vocabulary onto its own. Writers need to realize that these terms are not known to everybody. A good report should explain statistical terminology in words understandable by the intended readers. The following bulleted statements appeared in a management letter sent to 20,000 employees of a large company.

> Not Very Understandable
>
> - The C_{pK} for viscosity-enhancing synthetic chemical was improved to 2.35 (>6-sigma) from 0.56.
> - The DPU in the polyester yarn line was reduced 65%.
> - The Solutions Department MTBF improved 72% to 10.6 h.

This writer, no doubt, was of the opinion that all 20,000 employees should know what C_{pK}, 6-sigma, DPU, and MTBF mean. Some may; most will not. These statements could be written so that the average employee could understand them.

> Understandable
>
> - The specification conformance for a viscosity-enhancing synthetic chemical was improved from 2,000 units per million out of specification to less than 6 units per million out of specification.
> - The number of defects per spool on the polyester yarn line was reduced 65 percent.
> - The Solutions Department went from an operations shutdown every 17 h to 10.6 h.

Acronyms. The ease of understanding a technical document depends on many of the attributes listed in Fig. 6.1. An author needs to present information in a proper way and with a proper use of the language.

One of the most notable and easy ways to ease the understanding of a technical document is to limit or eliminate the use of acronyms. There is a tendency to use acronyms for everything. It is perfectly acceptable to write in technical jargon, but acronyms should be used sparingly. Never create an acronym. Even if an author must use acronyms that have been used in other documents, always define them when they first appear in a document. The only exception may be an informal document that is directed only to the writer's immediate peers. If there is any doubt about who may read the document, define the acronym or eliminate it. This is especially true of formal documents that may have long-term value.

RULE

Do not create your own acronyms; if you must use acronyms, define them.

The best approach is just to eliminate acronyms. It is possible to write in understandable terms without acronyms. Newspapers write with understandability as a prime concern. They only use selected acronyms published in agreed-to guidelines written for reporters. If and when U.S. newspapers go international, they must stop using common U.S. acronyms such as FBI for the Federal Bureau of Investigation and IRS for the Internal Revenue Service because these have a different meaning in other countries. [*In 1997, I had the harrowing experience of being detained by military police on trying to enter Belarus on a visit that was sponsored by the Belarus Academy of Sciences. Afterwards, my Belarus hosts suggested that my entry problems were related to my e-mail address KGB@xxxxx.com, even though they are my initials. Most of us older people remember that KGB was the acronym for the dreaded secret police, Komitet Gosudarstvennoi Bezopasnosti, in the former Union of Soviet Socialist Republics (USSR).*]

6.2 Presentation

The method of presenting information in a technical document is equally important as technical content. Presentation includes all factors listed in Fig. 6.1 as well as overall writing methodology. All these things taken together constitute your style of writing. Chapter 7, "Writing Style," is devoted to the topic. In this section, the critical attributes of presentation technique (or style), are briefly introduced.

Logical Sections. The most important aspect of a technical document is a logical structure for the reader. Many kinds of documents constitute technical writing. They look different and have different purposes, but all documents must be broken into distinct parts, which are placed in a logical order in a clear fashion. Letters start with a salutation; proposals start with background information. Quotation requests need to ask for price and delivery. There are recipes that work well for every similar writing situation. This is discussed later, but basically, a presentation needs to make sense.

Reports are a significant part of technical writing. As subsequent Chapters describe in more detail, formal reports should have specific sections. The basic elements of formal reports and technical papers are shown in Fig. 6.2. Less formal reports may have fewer sections.

Some word processing software has templates for reports that essentially contain the same elements. [*In 1998, I contributed a paper in a proceedings published by the American Chemical Society (ACS), a very large technical organization in the United States. The organization sent me a disk to put my paper on. It contained the format and the fields for all the report elements. This ensured that all contributors used exactly the same format, heading hierarchy, font, white space, and so forth to allow printing and editing the documents electronically as received. This is called a camera-ready document. Previously, camera-ready documents had to be laboriously typed on extra-size pieces of paper called mats. These came with about ten pages of instruction on presentation technique. The ACS system was a real pleasure to use. It even contained sample headings to overwrite with yours. The document fields were title, authors name/affiliation, introduction, experimental, results and discussion, conclusions, acknowledgments, and references.*]

Large firms may have similar templates for formal reports. They invariably have the basic sections in the preceding list or in Fig. 6.2. [*Many years ago, when I worked for General Motors, they used the sections of Fig. 6.2, but the conclusions were placed after the introduction. The rationale was that the readers were mostly managers, and they were too busy to read the whole report. They only needed to read the introduction and conclusion to have complete grasp of the content.*]

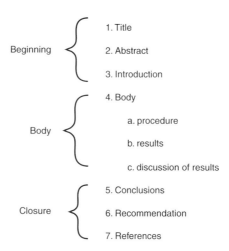

Fig. 6.2 Basic elements of a formal technical report

Most technical journals publish an author's guide that contains preferred presentation techniques. The journal editors want these techniques to be consistent among authors. Reviewers of papers are also supplied with a checklist of the basic elements that must be in the report. Figure 6.3 is an example of a typical reviewer's checklist. Be aware that these are considered mandatory requirements, and a reviewer will reject a paper that does not adhere.

In summary, most technical documents, even letters, need to be broken into parts. They need a beginning, a body, and an ending. Formal reports need specific sections. Most other reports need the same elements, but each element may only include one or two sentences. Figure 6.4 is an informal report with fewer sections, but the basic elements of a report are still included.

No Bias. Bias is imposing a personal opinion or proclivity in your writing. This is to be avoided at all costs. As stated in Chapter 1, "What is Technical Writing?" technical writing must be objective. State facts and formulate conclusions based on these facts. Bias is usually detected by a reader, and from that point on, credibility may be questioned. A writer may not even be aware that his writing shows a bias. A review by a trusted coworker or another party usually prevents opinionated writing, but authors should make conscious efforts to read and reread work to ferret out statements that reflect opinion rather than objective fact.

Interesting. Technical documents are not meant to entertain, but nobody likes dull reading either, including readers of technical documents. There is a tendency for technical documents to be on the dull side. How can you make an interesting report on the seizing of an edge-guide roller?

Is the purpose of the report and objective of the work clearly stated?

Is the procedure repeatable by others?

Is the work original?

Is the work free of commercialism and tradenames?

Are the test results clearly presented?

Does the discussion clarify and explain the results?

Are the graphs and figures clear, in SI units, and do they add value?

Are the references to the work of others adequate?

Is the abstract concise and does it summarize the work?

Is the spelling and grammar correct?

Fig. 6.3 Checklist for journal papers

A technical document can be made interesting by including facts that:

- Are new to the reader
- Demonstrate the importance of the work
- Define the reader's stake in the results or conclusions

Another technique to reduce reader boredom is by using attractive (but not distracting) page layouts and useful illustrations that stimulate interest. Attractive headings and fonts can help the reader. Writers should try to appreciate what will interest readers. This may not always be possible when standard templates are required, but interesting features in content and presentation can help maintain reader attention. [*My second version of this book was declared boring by two of five college-professor reviewers. These personal anecdotes are one response meant to increase reader interest. I hope that they are working.*]

BUD LABS

TO: Drabek Industries
SUBJECT: Bond strength of CCD glass covers
FROM: K.G. Budinski, Principal Investigator
CONTRIBUTORS: M. Kohler, C. Mroczek

PROBLEM:

(Introduction)   Bud Labs was requested by Drabek Industries to develop a test and measure the relative bond strength of three different adhesives used to attach glass covers to aluminum oxide charge-coupled device chip supports. They were experiencing failures in service and two new adhesives would be compared with the present for the glass-to-aluminum oxide bond strength.

It is the purpose of this report to present the results of laboratory tests to rate relative bond strength. The objective of this work is to find a stronger cover bond and no creep failures.

INVESTIGATION:

(Procedure) The configuration of the bonded assembly was such that a special test had to be developed. A fixture was fabricated to hold the CCD assembly to the base of a universal test machine (tensile tester). A throwaway aluminum tee was bonded to the horizontal glass cover with an ASTM A38 grade 4 epoxy adhesive (Figure 1), and the adhesive joint was loaded in tension until the glass cover separated from the CCD assembly. The failure load was recorded for thirty samples of each adhesive.

(Results) The test results which are shown in Figure 2 indicate that the current adhesive has a bond strength of less than both candidates. The No. 80 adhesive had significantly better bond strength than the others.

SUMMARY

(Conclusion) This study indicated that the current glass/aluminum adhesive is inferior in short term tensile properties to the other candidates. The No. 80 adhesive had the best strength. (Recommendations) It is recommended that additional tests be conducted to investigate loss of bond strength with aging and elevated temperature. These tests only assessed short term tensile properties at room temperature.

Fig. 6.4 Example of an informal report

Specific. It is generally accepted, in any kind of writing, that a good presentation progresses from the general to the specific. However, a good technical document must also contain specific statements rather than generalities and facts rather than opinions. Supporting data should be used to make points. Starting a discussion with a generality, such as a prevailing tendency for college freshman to write documents in an illogical manner, makes correcting such points extremely difficult. (This is the general problem.) Most formal reports received in the engineering library do not contain all of the mandatory sections that are needed in a formal technical report. Some lack introductions; some lack conclusions. This is a specific statement.

Report Mechanics. Another aspect of good presentation involves general report mechanics—that is, the general methods that affect the overall appearance of a document. For example, word processors make it easy to use just about any type font imaginable. However, some are almost impossible to read by most people, and some only belong on wedding invitations and graduation certificates. Figure 6.5 lists the presentation techniques (report mechanics) that are recommended by a technical society for papers submitted to their journals. Needless-to-say, these factors are important, and they constitute basic requirements for a technical document. They are the attributes that make a document clear and readable.

Report Mechanics	
Item	**Preferred form**
Page layout	left justified, no indent, skip line between pp, one-inch margins
Font	New York, 12 point, bold headings, upper and lower case
Headings	A = bold, 18 pt centered, B = 12 pt bold, one space above text, to left,
	C = bold, 12 pt, in line separated from text with dash.
Format	block spacing, skip line between pp., one-inch margin min. all sides
Pagination	centered at bottom, no number on first page
Equations	numbered, centered, two spaces from text
Reference citation	[2], author(s), title, journal, vol, issue, year, pages, etc.
Footnotes	discouraged
Acknowledgements	put at end of report
Tradenames	discouraged, use R for registered trademarks
Abstract	concise, at start of report

Fig. 6.5 Report mechanics

Hierarchical Headings. The details of various kinds of reports have not been described yet, but most technical documents need section headings. Any document, technical or not, containing just words with no paragraphs or breaks would be extremely difficult to read. Simple structural elements can improve readability. Putting in a skipped line between paragraphs helps readability. Putting in section and subsection headings makes reading easier still.

The brain works faster than the eye, and headings provide a visual road map for what is to come. A reader can fan through the pages of a long report and quickly assess the work by noting the section headings. A good report has logical sections, and each has a title. If a section is long, it is broken down into subsections. Writers should assign headings to them. The author of a technical document needs to use headings that define the hierarchy of contents within the document.

Headings can be designed to fit preferences, but they must stand out from text. The heading designs should clearly signify the hierarchy in a consistent way. For example, the top-level heading (or "A-head" in Fig. 6.6) may be bold, upper-case text with centered placement (i.e., centered text is placed one space above the next line of text). The next subdivision of headings (or "B-heads") may be also be bold text, but is distinguished from the A-head by being left-justified and placed one space above the next line of text. Another distinction of the B-head is that of using title-style capitalization, which means that only the first letter in each word is uppercased. The next heading in hierarchy (or "C-head") may be bold, title-capitalized text that is placed in line with the first line of text. This is just one example of how to assign a hierarchy to headings. Whatever design is used, the key is to be consistent throughout the document.

Good Page Layout. In addition to above items on presentation, reports should have an overall appearance that improves readability. Good page layout is discussed in more detail in Appendix 7, but one key attribute of a good report is ample white space. For example, one-inch margins mini-

A HEADS

An "A" head is the top-level head. In this example, an "A" head is bold, centered, upper-case text.

B-Heads

A "B" head is the next subdivision of headings. In this example, they are distinguished from an "A" head with its left-justified placement of text and the use of upper-case letters for just the first letter of each word in the heading. This format for capitalization is sometimes referred to as "Title Case."

C-Heads. The next level in the hierarchy is a "C" head, which is bold, title-case text that is placed in line with the first line of text.

Fig. 6.6 An example of hierarchy in the formats of headings

mum on top and bottom and both sides are recommended. Good design and consistent use of headings also enhances readability, as previously noted.

The most readable books, children's primers, are classic examples of ample white space. Often there are only one or two sentences on a page. Obviously, this is an extreme example, but its purpose is to make children's books the easiest to read. A proper visual appearance increases readability, and the effective use of white space is an easy way to improve readability.

6.3 Language Skills

Every language has its rules. By definition, a language is a particular type of verbal communication. Each is made up of letters or symbols, and there is a set of rules on how to use these letters and symbols. The language skills referred to in Fig. 6.1 have to do with conformance to the rules that have been established for a particular language. This text only deals with English. In the United States the Modern Language Association (MLA) is the keeper of the rules. Other languages have keepers of their rules.

As an example of the need for language "police," the Internet has become a widely used source of information in the technical community. It is used for obtaining property information, chemical symbols, manufacturers, safety information, and many types of information needed by technical professionals. How does a writer reference Internet sources in a document? Who decides what is right or wrong? By consensus, in the United States, it is the Modern Language Association.

This section reviews some of the more important document attributes that pertain to proper use of language in a technical document. Details on punctuation and grammar are described in Appendix 6. Only general aspects of language skills, which must constantly be considered in every writing task, are highlighted here.

No Jargon or Acronyms. As previously noted, acronyms can make documents difficult to read. The overuse of legitimate words or complex words can also obscure meaning, sometimes purposely. The excessive use of complex technical jargon, or "technobabble" (J.A. Barry, *Technobabble,* MIT Press, Cambridge, 1991) only makes the job of the reader more difficult. Even without the use of complex technical words, the use of jargon can obscure meaning:

Jargon

The Chairman's strategic initiative team has devised a path forward to quantify the leveraging of our assets and revenues to optimize shareholder delight.

Translation

The Chairman's staff has a plan to increase stock performance.

With more and more hype used in business affairs, it seems as if consultants and managers have created their own language to give the impression of importance. New meanings are created for common words such as:

- Strategic initiative—plan
- Leverage—use
- Shareholder delight—stock price

Maybe this type of writing could be called "businessbabble." In any case, the use of jargon in either technical or business communications just complicates and obscures meaning. Use words with concrete meaning and clear definitions, instead of complicating (or disguising) simple thoughts with unnecessary contortions of the English language.

No Misspelling. Spelling errors can destroy an otherwise perfect technical document. Right or wrong, many readers equate a misspelled word with the credibility of the entire work. If you cannot spell "differentiation," what do you know about the friction characteristics of film transport rolls? It is very easy to misspell a word. Readers do not object to the misspelled word, but rather that you did not proofread, and re-proofread to eliminate any misspelled words.

RULE

Never allow a misspelled word in a finished document.

No Punctuation Errors. Errors in punctuation are in the same category as misspelling. If any are allowed, they can negatively reflect on the author. Many rules exist on commas, fewer rules on periods, and fewer yet on semicolons, but overall proper punctuation requires knowing basic rules of application. Basic rules on proper punctuation are discussed in Appendix 6.

Punctuation can also be very important in terms of content. A misplaced comma, for example, can change the meaning of a sentence. A missing period can do the same. Technical documents require technical accuracy, and proper punctuation is no less important than the placement of a decimal point in a stress calculation. If it is wrong, there can be significant repercussions.

Mixed Sentence Length. Proper technical writing requires sentences that are neither too short nor too long. Long sentences are difficult to read, but short sentences present a choppy type of writing. A blend of sentences of varying length is preferred.

Written in the Active Voice. Voice in writing means if the subject in a sentence is doing the action implied by the verb or if the subject is acted on.

Active Voice

The construction division demolished the damaged powder-storage silo.

Passive Voice

The powder-storage silo was demolished.

Both sentences are grammatically correct; however, professional writers and English instructors believe that readers understand information written in the active voice better than in the passive. The author is perceived to be more assured of himself or herself when writing in the active voice. Therefore, authors should try to use verbs that connote action in the present tense. Some examples of active and passive voice are as follows:

Active	Passive
The field division surveyed the site.	The site was surveyed by the field division.
The pressure relief valves vented the system.	The system was vented by the pressure relief valve.
We purchased three lots of offshore steel.	Three lots of offshore steel were purchased by us.
Three chromium-plated rings prevent piston blowby.	Piston blowby is prevented by three chromium-plated rings.
The installation of R14 insulation has reduced heating costs 21%.	Heating costs were reduced 21% by the installation of R14 insulation.
Jack issued the change notice on November 3.	The change notice was issued on November 3 by Jack.

Reasonable Grammar. Most readers of technical documents are interested in the technical content of a document, and they may be tolerant of less-than-perfect grammar. However, the writing goal should be perfection. Punctuation and grammar are the basic rules of language, and a disregard of these rules often leads to poor readability or misinterpretation. Poor grammar can affect your credibility as a technical person. Readers may question why an author did not have a document proofread before sending it out.

One of the most common grammar errors in English writing is disagreement of verb and noun in a sentence.

Disagreement of Noun and Verb

Containment of the spill and notification of the environmental officer *is* essential in any chemical spill or leak.

Agreement

Containment of the spill and notification of the environmental officer *are* essential in any chemical spill or leak.

Proofreading by a trusted peer or support person is usually the best way to catch and correct grammar errors. Appendix 6 reviews basic grammar rules.

Most technical writing in the 1990s in the United States was performed on computers. In 1998, common word processing software programs started to include grammar checks along with spelling checks. Such software aides are extremely helpful in addressing proper use of the written language. These computer tools should be used. [*The program that was installed on my work computer highlights misspelled words in red and grammar errors in green as you type. It is uncannily accurate in finding noun-verb disagreement. It is like having a built-in English teacher. I love it.*]

Concision. A concise document contains no more words than necessary. Concision is an important attribute of technical writing because readers want to be informed in the fewest possible words. Readers determine the success of a document, and the planning and writing of a document must be directed toward readers. Everyone is busy, and readers do not need extra words, paragraphs, or unnecessary visual aids. If the reader thinks or does something, then the writing succeeded.

Subsequent chapters provide specific suggestions on how to achieve concision in writing. [*As an aside, I never heard the term "concision" before starting this book. I only knew the adjective concise. Some books that I reviewed on technical writing used the term conciseness as the noun meaning "no unnecessary words." But then I received a review on a draft of this book from a person who wrote a book "Dictionary of Concise Writing." I bought a copy of this book. It lists many words, phrases, and cliches that produce extra length in a document with no added value. This book contains a chapter on "Concision." I still am not sure that concision is a valid word, but I will use it in this text to mean compact writing. If a publisher accepted it for a chapter heading, this is enough evidence for me that it is a real word. Besides, I like the word.*]

Concision is discussed in more detail in Chapter 7, "Writing Style." The main point is to eliminate all unnecessary words by reviewing each document for concision. Figure 6.7 illustrates a marked up manuscript after copy editing. Note that many words were deleted without changing meaning. If you can eliminate words or phrases, do so.

RULE

Eliminate any unnecessary words, phrases, or sentences.

Summary

This Chapter introduces the basic criteria of good technical writing. The major elements include:

- Acceptable technical content
- Good presentation of information
- Proper use of English and its rules of grammar

The evaluation sheet in Fig. 6.8 is a useful checklist of elements for good technical writing. This type of sheet could be used by others to rate your document. Technical reviewers for journals frequently use this type of form. A more complete checklist for documents is given in "Getting It Done," the last Chapter of this book.

In summary, a good technical document must have good technical content that is presented in an easy-to-read format without technical or grammatical errors. Most of these issues are addressed in more detail in subsequent Chapters, but the key criteria discussed in this Chapter should give

reason for this is that when a designer goes to select a plastic for an application there is almost always a selection criterion that has already been established in this area. If the part under design is destined for high volume production, the designer will usually want the material to be thermoplastic so that the economies of thermoplastic processing processes can be obtained. Some designers, on the other hand, know that they will need thermosetting resins to obtain some specific properties. The other situation that exists is when a designer is objective and would consider candidate plastics from both categories; whatever the case, it simplifies plastic selection if this basic categorization is made. We have done this in Figure 4-1.

The other categorization that we have made in Figure 4-1 is that we divided thermoplastics and thermosetting plastics into commodity plastics and general-use plastics. We are using the term commodity to reflect plastics that are used primarily for consumer goods and routine plastic "things". These are the plastics that are used for garbage bags, for containers in fast food stands, for upholstery, for all sorts of ordinary applications. The ordinary applications for thermosetting plastics are things, like fiberglass boats, fiberglass roof panels, phenolic break linings for vehicles, and phenolic electrical boxes, switches and the like.

Fig. 6.7 Example of professional copy editing

Technical Writing Evaluation (10 is best)

<u>Content</u>

- Is the background material sufficient? _____
- Are the purpose and objective clearly stated? _____
- Is the information accurate? _____
- Is the information useful? _____
- Is the information timely? _____
- Is the message clear? _____
- Are sufficient data presented to justify the conclusions? _____
- Does the technical level of this document suit you? _____

<u>Presentation</u>

- Is this document as concise as it can be? _____
- Is the presentation of information logical? _____
- Is this work free of bias? _____
- Are the visual aides helpful? _____
- Is this document easy to read? _____

<u>Language Skills</u>

- Is the punctuation acceptable? _____
- Are there any misspelled words? _____
- Is the document free of jargon? _____
- Is the use of acronyms appropriate? _____
- Is the grammar acceptable? _____
- Is the composition to your liking? _____
  (page layout, font, headings, etc.)

Fig. 6.8 Technical writing evaluation sheet

you a general understanding of what is expected in writing and what writers should accomplish after reading this book.

Important Terms

- Readability
- Investigation
- Heading
- Factual
- Language Skills
- Bias
- Jargon
- Report Mechanics
- Grammar
- Plagiarism
- Credit
- Misspelling
- Concise
- Conclusion
- Active Voice
- Technobabble
- Recommendation
- Font
- Acronym
- Abstract
- Specific
- Timely
- Body
- Logical
- Procedure
- Caption
- Hierarchial

For Practice

1. Submit an example from a journal or newspaper of an article with good readability and an article with poor readability. State the reasons for your ratings.
2. Submit a published work containing jargon and rewrite the jargon portions in understandable terms.
3. Submit a published example of a poor or inadequate caption on a graph. Explain why it is inadequate.
4. Submit a paragraph or two from a published work that you believe contains unnecessary words. Edit it for concision.
5. What precedes and what follows the results section of a formal technical report?
6. List five attributes of a good report that pertain to technical content.
7. Submit an example of "technobabble" or businessbabble for class discussion. What words are meaningless?
8. You just finished a six-month investigation of the health hazards of methylene chloride. How soon should you publish your study for it to be considered timely?
9. State five presentation concerns of technical writing.
10. Describe a strategy to prevent language errors in technical writing.

To Dig Deeper

- R.A. Day, *How to Write and Publish a Scientific Paper,* 3rd ed., ORYX Press, New York, 1988
- R.H. Fiske, *Dictionary of Concise Writing,* Writers Digest Books, Cincinnati, OH, 1996
- J. Gibaldi, *MLA Handbook for Writers of Research Papers,* The Modern Language Association of America, New York, 1999
- M. Roze, *Technical Communication,* 3rd ed., Prentice Hall, Columbus, 1997

CHAPTER 7

Writing Style

CHAPTER GOALS

1. *Understand the elements of style in technical writing*
2. *Become familiar with the style that is most suitable for technical reports*

WRITING STYLE refers to mannerisms in word usage. It is the way that the writer consistently and unconsciously chooses words, forms sentences and paragraphs, and combines them with ideas and thoughts to create a document. Style is all of the attributes described in "Criteria for Good Technical Writing," Chapter 6, plus the product of the writer's persona, personality, who he or she is and how they feel. A document is a composition that integrates the elements of style into a finished work (Fig. 7.1). Like a work of music, each composition is different because each person is different. The work is a summation of technical knowledge combined with inner feelings and personal characteristics of the composer. This is the style of the composer, who does some things in certain ways most of the time.

Style can be important because the audience (readers) may be influenced by it. Style is usually what sells in the marketplace. Continuing with the analogy of music composers, the great ones like Mozart, Vivaldi, and Handel each had a style that set them apart from the millions of other musicians who composed music. People like the styles of these composers, and the listeners have become comfortable with particular styles by listening to certain composers' works for a very long time. Style makes a painter; style makes a singer. Everybody has access to the words and music of every published song. What makes some singers' renditions sell a million compact discs? It is his or her style—the words, combined with the notes, combined with the music, combined with the voice quality, combined with the voice inflections, combined with the tempo. It is the total composition that sells.

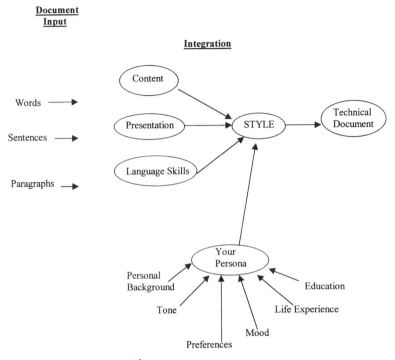

Fig. 7.1 Elements of style

Technical writing by definition is done for a reason. There is a purpose and objective for every document. You want to achieve something by writing this document. Writing style will help determine success in achieving the goals of a document.

The author of a document determines the style. It is the purpose of this Chapter to make the author aware of the factors that determine style so a writing style that is successful with readers and congruent with accepted technical writing practices can be developed. Chapter goals are addressed by discussing the elements of style and by giving examples of different writing styles. This Chapter then concludes with suggestions on developing a style that works in technical writing.

7.1 Elements of Style

Figure 7.1 is an attempt to illustrate the many things that make up style. Chapter 6, "Criteria for Good Technical Writing," already discusses document content, presentation, and language skills, but style starts earlier. It starts with writer choices of word usage, how words are put into sentences, how paragraphs are constructed, and then how they are presented to readership in the final work. These early choices encompass a variety of elements shown in Fig. 7.2.

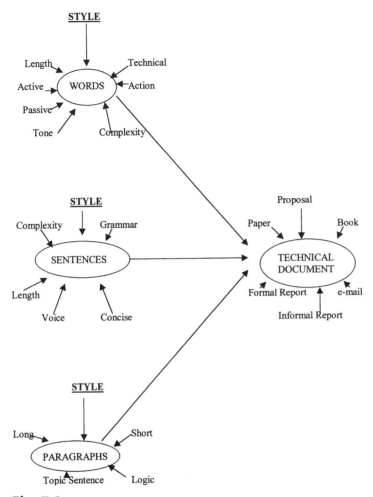

Fig. 7.2 Style is manifested in choice of words, sentences, and paragraphs.

Persona affects word choice and results in a particular style. If you are a gregarious person, you may be prone to use words like "bold, hearty, beautiful." An introverted person is likely to use words that do not draw attention to himself or herself. Words like "anticipate, therefore, follows" may be selected as a reflection of an introverted personality.

Likewise, writing style has a tone, which is a reflection of writer attitude or feelings. If you are angry at your manager at the time of writing, this may show in your writing as snipping remarks or angry words.

Angry Tone

The project would have been completed as scheduled if we had management co-operation.

Angry Words

Noncommittal

Procrastination

Disappointing

Uncooperative

Unqualified

Sloppy

Nonconforming

Malicious

Shoddy

Technical writing, by definition, should be objective and impersonal. The tone should be neutral. In other words, the tone should be dispassionate, impartial, and unbiased. Tone may be influenced by your feelings at the time of writing, how you feel about yourself, how you feel about your boss, company, mate, or health. Readers do not want to be distracted by these subjective connotations. If such a tone is evident, readers may question the objectivity of a work. [*I can personally attest to the importance of tone in writing. I had an unfortunate experience caused by my improper tone. Ten years ago, we built a new house. It turned into a disaster. Everything went wrong. After we moved in, I vowed to write a book about our misadventure so that others could avoid our mistakes. I spent a year writing the book, House From Hell. Because nontechnical books are generally submitted to publishers through an agent, I submitted a book proposal to 140 literary agents. Only two agreed to read it; both turned me down as a client because of the tone of the writing. One called the book sardonic; the other called it nasty. I was still angry at the general contractor, the architect, the mason, the plasterer, and the plumber, and it showed in my writing. One agent said that I would frighten people out of building a custom-built house and nobody would buy the book. I completely rewrote the book, but it is still unpublished. Tone can definitely have a negative impact on writing style.*]

RULE

Do not let personal feelings show in writing.

Word Choice

There are many types of words, and choice of words is a major part of writing style. The average person born in the United States knows about 21,000 English (U.S. modified) words. [*I was told this during a company psychological test on word knowledge.*] It takes about 500 words to fill a single-spaced typed page. Technical documents can be less than one page or more than 100 pages. However, most technical journals limit submissions to about ten pages or 5,000 words. The author must make a decision for each word included in a document.

Definition of some word categories can help define the concepts of good word selection. For example, a person's 21,000 word repertoire can be broken down by grammatical function into four categories:

- Utility words like conjunctions (and, or, but, if, and so forth)
- Substance words for action (verbs) or subjects (nouns)
- Descriptive words that modify other words (adverbs, adjectives, clauses)
- Contractions for people who do not have time to write out the real words

These basic categories are very general, but the usage of some types may have more priority than usage of other word types. Substance words may be more important than descriptive words, which are probably more important than utility words.

Many words also have a "personality" of their own, just as words may reflect the personality of the author. Figure 7.3 lists 15 categories of word types with unique characteristics, as described below in more detail. This

TYPES OF WORDS

Cliché	Foreign	Gender specific	Action	Archaic
Bottom line	Et cetera	His	Produce	Albeit
Done deal	Ibid	Her	Change	Herewith
Far flung	Et al	She	Exceed	Hence
Snug as a bug	Per se	He	Organize	Hereafter

Abbreviated	Negative	Technical	Slang	Euphemisms
Info	Mistake	Spectroscopy	Hosed	Passed away
Lab	Oversight	Martensite	Miffed	Incident
Dept	Accident	Fluoroscopy	Hoot	Pre-owned
Sub	Clumsy	Comminution	Bummer	Marketer

Long	Complex	Uncommon	Technobabble	Profanity
Predominant	Phraseology	Proclivity	Leverage	↑
Proximity	Diametrical	Unctuous	Boot	(Well known)
Ajudicate	Incongruent	Abrogate	Strategize	
Eventuality	Pernicious	Ameliorate	Overarching	↓

Active	Passive	Happy	Sad	Meaningless/ misleading
Achieved	Was	Joy	Death	Best
Finished	Were	Health	Disease	Highest
Built	Has	Smile	Sickness	Unsurpassed
Inspected	Is	Laugh	Cancer	Excel

Fig. 7.3 Spectra of words available for technical writing

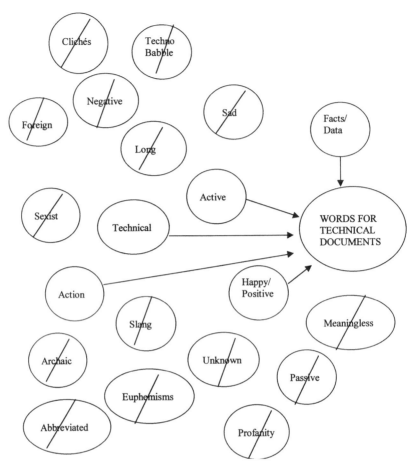

Fig. 7.4 Word choices for technical writing

breakdown is useful, because some word types are inappropriate for technical writing. It really is important not to use slang, cliches, "technobabble", or uncommon, angry, and negative words.

Figure 7.4 is an illustration showing preferred types of words for technical documents. There are few word types without a "do not use slash," but do not worry. There are enough active, technical, action, and positive words left to meet even the largest writing task. Remember, you know 20,000 or more.

Cliches are essentially figures of speech. Groups of words are taken together to have a contemporary meaning. They come and go in popularity.

Foreign words are also part of the English language, such as, restaurant (French), pizza (Italian), envoy (French), soprano (Italian), and resumé (French). There are many other examples. The foreign words listed in Fig. 7.3 are Latin terms widely used in English writing. You, the author, must decide if your readership will know their meanings:

- etc. = et cetera = and so forth
- Ibid = ibidem = in the same place (cited previously)

- et al. = and others
- per se = by itself
- i.e. = id est = that is
- e.g. = exempli gratia = for example

Words with a presumption of gender abound in the English language: chairman, congressman, salesman, meter maid, repairman, and so forth. These words suggest that only a man or a woman could do them, which is often an anachronism. Nowadays, professional writers avoid words with any arbitrary or presumptuous reference to a specific gender, religion, ethnic group, or special group. All readers should be approached in an inclusive manner without presumption of unnecessary distinctions. For example, twenty years ago, form letters in business often had the salutation of "Dear Sir." This presumption of gender reflects poorly on the author and should be avoided. Current salutations include "Dear Resident," "Dear Subscriber," and "Dear Reader." Salesperson is used in place of salesman. Chair is used instead of chairman; meter monitor is used instead of meter maid.

Action words are meant to evoke or elicit feelings of excitement or progress. It is commonplace in American business to develop vision and mission statements for every department as well as the overall company. The people who promote these exercises insist on using words of action in developing these statements. Some words by nature have that connotation. Race, fight, beat, and convert all imply action. When you use these words, you describe something that happened, or you try to evoke action or excitement from the reader. Conversely, verbs that do not suggest as much action are list, number, think, consider, study, and so forth. These words are simply not as bold.

Archaic or old fashioned words can be used, but possibly only old readers will know what they mean. Some biblical words like "thou shalt" fit in the archaic category. [*When I was a youth, all of the homes in the neighborhood were heated with coal. There were many words associated with coal furnaces that are probably unknown to people born after 1960. The coal man (gender specific, then) delivered either anthracite or coke (if you had a stoker) through a basement window into the coal bin. He (not he or she) used a wheelbarrow to unload the coal from the truck and dumped the coal down a coal chute. We had to stoke the fire several times a day, and, in the winter, we put the ashes and cinders on the driveway and sidewalks for traction. I suspect that many students in the second millennium will not be familiar with terms such as coal man, stoke, cinders, anthracite, and coal chute. They became archaic in only 50 or so years.*]

Contemporary abbreviations of words are very commonplace in the United States. They are not legitimate abbreviations but ones that simply evolve. The abbreviations listed in Fig. 7.3 are familiar to most English speaking people, but sometimes these shortened words are not generally known. [*I have noticed that public works departments have been using shortened words on road signs, such as Niagara Coll, thru traffic, City*

Cent, and 50 MPH Min. I assume that Coll is short for college, thru for through, cent for center, min for minimum, but I am not sure. These abbreviations will not be found in most approved lists of abbreviations.]

Negative words such as those listed in Fig. 7.3, suggest something undesirable, whether used as nouns, verbs, or modifiers. A mistake is almost always undesirable. An "oversight" and an "accident" are similar to a "mistake"; all imply a negative connotation. The last word on our negative word list, clumsy, is almost name-calling. Many English words almost always evoke negative reactions:

* Lazy
* Indolent
* Shiftless
* Ignorant
* Rude
* Crazy

* Loud
* Fat
* Boisterous
* Impolite
* Terse
* Dumb

Their use is usually avoided in technical writing.

Technical words are certainly to be expected in technical documents, but there can be an overdose.

> Too Many Technical Words
>
> The analysis of the spectral frequency output of the Johannsen-type goniophotometer was accomplished by embedding algorithms in a derived linear regression analysis computer program. Second order derivatives...

The technical words listed in Fig. 7.3 pertain to materials engineering, but there is a list of parochial words for every technical field. You know the ones in your field, but you may not know the ones used in other technologies. A technical document may miss its mark if the reader is not familiar with many of the words.

Slang abounds in our daily interactions with coworkers, service providers, and even bosses. Words like "hosed," "neat," "cool," and "bummer" have well-known meanings in the United States:

* Hosed means cheated.
* Neat means nice.
* Fine means attractive.
* Bummer means negative outcome.

They are used in most circles, but none of them read well in technical writing.

Slang in Technical Writing

We completed corrosion tests on eight types of cemented carbides. The cobalt binder was varied from 6 to 14 percent and the effect of binder content on corrosion rate was determined. Sample preparation was a *bummer*. Two grades of test materials, the 6 percent and 8 percent cobalt grades, could not be polished to the required surface roughness of 6 to 8 *micros*.

Euphemisms are words or phrases used in place of what really needs to be said. In many cases, euphemisms are used to make things sound better. For example, garbage collectors are called sanitation workers; it is more genteel to say "where is the restroom?" than "where is the toilet?" Euphemisms also are very commonly used in industry in the United States.

Industrial euphemisms	Meaning
Strategic framework	Plan
Downsizing	Layoff
Technology road maps	Plan
Operational excellence	Good work
Career enhancement	More work
Development opportunity	Problem
Diagonal slice	Opinions
Performance matrix	To-do list
Shareholder satisfaction	Quarterly results
Overarching goals	Objective

The user usually intends to make something sound nicer or grander than it is, but in reality the readers often miss the intended meaning or may be annoyed with the unnecessary words. In all of the above instances, the euphemism requires extra length.

Euphemisms are also commonplace with computer technology, and personal computers (PCs) have prompted almost a second language by assigning specialized meanings to ordinary words.

Personal computing term	Meaning
Boot	Turn on the computer
Memory	Storage capacity
Download	Transfer from another source
Upload	Transfer to others
On-line	Connected to other computers
Crash	Electrical problem
Program	Operating instruction for a computer
Software	Programs
Hardware	Electrical equipment associated with personal computers
Disk	A data storage device
Drive	An electrical device to read magnetic media

Long words and complex words, the next two types in Fig. 7.3, often conflict with the objective of concise writing. Good documents of any type are concise, and the following words and phrases are examples of wordy phrases and more concise alternatives:

Wordy phrase	Concise replacement
And so as a result	Hence
Day to day routine	Routine
On the rise	Growing
On the basis of	Based on
Not withholding the fact that	But
Is knowledgeable about	Knows
It is evident that	Clearly
It is important to note	Note

Concision can also be applied to individual words.

Big word	Concise replacement
Facilitate	Help
Contact	Call
Attempt	Try
Additionally	Also
Congregate	Gather
Significantly	Greatly
Springtime	Spring
Therefore	So
Utilization	Use

Simpler, easier to read words can usually replace complex words like those listed in Fig. 7.3, including "terms" for "phraseology," "opposing" for "incongruent," and "harmful" for "pernicious."

Uncommon words (such as proclivity, unctuous, abrogate, and ameliorate) are not widely known and make reading more difficult. As mentioned earlier, each English-speaking person knows a different quantity of the 40,000 or so words in the English language. Many times, writers use long, complex, and uncommon words to demonstrate intellectual authority. They may succeed at this but fail to make the point of the document. Newspapers are a good example of a proper level of word usage. Most professional writers in the United States possess a 30,000+ English vocabulary, but they write news articles using a 15,000 word vocabulary. They write for a readership with educations ranging from grade school through Ph.D. [*I have yet to hear a Ph.D. complain that newspapers are written at a level that is too simple. Simple writing is often the easiest reading.*]

Technobabble includes manufactured words and phrases similar to some industrial euphemisms. Every profession has words with special meanings, but overuse of specialized terms or big words (strategize, overarching, deliverable, and so forth) with simple meanings may only just obscure meaning and frustrate readers. These words have meaning within the profession, but the meaning outside may be very different. This must be considered in their use. Does your reader know what a deliverable is?

Profanity is, of course, totally unacceptable in any business or technical document. In U.S. business and industry, there are very strict rules about sexual harassment. Profane language often has sexual overtones that would violate most sexual harassment guidelines. One off-color remark could cost a person his or her job in spoken word or writing.

Active and passive words involve the reader in very different ways. As previously noted in Chapter 6, "Criteria for Good Technical Writing," writing in the active voice usually makes documents more readable. The active voice is a good attribute in any form of writing, because writing in the active voice makes your statements more forceful. The active voice simply means that the subject of the sentence does action. Passive voice, on the other hand, does not identify who is doing the action:

Passive Voice

Samples of each candidate plastic were sectioned and mounted.

Active Voice

The metallurgical laboratory sectioned and mounted a sample of each candidate plastic.

In the active voice, the reader knows who did the sectioning and mounting. Active voice generally requires the use of action verbs (like the ones in Fig. 7.3), while passive verbs tend to give readers observations. This is one reason why technical documents, which must present observational results, often use passive verbs. The passive voice is unavoidable; just try to eliminate passive verbs whenever you can. As shown in the following examples, the indirect result of using the active voice is to make clear who is doing the action.

Active Voice

- The Alpha Site Team achieved a tenfold improvement in knife life.
- The shop safety team finished instructing all employees on safe lifting practices.
- Cosgrove Construction Company built seven reinforced plastic storage tanks for Nocal Chemical Corp.
- The NDE technician from the Materials Engineering Lab inspected three hypo-solution tanks in B-58.

Passive Voice

- A noticeable difference in quality was observed between U.S. and Canadian products.
- Rehabilitation was discussed at the quarterly review meeting.
- Never has the subject of unionism been discussed.
- A production trial is scheduled for November 30.

Words that evoke happiness or sadness are two other categories listed in Fig. 7.3. Feelings of happiness or sadness are not objectives in report writing, but there may be occasions to use these kinds of words in business or day-to-day communications. For example, managers may write a holiday greeting to employees. Just keep in mind that some words can alter the mood and morale of a reader.

Meaningless or misleading words represent the final category of word types in Fig. 7.3. Many well-used words in the English language do not provide any meaningful information. For example, most superlatives (best, greatest, superior, fastest, and so forth) are meaningless or misleading words without factual foundation. These words (when you step back and think about them) are often unsubstantiated claims. The author did not and could not test every screwdriver in the world in order to claim that he or she sells the best. Never claim product superiority without test data to support the statement.

Not Credible

Cubic zirconia is superior to tool steel in wear and corrosion resistance.

Certainly the author did not test zirconia against all kinds of wear. It would last only seconds in a gyratory rock crusher. Similarly, there may be an environment that dissolves zirconia and not steel. The author has not tested all environments. Therefore, this statement is meaningless and, at the least, very misleading.

Words that cannot be backed with facts will lower the credibility of your document. Superlatives generally represent unsupportable claims and thus should be avoided in technical writing. Some other examples of meaningless or very misleading words include:

- outstanding
- biggest
- ultimate
- optimize
- maximize
- cheapest
- farthest
- robust
- highest
- friendliest
- fastest
- lowest

If you ask, "What do these words mean?" the answer may be "nothing."

Sentence Construction

Construction of good sentences follows directly from word choice. Build sentences with concise, vigorous, and substantive words in the active voice. Include valid facts and data. Suppress your inclination to use meaningless and negative words, euphemisms, and abbreviations.

What recommendations should you follow when forming sentences? In a way, readers are like voters in a political election. They do not want negative statements. Voters want to hear positive statements with substantive recommendations for the future. This same strategy applies in technical writing. Sentences need to be clear, concise, understandable, and based upon positive words.

The basic rules of sentence formation are taught in school, but bad habits often persist. Confusing clauses, slang, idioms, and errors in grammar or punctuation are the common causes of poor sentence construction. This section focuses on five poor "sentence habits" that commonly appear in technical writing. Elimination of these unsound habits can improve construction of sentences.

[Opinion plays a role in writing style. By definition, persona is a key factor in style. My suggestions on word use, sentence construction, and paragraphing are mostly based upon my experiences in reviewing the work of others. I review technician and junior engineer reports daily, one or two papers a month for technical journals, and periodically MS and Ph.D. theses and government research proposals. I am not picayune. I do not substitute my words for those of others. I only make suggestions for concision, clarity, and questionable facts. The following top-five sentence "problems" are my opinion based on reviewing experiences. They are from my Pareto chart. Others involved with technical writing may have another list, but we both have the same objective—a document that the reader understands.]

Problem One: Starting with Dependent Clause. It is not grammatically wrong to start a sentence with a dependent clause, but the resulting sentences often lack concision and are difficult to understand.

Starting with a Dependent Clause

At the start of the business year and at the beginning of the third quarter, the laboratory has a project review with the director of research.

The reader had to wade through 15 words to discover that you were talking about the laboratory. Why not tell the reader up front?

Starting with a Noun

The laboratory has a project review with the director of research each January and July.

The sentences in the preceding two examples have the same meaning, but the latter is more clear and concise. The modifying phrase with the data is still present, but it is easier to read when moved to the end of the sentence. Extra words were also removed to make the sentence more concise. The remodeled sentence has 15 words, reduced from 26.

When a sentence starts with a dependent clause, another problem may be confusion on what the clause refers to in the sentence.

Beginning with a Clause
Traveling to the western tank farm, he met three lab members near building 318.

Were the lab members traveling to the tank farm or was he? This is the kind of confusion that often occurs when a dependent clause is moved away from the word that it modifies.

Beginning with a Pronoun
He met three lab members on his way to the western tank farm. They were near building 318.

This second version is clearer. Many times clarity can easily be improved by breaking a long sentence into two or more sentences.

Starting sentences with clauses can be infectious. Some people do it on almost every sentence. It produces an indirect tone. Every statement seems to have a preamble. You can correct this bad habit by thinking about how you start each sentence. Do you really need to start with a modifying clause, or can you say what you want directly without qualifiers? If you can, then do it. It helps the reader.

Problem Two: Subject and Verb Do Not Agree. As we all know, the rules of English require agreement of subject and verb. If the subject is singular, the verb must also be singular. Long sentences with many clauses often create a significant physical separation of the subject and verb, and this often leads to subject-verb disagreement.

Subject and Verb Do Not Agree
Metallography of failed balls and races are key aspects of failure analysis.

The subject, metallography, is singular, but the sentence uses a plural verb, because the author has confused the modifying words "of balls and races" as the subject. Even the grammar checks in word processing programs sometimes miss this subtle mistake.

This type of mistake also occurs when an author makes the verb agree with words that immediately precede it:

Subject and Verb Do Not Agree
The output of three wear testers, the block-on-ring, the ball-on-plane, and the microtribometer are directed to a single computer.

> Agreement
>
> The output of three wear testers, the block-on-ring, the ball-on-plane, and the microtribometer is directed to a single computer.

The subject of this sentence is output, which is singular.

Quirks of the English language may also cause problems with subject and verb agreement. For example "plural-sounding" words like series, group, flock, gathering are really singular.

> Agreement
>
> A group of distinguished metallurgists is to convene for the selection process.

Other words (for example, none, some, all, or more) can be used properly in either singular or plural form:

- All of the literature is complete.
- All were complete.
- Some of the literature has been delivered.
- Some are to be delivered.

In summary, watch for subject-verb agreement. It can be tricky. This mistake is often noticed and can distract the reader from a message.

Problem Three: Incoherent Sentences. Sometimes during review of a document, a sentence just does not make any sense. What is the author trying to say? Very often this occurs when writers are "on a roll" or in a "stream of consciousness," where their mind is going faster than their fingers can type. Parts of the next sentence get merged with the sentence in progress. Sometimes a key word is inadvertently omitted.

Incoherent sentences usually result from the omission of a word.

> Incoherent
>
> The procedure for cutting circuit boards for failure study begins stripping the devices.

In this example, the writer left out "with" between begins and stripping. Another example can be a missing verb.

> Incoherent
>
> Differential scanning calorimetry (DSC) and thermogravimetric analysis (TGA) to determine if any significant differences existed between the samples identified as A and D.

It appears that a "were used" or "were employed" was intended after TGA.

Incoherent sentences may also occur when two sentences are inadvertantly merged:

> Incoherent
>
> It is typical for polypropylene-polyethylene blends to have a melt flow index and the past processing cooling on a mandrel reduces distortion.

Whatever the cause, incoherent sentences can easily be prevented by having peer review of every document.

Problem Four: Same-Sounding Word and Wrong Usage. A significant number of English words have the same pronounciation but very different meanings, such as:

- Two, to, and too
- Coarse and course
- Principal and principle
- Loose and lose
- Plane and plain
- Their and there
- See and sea
- Affect and effect
- Brake and break

The words in the preceding list are frequently misused. The root cause can be reliance on only computer spelling checking applications. Sometimes grammar checkers can pick up misuse, but even here, programming misses some quirks of the English language. The machines do not know what the writer is trying to say, and they often cannot detect misuse of words in technical phrases.

The following is an example of a document that made it through a spelling checker with just about every word misused:

> The problem width a spell checker
>
> Eye wood like to show that spell check hers will knot all ways fined words that are used inn correctly. The rite spelling is only pro deuced by reeding the work over and by watching for words with multi pull spellings. All duel spellings caught bye soft wear cannot be assummed.

The solution to this problem is to reread your work after using the spelling checker.

Problem Five: Sentences Containing Slang or Idioms. Slang is never proper in writing. The readers may assume that the writer does not know the proper terms and question author credibility. Slang and idioms are particularly inappropriate in communications. Readers who have English as a second language may not be familiar with slang or idiomatic expressions and may completely misinterpret your document.

Use of an Idiomatic Expression

The rejection rate is higher than normal for the delta station chassis that we have received from France. Please get on top of the situation.

A French-speaking person may have no idea what is meant by "get on top of the situation." Similarly, that person may not recognize slang words.

Use of Slang

The transport roller bearings were hammered. There was reddish goop coming out of the seal. We decided that this baby had to go, and we served up a new issue.

Translation

The transport bearing failed. Wear debris was apparent in the seal area, and a decision was made to replace it.

Summary. In summary of this section, sentences need to be clear, concise, understandable, and based on positive words. Keep in mind some of the examples of common sentence problems. A writing style should use sentences free of clauses that can confuse, slang, idioms, and errors in grammar or punctuation. These mistakes can easily be prevented by acquiring peer review of every document.

Paragraphs

Paragraphs make reading easier by breaking a document into pieces. It would be difficult to read any document without paragraphs, as the human eye and brain need to recognize when an author finishes one thought and moves to another. Reading a book without paragraphs would be comparable to eating an eight-ounce steak without cutting it. A paragraph break (denoted by an indentation or skipped line of text) provides a necessary signal that a new thought, idea, or piece of information is coming.

Paragraphs are essential in good writing, and most people take paragraphs for granted. Good paragraphs are unnoticeable. We have always read in paragraphs; everything is broken into paragraphs in newspapers, books, and letters. Good paragraphs are inconspicuous, because they just make reading easier by grouping related sentences together. This basic objective depends on the choice of paragraph content and length.

Paragraph Content. The key part of any paragraph is its topic sentence. Paragraphs must have a topic sentence, which is supported by the other sentences in the paragraph. Most often, the paragraph begins with the topic sentence. It continues with supporting sentences and then ends with some form of thesis restatement and a transition to the next paragraph.

For technical documents, the purpose of a paragraph is either to persuade or to share information. These purposes lend themselves to specific strategies for development of paragraph contents. To persuade someone of something, first assert what you want the readers to think or do. This is done with a topic sentence.

RULE

Every paragraph must have a topic sentence.

This is the topic sentence for the next section, where the purpose is to persuade you about the importance of paragraph length. In the next step, the paragraph is built with examples and arguments that convince readers of the importance and validity of the topic sentence. Readers may be persuaded in many ways, and so a strategy or plan is needed. Besides examples of long and short paragraphs, one could try to make paragraphs the right length by listing grammar rules on paragraphs. Another persuasion technique is the use of analogies (like the steak one).

Most persuasion strategies involve a presentation that progresses from general to specific statements. Start with your topic sentence and some general related sentences. Then, move on to specific examples, facts, arguments—whatever it takes to make the point. End the paragraph by restating the topic sentence in a modified form. This is the classic approach: tell them what you want to say, say it, and tell them what you said.

The strategy for information sharing is similar. You are not sharing information without reason. Publishing a paper in a public journal is the ultimate example of sharing, but the work is out there for all to criticize. Therefore, your purpose is to be convincing as well. In writing a trip report, where one simply describes what he or she saw at the Atlas Welding Company, intentionally or not, the writer's words are producing a reader result. The reader forms a favorable or unfavorable opinion of Atlas Welding, depending on writing style. The writer ends up persuading the reader.

RULE

Paragraphs contain a complete thought.

Paragraph Length. As pointed out in the previous section, paragraphs need to be the right length. Is the correct length one sentence, ten sentences, 250 words, or 500 words? Unfortunately (for us engineers who like to think by the numbers), there is no numerical rule on the number of sentences or words in a paragraph. If a complete thought can be expressed in one sentence, and if that is all that you want the reader to know, then a paragraph can be one sentence.

> One Sentence Paragraph
>
> In developing this standard, other standard recommended practices from some of the largest manufacturers (Polk, LTC, and Toledo Gear) of gear reducers were compiled along with Tribology's knowledge of company applications to determine a common procedure that would encompass the vast majority of gear reducers in the company.

[This paragraph has other problems, but it was still published in a company newsletter.]

On the other extreme, a paragraph that goes on for several pages is very difficult to swallow and digest. A page or screen full of words just does not look right. Long paragraphs often reflect poorly on the author as a "windbag," which is a likely perception of readers subjected to an uninterrupted page of words. [*I would never even consider reading any document that contained even one page without paragraph breaks. This is a personality fault. It was probably brought on by having to read Milton's "Paradise Lost" in a high school English course. Boring!*] Some people never stop talking. Those are the people who write long paragraphs.

One rule of thumb for paragraph length is that good paragraphs are visually appealing with easily discernible and meaningful paragraph breaks on each page. Each page (or a screen view) needs at least one paragraph break, but more than three may be too many. This is about the extent of any numerical rule on the length of paragraphs, but it can help in identifying places where writing can be improved.

If a complete thought takes more than a full screen or printed page of words, then alter writing strategy to yield more bit-sized arguments. If you come up with "too small paragraphs," the same is recommended. Change your writing strategy to add more support to the thesis. One-sentence paragraphs are acceptable, but that should be all that needs to be said. As you adjust your writing strategy, strive to keep a balanced flow of paragraph breaks. The breaks just have to coincide with the "complete thought" rule. Each paragraph contains related sentences, an ending, and a transition that do not go on forever.

RULE

Paragraph length needs to be just right; just give enough information for a thorough discussion of the topic sentence.

Another writing concept that may help determine the length of paragraphs is the idea of parallel construction for lists. When authors make lists, each item in the list is treated in the same way as the other list items. For example, if the first item in a list begins with an adjective, then all the items in the list should begin with an adjective. If the first item is expressed in one sentence, then all the items in the list should be expressed in one sentence, if possible. This is the idea of parallel construction for lists. An example of this is the list of "Chapter Goals" at the beginning of each chapter in this book. These lists have parallel construction, as each list item begins with a verb.

The concept of parallel construction can also help one shape the length of paragraphs. Just think of all paragraphs as a list of thoughts or topic sentences. Like any list, you want a balanced presentation of each item. If you treat each topic sentence as parallel ideas with roughly the same amount of support, then you can achieve a good balance of paragraph length. Obviously some paragraphs may be more important than others. However, this does not mean they have to be longer.

The headings in a document are also useful tools in shaping paragraphs. Important paragraphs should be appropriately placed in your heading hierarchy. Important ideas do not require long paragraphs; they do deserve prominent positioning in heading structure. Likewise, a series of less-important but parallel topics can be broken down by headings. Good use of headings is thus another way to achieve an even balance and flow of paragraph breaks in documents.

Summary of the Elements in Style

Writing style is like personal lifestyles. Everybody is different in the use of words, sentences, and paragraphs. However, technical writing requires a style different from creative writing. It requires a style that conveys information rather than art.

Words must be chosen carefully. There are many types of words available to the technical writer, but only particular types of words should be used (active, factual, positive, and so forth). Be very aware of tone. Personal feelings (biases) can be determined from word choice.

Sentences and paragraphs also need good structure, length, and focus. Sentences should be concise and direct statements with a complete thought. All paragraphs need a topic sentence that is supported by the other sentences in the paragraph. These are just some of the elements of style. Other elements are described in subsequent Chapters.

7.2 Examples of Writing Styles

The writing style of an author often becomes apparent the second time that you read something by the author. Mannerisms become apparent. A formal style of technical writing by an accomplished author looks like the following:

Formal Technical Writing Style (P.J. Blau, *Friction Science and Technology*, Marcel Dekker Inc., 1995, p 204, used with permission)

The process of lubrication is one of supporting the contact pressure between opposing surfaces, helping to separate them, and at the same time, reducing the slid-

> ing or rolling resistance in the interface. There are several ways to accomplish this. One way is to create in the gap between the bodies geometric conditions that produce a fluid pressure sufficient to prevent the opposing asperities from touching while still permitting shear to be fully accommodated within the fluid. That method relies on fluid mechanics and modifications of the lubricant chemistry to tailor the liquid's properties—especially its viscosity and the dependence of viscosity on temperature and pressure . . .

This style is impersonal. It presents facts, and it uses enumeration in the paragraph to make a point. There are no personal pronouns, slang, idioms, or any of the "less-preferred" words. The grammar may not meet every English teacher's expectations, but the message is clear. This example mirrors a basic lesson advocated in this book: a clear message should be your goal. Strive for perfection, but do not make pursuit of good grammar the primary goal. Reasonable grammar, which obeys the rules of English, is adequate when your primary goal is the clarity of content.

Technical documents can have a less formal style than the previous example. Reports to peers or within a department may have a more informal style with some limited use of personal pronouns:

> **Informal Style**
>
> The Materials Engineering Laboratory has an on-going program to promote the use of ceramics and cermets for tools subject to corrosion. A limitation of the use of most ceramics is their inability to be machined in the sintered state. There are some ceramics that have been modified by addition of a conductive phase (like graphite) to allow electrical discharge machining, but to-date, these additions have had a detrimental effect on mechanical properties.
>
> It is the purpose of this report to present experimental results on machining of zirconia with conventional machine tools. We attempted milling with diamond-coated cemented carbide end mills. We will show surface finish and dimensional results in machining dies for edge trimming of polyester film support.

A descriptive style pertains to documents that need to explain something. The document may be an instruction manual or a report that explains a process or a design concept. These types of descriptive documents can be written in different ways depending on author style. For example, a description in the active voice has a different connotation than the passive voice:

> **Active voice**
>
> Put the oil indicator into the sump and connect it to the data logger.
>
> **Passive voice**
>
> The oil indicator is put into the sump and connected to the data logger.

The statement in the active voice is written as an imperative. It is telling you to do it like this.

In general, machine or process descriptions first explain the purpose of the device or process, the way it is used, and any essential background information on its importance. Reasons for the reader to continue or put the document aside are given. The overall size or dimensions of the device or the operating specifications and capabilities of a process are explained or described. Next component parts may be listed—what they look like and what they do. Then how the parts are put together and how they work as a unit are described.

> Descriptive Writing
>
> The paper-clip friction tester has been used since 1987 to measure the presence of wax on photographic films. It is basically an inclined-plane friction tester like those used in high school physics labs. In this version, a strip of film (usually 35 mm wide) is fastened with clips to the plane that is raised to produce motion of a rider. The rider is the radiused end of a paper clip. The end of the paper clip is placed on the web (vertically with its flatwise plane aligned with the long axis of the inclined plane). The plane is hinged at one end and the free end is raised until the rider motion is produced. The tangent of the angle of inclination of the plane when the paper clip starts to slide is the static coefficient of friction.

It is probably apparent at this point that this description needs an illustration. How does one slide a paper clip on-end down a plane? A sketch and a list of key parts would clarify things.

> Description (continued)
>
> As shown in Fig. 7.5, the paper clip is cut so that one loop extends about 0.5 inches from an acrylic sheet that is cut in a "U" shape with the paper clip at the valley of the "U." The "U" is inserted, and it straddles the plane that holds the film to be

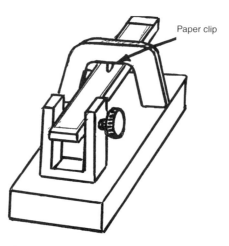

Paper clip

Fig. 7.5 Schematic of paper clip test

tested. The dimensions of all components and the placement of the inclined plane hinge are given in diagram A.

The details of the test procedure and reporting of data are available in the ASTM G 164 Standard Test Method. The output of the test is the breakaway friction co-efficient, which, in turn, will indicate if wax or another lubricant is present. If the value is less than 0.2, a lubricant is present.

It would take considerably more text to completely describe this device in words. The sketch probably saves three pages. This example illustrates the value of graphics in descriptive writing.

Some technical writing is done strictly to inform. There is no subliminal message, no attempt to persuade the reader. The goal is simply to inform on a subject that may be of general interest. Professionally written pieces in technical magazines often use this style.

Informational Technical Writing (Steel, Material for the 21st Century, *Advanced Materials & Processes*, ASM International, Jan 1996, p 29, used with permission)

In spite of inroads by a range of competing materials, steel is still the primary structural material because of its outstanding strength, ductility, fracture toughness, repairability, and recyclability. Over the past ten years, advances in steelmaking and processing technologies have enabled the development of a wide range of new steel products with improved properties. For example, combinations of closely controlled chemical composition, rolling practices, and cooling rates now permit the production of steels with enhanced fracture toughness and lower susceptibility to hydrogen cracking.

The expanded use of vacuum degassing, ladle treatment, and continuous casting has led to the introduction of ultra-formable steels for complex automotive parts and has provided improved electrical steels for the electric motor and transformer industries. The higher ductility of extra-deep-drawing steels has allowed auto designers to consolidate parts. At the same time, stamping reliability has been improved, as demonstrated by lower rates of breakage and splitting.

Improved cleanliness in bar products has led to longer fatigue life and opportunities for bearing downsizing. The expanded use of ladle treatment has also allowed the development of superior seamless pipe grades with improved toughness through closer control of composition.

Eloquent use of words, proper grammar, proper structure, tone, and presentation usually typify professional technical writing. Professionalism shows through. These articles also tend to be shallow in the technical aspects of what they concern. This is intentional. They are writing to a broad readership. [*Most of us practicing engineers can only aspire to this style of writing. It is a talent, and they have more than many of us. However, having a lesser command of language should not discourage you. Remember that technical writing is predominately task oriented. You can still obtain your writing objectives with ordinary language skills by following our technical writing suggestions for ordinary technical professionals.*]

A style of writing common among academicians is referenced writing. The author cites a reference every few sentences. The work of others needs to be acknowledged, but too many references can make the reader think that the writer has no personal knowledge of the subject and he or she is simply parroting others.

Referenced Writing Style

Attempts at lubricant pressurization have been repeatedly made. Direct supply to the tool interface was abandoned because the orifice became blocked with detritus [57]. Milford [58] proposed a powdered metal compact die through which oil could be forced. The concept of supplying oil through radial holes in a die was first proposed by Turner [59].

The author's intentions are noble; he or she wants to be thorough. Some readers prefer this writing style. Others [*moi*] are distracted by frequent interruptions to say what others have done. It is not necessary to cite everything ever written about a subject; it is appropriate to cite papers that support or challenge a thesis. References to the work of others should be included in background information and in discussing results. The remainder of the paper should be the author's words, not the words of others. In books, some authors do reference writing; others provide general references that readers can use if they want more information. Neither is wrong. They are different styles.

Another often-used style of technical writing is "mathematical" writing. The object of some research is to derive a model that can be used to predict some phenomenon. The article consists mostly of derivations and mathematical computations. Of course, mathematics, physics, and chemistry tests and textbooks are heavy in mathematical writing, but often the instructor helps students and provides ample time to digest derivations and models. Most journal readers do not have the time to wade through mathematical writing.

Mathematical Writing

For the stated conditions, monotonic plastic deformation occurs and equation 8 applies.

$$H > Q\left[\frac{PR + P}{Y(P^2 + Q^2)(1 - P^2)}\right] \quad (10)$$

This equation can be further simplified. Assuming $P = A_1 B$ and $R = A_2 Y$

$$H > Q\left[1 - \frac{\mu A_2 - A_1}{\sqrt{(A_1^2 + A_2^2)(1 - P^2)}}\right]$$

The coefficients A_1 and A_2 can be obtained from equation 11. And so on...

The writing style in the mathematical example only conveys a message to those who have the interest and time to follow the mathematics. If this is the goal, then let this be the style. If you wish to convey a message to a larger readership, limit the math and explain the thesis in words and illustrations.

On the other extreme, some technical documents are written in a conversational style. They are written like you would talk to a coworker.

Conversational Writing

I made forty microhardness indentations in the coated surface and gave the results to John Jackson for profilometer tracing. He traced them after lunch, and I got them back. Next, I cut the bar with my abrasive cutoff saw so that I could fit the samples in the SEM. Pete Fong said that he would do the job as soon as a hole opens in his schedule.

A conversational style is inappropriate for most technical documents. It reads awkwardly. The words used to communicate with coworkers sound fine because of accents and intonations that occur in speaking but do not read well. For example, you may say to your officemate, "I found the answer to shaft failure on the photomics I took." Stated properly in a report this becomes, "Metallographic examination of the shaft failure indicated that the failure was caused by low cycle bending fatigue." It is acceptable to be somewhat informal in writing, but do not write like you talk.

RULE

Do not write like you talk.

The final example of style is shallow writing. It is characterized by unsubstantiated statements and must be avoided at all times.

Shallow Writing

Polycrystalline zirconia is outstanding in wear applications. It is widely used in all industries. It is harder than steels and has low friction. It is also very corrosion resistant and will not rust like metals. It is proposed that this material be used to replace all machine parts that wear out. We can set up a zirconia manufacturing operation to supply the parts.

The author sounds like a used car salesperson. The use of superlatives like outstanding, widely, very highest, superior, fastest, and cheapest immediately raise a flag. These terms cannot be substantiated. This type of writing is the extreme opposite to referenced writing. When making a statement that something is harder or more resistant, present facts to support the claim. The style of writing in the above example will probably have low credibility.

Summary. Examples for eight styles of technical writing illustrate how a particular style is manifested in words. There are as many styles of technical writing as there are people writing. Each person writes differently. However, as shown by these examples, style has an effect on the reader. Just as you reacted differently to these examples, so will your readers. Use a style that will appeal to the intended readership.

7.3 Recommended Style

How should you write? What styles should you try to have? For good or bad, your style is somewhat predetermined by your persona—who you are, your personality, characteristics, likes, and dislikes. Previous sections describe the key elements of style: word choice, sentence and paragraph construction, and examples of different styles in technical writing.

The purpose of this section is to recommend a writing style consistent with the best practices for technical writing. [*The term "best practices" is used in U.S. industry for what most people think is the way things should be done to get good results. Pardon the euphemism, but in this case it aids concision.*] First the attributes of technical writing (summarized from discussions in Chapter 1, "What Is Technical Writing?") are described briefly, and then suggestions are given for some of the major components of style: content, presentation, format, and language skills.

Attributes of Technical Writing—Again. Technical writing style pertains to technical subjects. No need to use technical writing style in an e-mail to grandmother. Your writing should have a purpose and objective. Decide on both. What is your purpose? What is your objective? The purpose is what you hope to achieve with your technical document. The objective is the overall goal of the body of work. Style should be such that one conveys information in the form of test results, statistics, budget numbers, and so forth—facts. This is a key aspect of technical writing. Saying that you improved the slitting operation is insufficient. You should state that tool life has increased from one to twelve weeks and then calculate the savings.

Readers appreciate quantitative information. Technical writing is impersonal. You should not thank Henry Washiski for his help in completing your heat treating experiments. There are special places for this type of acknowledgment; these places are not in the text. Technical writing is concise. Every word and sentence is scrutinized to determine if each adds value to the document. If not, delete it; be brutal. Technical writing style is directed at a particular readership. Write to somebody. Determine who your priority readers are and write to them. Adjust the technical level to suit them. Technical writing is archival. The document should be written in such a manner for archiving and possible use five or ten years from when you wrote it. Finally, when you refer to the work of others, use appropriate attribution.

In summary, as stated in Chapter 1, "What Is Technical Writing," technical writing has certain attributes, and your writing style should give your readers technical documents with these attributes.

Content. The content of a technical document should warrant sharing. It should have some technical value with meaningful information. If you send out a technical document on a new piece of equipment that you saw demonstrated [*but have a zero chance of buying*], you may earn yourself the reputation as a purveyor of trite. Only write technical documents on matters that need to be shared, that have value in retaining, and that will produce some tangible savings or benefit to your employer. If you develop an innovative etching process for stainless steels, there may be no company value in sharing this information if your department is the only company unit that would ever use this process.

A technical report on a new chemical process that you developed may be suitable for sharing with others. What kinds of information/data should be included? Include only what is necessary to meet the purpose and objective of the document. Do not include details of every blend alloy encountered. Only report on things that matter, things that would produce value if shared. If you are a researcher writing about significant work that you completed, give sufficient details so that the work can be reproduced by others. A researcher may challenge your thesis and want to do what you did and see if the same results are obtained.

RULE

Be concise, but be complete, too.

Presentation (Format). Preferred formats have already been mentioned for various technical documents, but their importance is worth repeating. Format is an integral part of writing style, and Fig. 7.6 illustrates recommended formats for informal reports, formal reports, and papers. All technical reports need the same elements.

- Introduction
- Body
- Conclusion
- Recommendations

Informal reports can be one page followed by supporting data in graphs or photographs. Formal reports do not need to be very long, but they should have an abstract. Papers require an abstract and distinct requirements in the introduction, body, and ending matter.

Most engineering projects require the preparation of a formal report, and so subsequent Chapters focus on the preparation of formal reports. At this point, the main intention is to convince you, as an author, that your report should follow the proposed formats (unless there is a mandate from some higher authority or some other compelling reason).

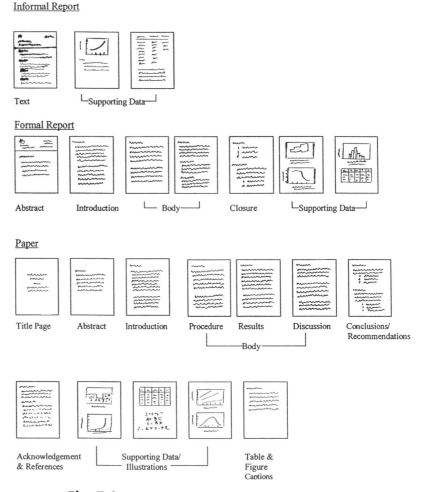

Fig. 7.6 Typical formats for various types of technical reports

[*One of the reasons for writing this book is my frustration with unintelligible reports from coworkers. In the 1990s, large companies continually experimented with various organizational schemes to improve profits. We now have teams for everything. I work with team members from at least a dozen different divisions. Each person is performing assignments, mostly laboratory experiments, and we report to each other in periodic reports. Some reports have the format of an abstract, conclusions, graphs without captions, and assorted spreadsheets. The conclusions are not really conclusions. What is written under the conclusions heading is really discussion.*

Another uses a different format: title, experiment name, purpose of experiment, list of variables, and levels for this experiment, list of responses for this experiment, summary of findings, and statistical program graphic output (several pages) with no captions and only an acronym title. The text under all six headings is usually less than one-half page. Yet another person uses another format: title, experiment name, contributors, data, experiment, purpose, objectives, experiment design, process conditions, test results, data summary, conclusion, path forward, graphs and photographs with no captions.

There is no consistency in reporting. Many documents also have little value, because they are not properly written and data are not adequately identified. Essentially, we are spending millions in research and development, sometimes without adequate documentation.]

Many engineers, researchers, and managers write reports without an established format. There is no consistency; important details are omitted. The net effect is poor communication. The recommended formats of technical reports (Fig. 7.6) are proposed as an effective way to instill some quality control in report writing. Avoid illogical formats, tomes, micro-reports with too few words, and "theatrical reports" with overwhelming section headings (bold, 18 point, underlined, and italicized).

RULE

Use accepted formats for report writing.

Standard formats are essential, and the format of the formal report has been accepted worldwide by the international technical community. It is even duplicated as templates in many word processing programs. You simply put words into the standard section heads dictated by the software. Most programs use the standard format similar to the one advocated here.

Language Skills. Make every effort to follow accepted rules of language. Spelling and punctuation must be correct. Grammar should be reasonable if clear communication is the goal, but make every effort to eliminate common mistakes and repetitive errors in grammar. Choose words and construct sentences so that ideas are unambiguous. Try to use the active voice, and avoid jargon or phrases that are unfamiliar or inappropriate for readers. Write to readers at a level meant for the intended readers.

Do not mix thoughts and write complex sentences. The length of paragraphs should be balanced with convincing support of the topic sentence in each paragraph. Conclude sections with transitions to the next section. Make sure there is continuity of ideas and coherence throughout the text.

Summary

Writing style is a reflection of your personality, but technical writing requires a style that achieves the attributes of good technical documents. Use a standard report format as a basic skeleton, and organize thoughts and information into a series of balanced paragraphs with clear and concise topic sentences. Do not make the reader wonder where you are going. Make the content something that the readership wants to know. Make sure sentences are clear, and be extremely careful of tone. Do not let feelings or mood show through your writing. Finally, have a trusted person read your work to make sure that your content, presentation, language skills, and style meet expectations.

In summary, the following concepts reflect the style requirements of good technical writing:

- Use an accepted report format.
- Choose concise words.
- Be unbiased and maintain a neutral, objective tone.
- Construct clear sentences of mixed length.
- Develop a balanced flow of paragraphs with clear and concise topic sentences.
- Avoid the use of unfamiliar technical words.
- Avoid acronyms. Many readers may not know what they mean.
- Use an impersonal style, especially for formal reports or papers.
- Adopt a positive style of writing, and write in the active voice.
- Avoid euphemisms and words with contemporary meanings that are different from dictionary meanings.
- Write with credibility; avoid the use of superlatives like superior, fastest, and so forth that are often just unsupported exaggerations.
- Avoid conversational writing in technical reports.
- Adopt the concept of parallel construction when listing items or shaping the length and content of many paragraphs
- Use topic headings to help organize and prioritize the flow of paragraphs.

Important Terms

- Tone
- Passive
- General to specific
- Enumeration
- Style
- Euphemism
- Superlatives
- Analogy
- Format
- Conversational

- Credibility
- Coherence
- Impersonal
- Jargon
- Word choice
- Cause and effect
- Active
- Technobabble
- Tome
- Persuasive

For Practice

1. Define writing style and explain why it is important.
2. List the personality traits that you think show in your writing. Are there any that would have a negative effect on technical writing?
3. What is the difference between style and tone? Why are they important?

4. Analyze the writing style in a newspaper front-page story. What tense is used?
5. What is wrong with euphemisms?
6. Write a procedure for changing an automobile tire. Review it for use of undesirable words (slang, euphemisms, passive words, and so forth), and correct them.
7. Write a paragraph on a recent vacation. Now edit it to be more concise.
8. State five negative words and five positive words.
9. Write a paragraph with an angry tone. Now rewrite it so it is impersonal and objective (no tone).
10. Write a paragraph in the passive voice. Convert it to the active voice and compare the two from the readers' viewpoint.
11. What is meant by parallel construction?

To Dig Deeper

- D. Beer and D. McMurrey, *A Guide to Writing as an Engineer,* John Wiley & Sons, 1997
- R.H. Fiske, *Dictionary of Concise Writing,* Writers Digest Books, Cincinnati, OH, 1996
- D. Jones, *Technical Writing Style,* Allyn and Bacon, Boston, 1998
- P. Ruben, Ed., *Science and Technical Writing: A Manual of Style,* Henry Holt and Company Inc., New York, 1992
- W. Strunk, Jr. and E.B. White, *The Elements of Style,* 3rd ed., Allyn and Bacon, Boston, 1979

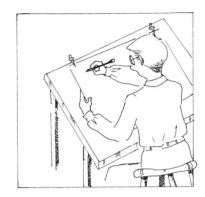

CHAPTER 8

Using Illustrations

CHAPTER GOALS

1. *Understand when to use illustrations*
2. *Understand what constitutes a helpful illustration*
3. *Know how to cite a referenced illustration in the text*
4. *Know how to write a proper caption*

AN ILLUSTRATION is a graphic or an image that conveys information or a message. There are basically three categories of illustrations used in technical documents:

- Photographs from digital recordings or silver-halide film
- Line art such as graphs, charts, or drawings
- Tables with numbers or words organized by row and column

Illustrations usually make a document more interesting, and they help clarify content by reducing words and enhancing the transfer, presentation, and interpretation of information. They help the author meet his or her objective, whatever it may be.

This Chapter describes various types of illustrations and how to properly use them. The overall objective of the Chapter is to gain an understanding of successful illustrations in technical documents. The Chapter discusses reasons for using illustrations, how to prepare various types of illustrations, how to write captions, and how to place and cite illustrations in a technical document.

8.1 Reasons for Using Illustrations

The answer to the question, "When should I use an illustration?" is best answered by explaining the benefits and functions of illustrations. Illustrations should be considered in every technical document to clarify points, help interpret information, and make the document more interesting and easy to read. In other words, try to use them effectively in all technical documents.

Interest. Illustrations increase interest and readability by giving readers a visual break. An uninterrupted page of words can be boring. Newspapers and magazines are full of illustrations. They create visual interest, and often photos are used to lure readers to particular pieces. The opposite extreme is a legal document like an insurance policy, which contains only words (and often long complex sentences). Which type of document do you like to read? Most people prefer a document with some photos, graphs, or other illustrations.

An author can use illustrations to increase visual appeal. A verbal description of a machine can be made more interesting with a good photo of the overall machine. [*It is standard practice in my lab to include a schematic or photo of test rigs used in studies. In fact, now that we have two digital cameras, this type of illustration is easier to do than ever before.*]

Another way to increase interest is use photos with people in them. A photo of equipment with an operator is more interesting than just a photo of machinery. People in photos give a sense of action and scale. They help animate photos of equipment and machinery, and this works for any type of document. However, be sure to get permission to use a person's photo in a document.

Clarification/Simplification. The classic use of an illustration is to clarify the assembly of a device or appliance. If you have ever bought a tool, or piece of furniture with "some assembly required," you recognize the benefit of an assembly drawing. Directions of any sort can usually benefit from an illustration.

Numerical data and calculations also can be simplified by the use of tables or graphs. Hardness conversions and temperature conversions are examples. Conversion can be accomplished without calculations when the information is presented in a graph or a table. Just look up 70 °F in a table in one column, and you will find 21 °C as the metric conversion. Another example is the tax computation table, which makes calculations easier when determining taxes for the government [*bless their hearts*].

Spreadsheet software on computers makes it very easy for authors to enhance documents with graphs and tables. Mathematical computations are also simplified by computer spreadsheets. [*We used to have to manipulate some very complicated formulas to calculate volume losses on balls used in wear tests. The ball would have a flat worn on it. We would measure the scar diameter and then use the formulas for a sector of a sphere to calculate volume lost. We now have a computer-generated table that converts scar diameter to volume—no more calculations and potential mistakes from a misplaced decimal point.*] In a similar way, tables can be used to include a very large amount of information, numerical or verbal, in a technical document.

Concision. "A picture is worth a thousand words" is an old and worn cliche, but it is still true. Illustrations help make a document more concise. This is always desirable. The following description has 150 words, and you probably still cannot visualize what the house looks like.

> The house is 36 feet wide and 62 feet long at the base. It has two garage doors, an entry door, and a barn door on the second floor facing the street. The second floor is smaller than the first floor with dimensions of 36 feet by 34 feet over the main part of the house, and there is a 26 foot by 26 foot second story over the two car garage. There is a 14-foot square tower in the center of the house with an attached deck. The tower has a peaked roof, and the remainder of the house has gabled roofs—one aligned north-south and one east-west. The house is gray with white trim. The roof is black, and the siding is cedar shingles.

In contrast, see Fig. 8.1. This photo replaces the 150 word paragraph, and it captures the content in a more memorable way. This is why almost all realtors use photos to describe properties.

Photos of people are even a better example. Pictures of people are interesting and descriptive. Can you visualize the words it would require to describe what you look like to a pen-pal who has never seen you? In this instance, a photo certainly may replace a thousand words.

Speeding Up Communication. Illustrations speed up the communication process by condensing information into a more useful form. In fact, a

Fig. 8.1 A hard-to-describe house

Table 8.1   1997 home sales on Edgemere Drive

Sale date	Address	Assessed at	Sale price	Delta
3/22/97	85 Edgemere Dr.	$105,100	$114,000	10.4K
8/30/97	131 Edgemere Dr.	107,300	84,000	−23.3
5/10/97	288 Edgemere Dr.	260,000	260,000	−0.6
3/22/97	660 Edgemere Dr.	89,300	87,000	−2
9/6/97	1112 Edgemere Dr.	117,600	152,000	−25.6
10/11/97	1272 Edgemere Dr.	373,500	386,000	12.5
9/20/97	1338 Edgemere Dr.	257,600	275,000	17.4
10/11/97	1450 Edgemere Dr.	153,500	114,900	−38.6
3/1/97	1545 Edgemere Dr.	27,000	26,000	−1
10/25/97	1590 Edgemere Dr.	229,000	200,000	−29
10/25/97	1593 Edgemere Dr.	20,000	20,000	0
9/13/97	1718 Edgemere Dr.	142,600	100,000	−42.6
12/20/97	1766 Edgemere Dr.	139,500	138,000	−1.5
4/12/97	1834 Edgemere Dr.	123,000	105,000	−18
10/11/97	1971 Edgemere Dr.	135,000	29,000	−104
8/16/97	1980 Edgemere Dr.	121,500	85,000	−36.5
1/4/97	2171 Edgemere Dr.	57,200	71,000	13.8
9/27/97	2223 Edgemere Dr.	78,500	33,500	−45
8/30/97	2337 Edgemere Dr.	68,000	54,750	−13.25
8/30/97	2346 Edgemere Dr.	209,900	193,000	−16.9
8/30/97	2547 Edgemere Dr.	55,000	53,000	−2
11/8/97	2572 Edgemere Dr.	112,300	80,500	−31.8
4/12/97	2644 Edgemere Dr.	110,000	65,000	−45
4/12/97	2650 Edgemere Dr.	86,000	40,000	−16.7
12/27/97	2704 Edgemere Dr.	148,700	132,000	−16.7
May-97	3191 Edgemere Dr.	186,700	185,000	−1.7
		$3,573,800	$3,083,650	$457,650

well-designed illustration can communicate content without any words. [*I subscribe to a do-it-yourself magazine that runs a feature in every issue where one page of cartoon-type illustrations with no text shows the reader how to build a simple project. The cartoon characters show the steps and materials needed to execute the project.*]

An organization chart is a good example of speeding up communication with an illustration. You can convey how your department works very quickly by simply including an organizational chart in your document. There are numerous other examples, but illustrations can be very effective in making the communication process quicker and more effective for your readers.

Easier Interpretation of Information. Graphs, charts, and tables are often indispensable when large amounts of information are presented. Well-designed tables or graphs can communicate information on one page that may otherwise require ten pages of words and/or numbers. A well-designed illustration also gives a visual structure that allows scanning and comparison of numerical or verbal content.

Tables are useful tools for communicating large amounts of repetitive numerical or verbal information, such as the examples in Tables 8.1 and 8.2. They allow easy scanning of the contents for comparison or quick lookup. However, tables can be boring, as in the case of logarithm tables or the phone book. Sometimes graphs and charts, which can be formatted in various ways, are more effective.

Table 8.2 Example of table that organizes verbal information regarding external laboratory facilities

Facility type	Disadvantages	Advantages
National or government laboratories	Difficult to own technology (public domain) Can be slow Limited resources Not integrated (labs do not work together) Significant infrastructure Generally for larger projects for programs	World class researchers Unique equipment Multiple funding sources to spread cost of work Research available to public domain
Independent laboratories	Expensive For larger projects or programs	World class researchers Unique equipment Specialized facilities or services Unbiased results/opinions Minimal corporate infrastructure
University facilities	Very specialized facilities Expensive Not well integrated (labs do not work together) Very slow response Difficult to own technology (public domain) Teach students technology who then work for competitor	Very specialized facilities Tax deductible World class researchers Graduate students are cheap Multiple funding sources to spread cost of work
Consortia	Very specialized focus Very slow response Committee driven Difficult to own technology (public domain or available to all participants) Technology available to competitors Expensive to join, must belong for life of program Results may not be exactly what you need (must benefit all participants)	Very specialized focus Multiple funding sources to spread cost of work
Testing laboratories	Specialized focus (metals, polymers, adhesives, surface analysis, analytical, environmental...) Expensive for specialized tests Nondisclosure agreements required for intellectual property protection Limited specialized equipment Task oriented/limited scope Slow response for specialized/nonstandard test and investigations Not well integrated (labs do not work together); not one-stop-shopping	Specialized focus (metals, plastics, adhesives, surface analysis, analytical, environmental...) Fast response for standard tests Cost effective for standard tests For a price, they will do whatever customer asks

Unlike tables, graphs and charts provide visuals that enhance the interpretation of data. The reader must compare data in a table. Graphs and charts can show comparisons or trends (Fig. 8.2), which definitely helps the reader. They catch reader interest, and they can be very effective in persuading the reader to your point of view. Most readers would glance at Fig. 8.2 and draw the conclusion that this organization has a severe safety problem. The transfer of this information is almost instantaneous—more effective than a word description or a table with numbers.

8.2 How to Prepare Effective Illustrations

The three basic categories of illustrations are:

- Photos
- Line art (graphs, charts, schematics)
- Tables

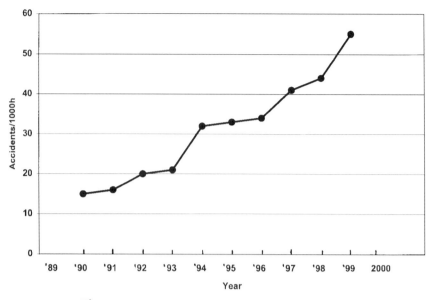

Fig. 8.2 Example of a trend shown visually with a graph

There are some general guidelines that can help make these types of illustrations more reader-friendly and effective. That is the purpose of this section—to present suggestions on how to prepare effective illustrations. A poor illustration can have the same negative effect on your readers as grammar or spelling errors. The objective is to produce illustrations that work for your reader.

Photographs

You only have two choices of media for capturing images as photographs: silver-halide film or digital files. Both can be either black-and-white or in color. The choice between digital image files and film often depends on the availability of equipment. Some people do not have both digital and film cameras. However, when one has the option, the choice between the two can become a question of speed and resolution. Speed depends on the specific type of equipment and/or film.

The resolution of two-dimensional image recorders (that is, cameras) is defined in terms of dimensional resolution and contrast resolution. Dimensional resolution is expressed in terms of pixels, which is just a shortened term for "picture element." Any two-dimensional image can be divided into any array of many small pixels. This pixel array defines the dimensional resolution and the size of the smallest feature that can be discerned. Ordinary televisions have a resolution of about 300,000 pixels, whereas digital high-definition televisions have a resolution of about 2×10^6 pixels. The dimensional resolution of a good digital camera (in 2000) may have a capability of 3×10^6 pixels in an array. Typical resolution with average 35 mm silver-halide film is about 20×10^6 pixels.

Contrast resolution refers to the level of shades that can be recorded or displayed. For example, digital cameras in the 1980s had 62 (2^6) gray levels for each pixel. Currently, most commercial digital cameras offer at least 256 (2^8) gray levels, although there are 12 bit cameras with 4,096 (2^{12}) gray levels and 16 bit cameras with 65,536 (2^{16}) gray levels (J.C. Grande, Principles of Image Analysis, *Practical Guide to Image Analysis,* ASM International, 2000). The contrast sensitivity of a recording device is important, because contrast is the underlying signal strength for observing images. If contrast sensitivity is low, then changes in gray scales may become less distinct, even with a high degree of dimensional resolution.

If image resolution is a priority, a silver halide photo can be better, although digital cameras are becoming more and more capable. Digital images may also offer advantages in handling, reuse, and image analysis. However, when a digital image is incorporated into a printed document or viewed on a monitor, the output device determines the final resolution. This final resolution must be considered when choosing between digital and film recording of images.

The resolution of printers and monitors is defined in units of dots per inch (dpi). The required dpi resolution of an output device (such as a printer or a monitor) depends on the required number of gray levels in a pixel. If an image is digitized into a pixel array, the output dpi can be virtually any number depending on the viewing device. Printers typically require a 4×4 dot array for each pixel when 16 shades of gray are needed. If correct reproduction of a 300 dpi image having 64 (8×8) gray levels is required, then a 600 dpi printer would be needed. A resolution of 600 dpi for line printers is suitable for most applications. A printer resolution of 1200 dpi is for very high-quality images.

Electronic images can also be produced from the scanning of photographs that are developed from film. This allows electronic manipulation like a digital photo. However, the resolution of the scanned image becomes limited by the resolution of the scanner, not the silver halide film. Generally, scanning at 300 dpi is sufficient for most photographs. The human eye cannot discern differences when scan resolutions are higher than this. Higher scan resolutions may also result in extremely large electronic files. It is also advisable to only scan from a continuous-tone (glossy) print. This assures good reproduction with a scan at 300 dpi.

It is poor practice to scan photographs from printed books. First, scanning from books may require a permission request for copyright clearance. Secondly, photographs in printed books are not like the continuous-tone (glossy) prints produced from film. Photographs in printed books are halftone images—that is, they are reproductions based on copy-dot patterns. These copy-dot patterns, which are like a screen covering the original image, degrade the image and can produce interference patterns (known technically as Moire patterns) after reproduction or scanning. Sophisticated scanning methods can apply reverse screens to eliminate the copy-dot patterns. However, this type of scanning is beyond the capability of most authors. Therefore, if you are scanning a photograph, it is advisable to scan from a continuous-tone (glossy) print.

When you generate electronic files for images, it is also important to understand the main differences between some common formats of electronic images. Electronic formats for current computer applications are defined by various acronyms known as file extensions (such as .tif, .gif, .jpg, or .eps). File extensions for some common types of image formats include:

- .tif for tagged images file format (TIFF)
- .jpg for joint photographic experts group (JPEG)
- .gif for graphics interchange format (GIF)
- .eps for encapsulated postscript file
- .doc for any document in Microsoft Word

There are some very important differences in these formats. Images in the JPEG format are intended for use on the Internet, and the files are compressed to a resolution of 72 dpi for quicker Internet access and viewing on screen (as most computer monitors typically have resolutions of 72 dpi). Images embedded in a Microsoft Word document are also compressed to 72 dpi. The resolution of 72 dpi may be suitable if you want to print just a few copies on your printer, but it is generally inadequate if you plan to use the files for print reproduction. If you plan to reproduce the photos (or submit them to a publisher for print or electronic publication), then the best format is a ".tif" file with a resolution of 300 dpi. This is the most adaptable format, and it ensures sufficient resolution for good reproduction. The other formats are for special applications (or one-time use in the case of .doc files).

Finally, the last question is the use of color. Should any photos be in color? The answer to this question hinges on the type of document being written and the reproduction facilities available. In the year 2000, most technical journals (without advertising) in materials engineering only accept black and white images. Journals in other fields may be different. For published papers, the use of color can be determined by reviewing the author instructions for a journal. For internal reports, the choice of color depends on reproduction capabilities. Many organizations do not have color copiers, and the use of color illustrations often means that the message intended by the color is lost. In fact, color should not be used in illustrations just because it is available. It should have a purpose. A color photo of a storage silo usually offers no value. On the other extreme, a color photo of a hot rolling mill will show what is hot and what is not.

RULES

Only use color in illustrations when color adds value.

Only scan photographs from continuous-tone prints.

The final consideration in photographs is composition. The best photograph is one made by a professional photographer in a studio. Professional photographers arrange lighting to prevent unsightly shadows and reflections. Studios have backdrops to prevent messy backgrounds. If you are taking photos yourself, consider lighting and background, and do your best to make the object of your photo prominent. Do what you can to remove "non-pertinent" items from the background (Fig. 8.3).

In summary, photos should be used when it is very difficult to describe something in words or schematics. They are often more costly to reproduce than tables or line art, but electronic photos are becoming more and more convenient for authors, publishers, and printers. When photos are necessary, decide on the needed resolution, and then take the photo with proper lighting and background. Make a point with the photo, and do not use color without a good reason.

(a)

(b)

Fig. 8.3   Examples of good and poor background in photos. (a) Photo with uncluttered background emphasizes the subject. (b) A cluttered background is distracting.

Line Art (Graphs, Charts, Schematics)

Line art refers to illustrations that can be drawn with lines, text, and lines formed into letters, words, and sentences. In the past, artists with drawing instruments and a drawing table prepared line art. Today, computer software can do the job. The most popular types of line art are illustrated in Fig. 8.4. Like photos, line-art illustrations can be produced in print or electronic form. If you plan to reproduce illustrations from electronic files (or submit them to a publisher for print or electronic publication), then the best electronic format for line art is a ".tif" file with a resolution of 600 dpi. Scanning or saving line-art illustrations at 600 dpi gives a crisp image without excessively large files (unlike photos, which should be scanned or saved at 300 dpi).

Schematics are simplified sketches of a process or object. The schematic in Fig. 8.5 illustrates how a particular test rig works. A block of test material rubs against a continuous web of film. It is not dimensionally accurate. It was made to present the concept of the device. It is easier to see what is happening in a schematic than if a photo was taken of the device. The extraneous items have been removed. This is how object/machine schematics should be made. They can be made with computer drawing programs, some word-processing software, or by hand.

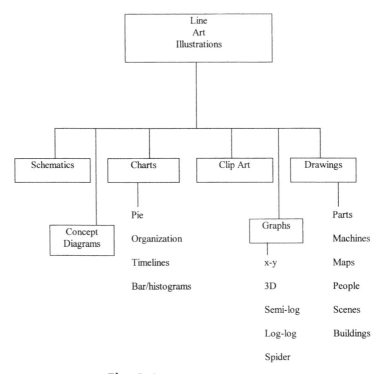

Fig. 8.4 Principal types of line art

Figure 8.4 is an example of a concept diagram, where the types of line art are grouped in specific categories. There are many concept diagrams used in this book. They organize concepts into groups with boxes, circles, and so forth, and then show how the concepts interrelate. Concept diagrams can show where plans, processes, concepts, or documents are going. Done properly, they help the reader to follow your thought process.

Charts. Line-art illustrations also include charts, which are visual presentations of numerical or verbal information. These are a type of informal line art that has become prevalent with computer spreadsheet software and tabular data. Charts differ from graphs in that they are less mathematical than graphs, and they may not be appropriate for showing trends. In simpler terms, charts are usually applied to business information. In contrast, graphs are used to present trends and scientific information. Their use, however, can overlap.

Pie charts (Fig. 8.6) are a classic way to show the relative portions of a whole. They are favorites for showing the sources of revenue and expenditures for government organization, schools, and businesses. Their weakness is that they only work well if the pie is divided into a relatively small number of pieces and if the pieces are significantly different in size. A pie of five pieces—20, 16, 19, 14, and 18%—will challenge the reader to visually discriminate differences in size. The labels on the pieces may need to display the percentage figures.

Organization charts (Fig. 8.7) are extremely helpful in technical communications for showing department personnel or functions. They are also

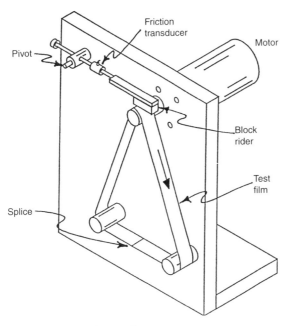

Fig. 8.5 Schematic of a device

extremely helpful to authors in developing document distribution lists. Blocks denote positions or functions, and lines between blocks show responsibilities. Most desktop software packages have the capability of generating these kinds of illustrations, and many company directories contain embedded organization charts for use in developing distribution lists.

Timelines (Fig. 8.8) are helpful illustrations for documents relating to project management. They detail tasks, who is to do them, and by when. When properly executed, they contain a significant amount of information and can guide the execution of a significant project. Similar to applications for creating organization charts, computer software programs can facilitate preparation of project management tasks. Some of these programs allow enhancing the data with shading, cross-hatching, and other types of adornment. Many times these features make the illustration so busy it loses its readability.

RULE

Keep timelines simple.

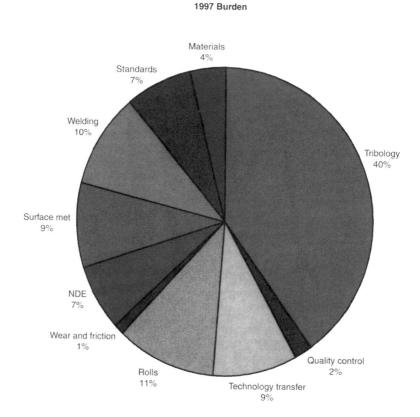

1997 Burden

Fig. 8.6 Example of a pie chart

Bar charts (Fig. 8.9) are well suited to illustrate relative properties of a number of items. They are interpreted easily by readers, but there are precautions to keep in mind in their use. Computer software offers many ways to make bars. You can have bars that are cumulative of two, three, or four quantities. The bars can be horizontal or vertical, shaded or not; there can be three or four bars for each item. Bar charts can look strange if there are either too few items (Fig. 8.10a) or too many items (Fig. 8.10b) on display.

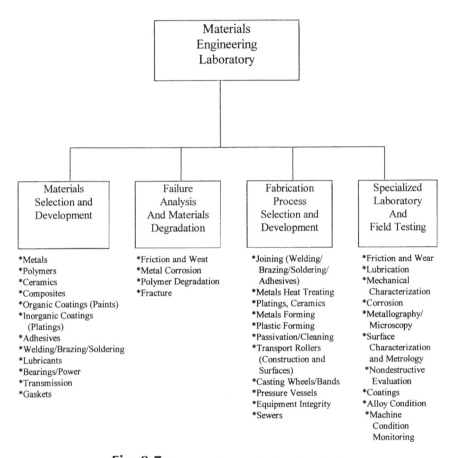

Fig. 8.7 Example of an organization-type chart

	Start	Finish	10/4	10/11	10/18	10/25	11/01	11/08	11/15	11/22	11/29
Task 1	10/4	10/18	Δ	→	♦						
Task 2	10/18	11/01			Δ	→	♦				
Task 3	10/18	11/15			Δ	→	→	→	♦		
Task 4	11/01	11/28					Δ	→	→	→	♦
			Δ	= start							
			♦	= finish							
			→	= progress							

Fig. 8.8 Example of a timeline

Do not use "monster" bars if you only have a few items to compare. Make the chart small. Do not put too many items on a single bar chart. Use several charts if necessary. Try to keep the background uncluttered with lines and chose shading and colors that do not overwhelm the message intended by the chart.

Histograms (Fig. 8.11) are very useful forms of bar charts that show the distribution of a large amount of data. For example, they can be used to show which age groups vote in an election or the distribution of hole sizes in a thousand "identical" parts. They are great tools for interpreting data and determining if data are normally distributed.

Clip art is available on most personal computers and is intended to add interest to a document. There are various opinions on the appropriateness of this computer product in technical writing, but it is recommended that you not use clip art in technical documents for decoration. The reason for this opinion is that it may impugn credibility. Clip art images are usually frivolous, whereas technical writing is purposeful—not entertaining.

Clip art is more appropriate on newsletters and department notices but not in formal or informal technical reports. The one exception for a report may be a department or business logo, which can be tastefully included on the report cover sheet as clip art. This type of use is acceptable in technical documents that start with a standard format. Clip art is in the public domain. It can be used without approvals or attribution but is not permitted in papers for most technical journals. Restrict its use to printing for department parties, golf outings, and club activities.

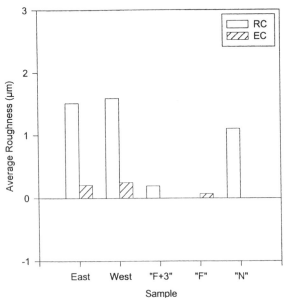

Fig. 8.9 Typical bar chart

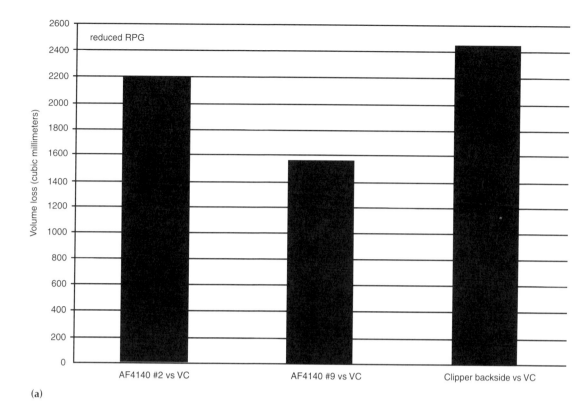

(a)

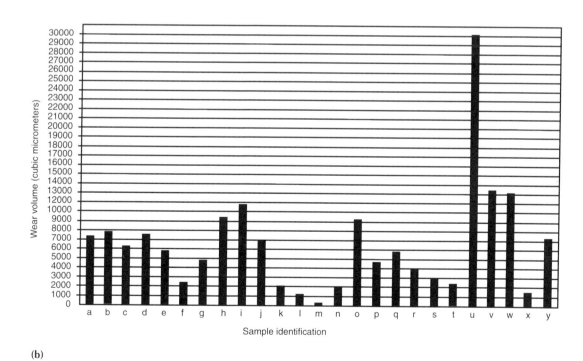

(b)

Fig. 8.10 Bar charts with (a) oversized and (b) undersized bars

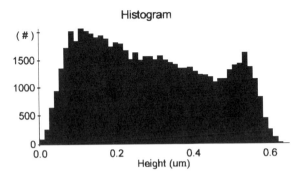

Fig. 8.11 Example of a histogram showing the distribution of a population

Graphs are an indispensable part of many technical documents. They interpret data and serve as the basis for models and theories. Figure 8.12 shows some of the graph options available. The independent variable is usually plotted horizontally; the dependent variable is usually plotted on the vertical axis.

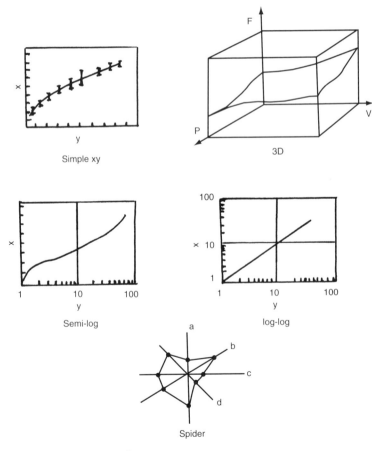

Fig. 8.12 Types of line graphs

As an author of a technical document, you have the responsibility for selecting a graph type (if it is felt to be necessary and to the point). The simple *x-y* graph is the easiest to interpret. These types of graphs should be enclosed with tick marks at logical increments. The plotted points are denoted by a symbol such as a box or triangle, and you may or may not fit a line to the points—your choice. Some suggestions for readability and clarity are:

- Do not plot too many variables on the same graph, where the plot lines obscure one another or are difficult compare (Fig. 8.13).
- Do not use so many graphs that they distract from reading the document.
- Do not clutter the plot area with horizontal or vertical lines associated with the graph increments.
- Clearly label the vertical axis vertically and the horizontal axis horizontally (Fig. 8.14).
- Avoid plotting complex functions that are difficult for ordinary readers to interpret (Fig. 8.15).
- Do not use odd-scale increments (as many computers with auto-scaling often do).
- Enclose the graph within a border.
- Use error bars to denote statistics on plotted data (plot the mean with error bars of plus and minus two standard deviations).
- Select scales such that the curve is centered in the graph.
- Use mathematics (least squares, and so forth) to fit lines to the data.

Three-dimensional graphs are fun to make, but they usually miss the mark in readability. Not many readers grasp your point. They should only be used for limited readers.

Semilog graphs are useful when one of the plotted quantities varies over decades. Log-log graphs are somewhat like three-dimensional (3D) graphs in applicability. The average reader has difficulty with them. They tend to make a linear relationship out of just about any variable. Spider graphs also fit into the 3D/log-log category. They are hard to interpret.

Videos. Computers have made it possible to insert animation into computer documents. [*I have been working with finite element specialists on some metal forming problems. They have sent me reports where you click on an application icon to show a moving punch. The part is blanked and formed in the simulation sequence, and the resultant shape of the part is displayed. It is really quite impressive.*] These types of visual aids produced by simulation are not trivial (or cheap) to generate. It is not necessary to use them unless the animation offers value. Sponsors of engineering projects only want to pay for activities that have the potential for either a technical or financial return. In some cases, video recordings are part of an experiment or test. In this case, treat the video like other data—put it in the report as an appendix.

Drawings. Engineering drawings like the one in Fig. 8.16 are the last type of line art given in Fig. 8.4. Engineering drawings are often included in edited form in technical documents. Most engineering drawings contain extra "boilerplates" that need to be deleted for concision when drawings are part of a text document. Machines, assemblies, or test setups are usually drawn in schematic form (for example, Fig. 8.5), because a normal assembly drawing may be too complex for average readers. Maps are

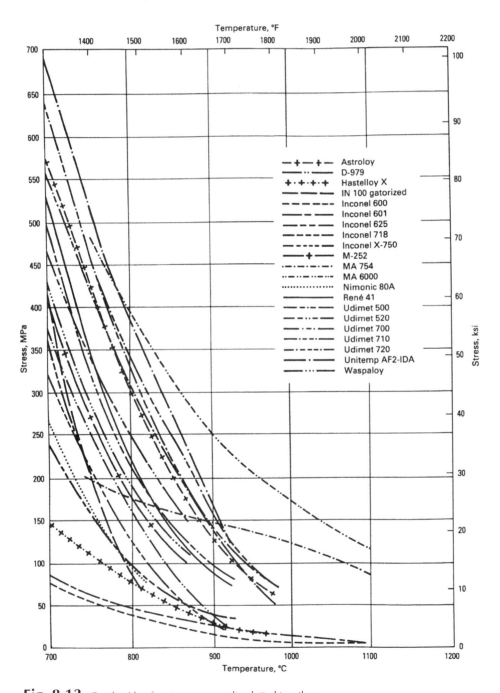

Fig. 8.13 Graph with a few too many results plotted together

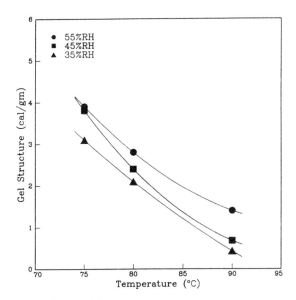

Fig. 8.14 Preferred appearance of a graph

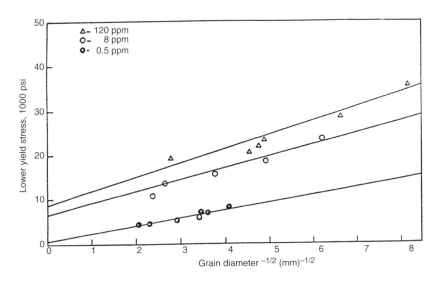

Fig. 8.15 Graph that plots a complex quantity (the inverse of the square root of the grain diameter)

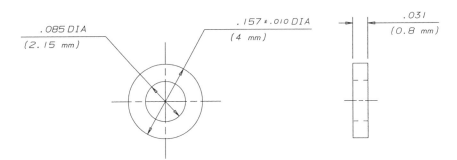

Fig. 8.16 Typical engineering drawing

appropriate illustrations for surveying and site plan documents. People drawings may be appropriate in biology or medical documents. Scenes and buildings are typical illustrations in architectural documents.

Summary of Illustration Preparation

All types of illustrations can be useful tools in technical documents. Use the types that make your point and remember proper ways to use them, as suggested in this section. If you are using data from others, use an attribution in the caption in the same format as that used for text reference. Seek permission from the original publisher when a published illustration is being reproduced in another published work. Permission to reprint materials is requested in writing by a form that varies from publisher to publisher. If you are seeking permission to reprint published material, the publisher of your work can provide permission forms to submit to the original publisher of the materials.

RULE

Only use illustrations when they add value (understanding, clarity, and so forth).

8.3 Captions for Illustrations

As described in the previous section, the value and readability of technical documents can be enhanced by the use of properly prepared graphs, charts, photos, and tables. However, the benefits of an illustration can be negated by an inadequate caption. This is a common problem with illustrations. For example, symbols, measurement units, abbreviations, or legends are sometimes inadequately defined in an illustration. This nullifies its usefulness.

Part of the problem with improper captioning of illustrations is that computer software may not allow proper positioning of a caption above or below an illustration. Everyone has a different style regarding figure captions and table titles. Sometimes captions are placed above a chart or graph, or they may be placed below the graph. Generally captions are placed below figures, while table titles are placed on the top. This is a matter of style and convenience; just be consistent within a given document or report format.

The key issue is the contents of a figure caption or table title. The contents must explain what is shown in the figure, and the caption must give readers all pertinent facts to interpret it. Figure captions must cite the source of the data and give any attribution, if the figure is taken from the work of others. In the case of tables, background information and attribution are often cited as footnotes at the bottom of the table. This is one reason why table titles are typically placed on the top in large type, while footnotes are in small type at the bottom. This is the basic style for this book, at least.

All figure captions and table titles should pass the person-on-the-street test. The meaning of the illustrations or table should be evident on reading its caption or title. It is important to use phrases for the beginning of any table title or figure caption. Do not start with a full sentence! The caption should be phrased like a title for a paper. Just state in a clear, concise phrase *what* the illustration shows. If any further explanations are needed, include this information as supporting sentences after the caption or as footnotes in a table. Use this book or a teaching textbook on any technical subject for other examples of how to write a proper caption. A proper caption should look like those in traditional textbooks.

With most computer programs, table titles and footnotes are easily produced along with the body of a table. Computer software for graphs, charts, and drawings is another matter. If your computer software does not allow you to put a proper caption on a graph, then it must be composed separately and put in place. It is wise to include a list of figure captions at the end of a paper or report. This helps you keep your figures and photos organized.

RULE

**Every figure caption and table title leads with a
clear and concise phrase of *what* is shown.**

8.4 Referring to Illustrations

All numbered illustrations must be referred to in the text body, and the following are basic recommendations:

- Use the term "figure" to refer to any graphic, photo, or piece of artwork.
- Use the term "table" to refer to any type of tabular format that can be typeset or presented with spreadsheet or tabular programs.
- Number the figures and tables in sequence with their mentioning in text.
- Put the figure or table number preceding the caption or title (Figure 1, 2, 3, and Table 1, 2, 3, etc.).
- Include attribution if the illustration belongs to another.

Placement of Illustrations. There are two options on where you want to place illustrations within a document. In some organizations, it is acceptable to put illustrations at the end of reports. In fact, it is usually done this way on informal reports. Papers submitted to archival journals also require illustrations placed at the end of the text. Figures may have no captions but only penciled identifications such as "Fig. 7." The figures are preceded by a typed list of figure captions and table titles. Most book publishers also want illustrations submitted as separate addenda that immediately follow a list of captions. The reason for this is that illustrations usually require distinct processing that is different from the text body.

Imbedding figures within the text of some word processing programs may also reduce resolution, as previously described in the section on photographs.

Nonetheless, embedded illustrations look more professional, and these can be provided if your computer skills and equipment allow cutting and pasting, scanning, and file transfers. Each illustration should be placed at the first appropriate break after its first reference in text. If a formal report is intended only for your organization, the illustrations can be embedded or placed at the end of the report. However, embedded graphic or art files may not reproduce with the same resolution as original files. In either case, complete captions are always required and placed under the figure and/or in a caption list.

Summary

Graphs, tables, and photographs add credibility, clarification, concision, and interest to a report. If you make the statement "accident rate has risen dramatically in the last quarter," most readers expect a graph to follow with data that illustrate how the number of accidents varied over the last year or two. Good graphs include error bars on data points reflecting the application of statistics to the test results. The use of illustrations in technical reports is strongly encouraged, and the following guidelines summarize some preferred practices for their effective use:

- Keep graphs simple; 3D graphs are discouraged.
- Avoid plotting complex mathematical functions as a scale.
- Avoid acronyms in captions, titles, and axis labels.
- The caption should describe what the graph shows and cite sources of data.
- Show all data points on graphs and show any curve fitting line and/or error bars.
- When using multiple graphs to compare items, use the same vertical and horizontal range. Beware of self-scaling computer programs.
- Wherever possible, start the scales for numerical data at zero.
- Use a key to identify multiple data sets.
- Graphs and photos should have a surrounding border, only if they are embedded in the text body. Borders are not needed if illustrations are placed on separate pages at the end of the article.
- Tick marks on graphs should have whole number increments (not $22\frac{1}{2}$, $24\frac{1}{2}$, $26\frac{1}{2}$, etc.).
- Make it clear in the text why you are showing an illustration.
- Consider the readership in designing illustrations.
- Be aware that the preparation of photos and artwork may add cost but that the handling and duplication costs are reduced when electronic files with sufficient resolution are used.

- When scanning photographs, scan from continuous-tone prints and scan to a resolution of 300 dpi. Do not scan photographs from printed books.
- When scanning line art, scan from good prints, and scan to a resolution of 600 dpi.
- The preferred format of electronic art for reproduction is a TIFF (.tif) file.
- Embedding art into a Microsoft Word document causes compression to a resolution of 72 dpi. This is generally unsuitable for good-quality reproduction, but may be suitable for internal reports.
- Horizontal and vertical axes should be identified by parallel lettering followed by the units.
- Attribute the source if data or illustrations are not yours.
- Use color judiciously. Avoid the use of color if color reproduction is not available.
- Graph titles generated from art software usually do not constitute an adequate caption.
- Use decimals in numeric tables and line up the decimal points.
- Tables with more than twelve rows should be put in an appendix.
- Label each row and column in a table.
- Avoid putting font sizes smaller than 10 points in tables (no "funny" fonts).
- Use as few lines as possible in tables (horizontal and vertical).
- Put lengthy derivations and calculations in an appendix.
- Do not mix lettering styles in an axis.
- Do not interrupt text with large numbers of illustrations; put them in an appendix.
- Surround illustrations with white space.

Important Terms

- Graph
- Timeline
- Pie chart
- Caption
- Line art
- 3D graph
- Axis label
- Digital photo
- Line graph
- Schematic
- Appendix
- Bar graph
- Organization chart
- Table
- Spider graph
- Histogram
- Credibility
- Log graph
- Resolution
- dpi

For Practice

1. Present data from an experiment in which you measured the density of a plastic foam at 50° increments from room temperature to 325 °F. Density values were 1.2, 1.4, 1.7, 1.7, and 1.7 g/cc.
2. Make a word table showing what you ate for a week.
3. Draw a schematic of a paper cutter.

4. You measured the morning temperature outside your house/apartment every day for two weeks. Plot the data and write an appropriate caption.
5. How do you put a digital photograph into a report?
6. What is a continuous-tone print?
7. What is a half-tone print?
8. Critique a graph taken from a published journal or newspaper. (What is done right or wrong?)
9. What is wrong with using a spreadsheet full of numbers as an illustration?
10. State five requirements of a good graph.
11. Describe the proper way of labeling the axis on a graph of temperature versus time of day.
12. Draw the organization chart for the organization where you are currently employed or at your last place of employment.

To Dig Deeper

- L.V. Anderson, *Technical Writing—A Reader-Centered Approach,* 3rd ed., Harcourt Co., 1995
- M. Markel, *Writing in the Technical Fields,* Institute of Electrical and Electronic Engineers, Inc., 1994
- S.E. Pauley, *Technical Report Writing Today,* Houghton Mifflin Co., 1979
- M. Rose, *Technical Communication,* Prentice Hall, 1997
- R.E. Wileman, *Visual Communicating,* Education Technology Publications, Englewood Cliffs, NJ, 1993

CHAPTER 9

Formal Reports: The Outline and Introduction

CHAPTER GOALS

1. *Understand the importance of an outline*
2. *Understand how to choose a title to invite reading*
3. *Know what constitutes a proper introduction*

REPORTS are the most frequent writing situations for most technical professionals. They are appropriate to propose and conclude a project, request a capital expenditure, a process change, a personnel change, and so forth.

Because reports are so important in technical writing, this book devotes three Chapters to them. It is the purpose of this Chapter to discuss how to start a report. It addresses the outline, title, and introduction. Subsequent Chapters address other portions of a formal report.

9.1 Outline

Sometimes the decision on whether to write a report (or not) is made by your boss or client. You are requested to write a report on a test, project, or whatever. The very first step in technical writing is to decide to write a document on some aspect of your work. If you just completed a project, a report is the logical way to present results.

After the decision to write a report, the next step is to decide on readership. Who do you want to share results with? Certainly the person, department, division, or company that paid for your study deserves to be listed as primary reader, but there may be others who are also interested. They should be noted.

Next, think about the reasons for writing the report. Are you trying to get funding for a project? Are you researching a mechanism? Whatever the purpose or purposes, state them in the outline and focus on them in the outline.

Sample Outline

Title

Fretting Corrosion/Resistance of Materials to Fretting Damage

Readership

1. Company machine designers
2. Publish paper in STLE proceedings

Purpose

1. Summarize lab studies
2. Make recommendations on material selection

At this time, you can start with just a preliminary working title. The exact title can be finalized later.

Next, it is necessary to state the objective of the work.

Objective

Prevent fretting corrosion damage of tool inserts in capping presses and reduce tool replacement costs 20%.

At some point in this initial stage of drafting an outline, deciding on the type of document is also necessary. You can mentally scan the options that were discussed in "Writing Strategy," Chapter 4. This example is a formal report that will be submitted as an article to a journal. Often, the only difference between the two is that the former may contain some proprietary information, and the latter may have some format conventions that are different from the accepted format of your organization. As mentioned a number of times [*and there will be more*], a formal report contains an introduction, a body made up of an investigation, results and discussion, and a concluding section. You could list these section heads in your outline and start to put in details, but some writing professionals prefer to start by simply listing the major points that they want to make in their report.

Major Points

1. Fretting corrosion is a problem.
2. Type of test used—not a normal test
3. Range of materials and coatings tested

4. Carbide against other metals produces low damage.
5. Dry film lubricants did not help.
6. Unique system for measuring damage
7. Application of results
8. No liquid lubricant allowed
9. Others try to solve problems with reduced motion or oils

These major points are not necessarily in the order that you will put them into the report. They are reminders that you want to make these points some place in your document.

Now you are ready to work on a more formal outline. It is usually helpful in the formal outline to assign a hierarchy to sections. There should be A, B, C, and maybe D heads for sections. Capital letters can be used for A-heads, numbers for B-heads, lower case letters for C-heads, and italics or Roman numerals or some other designation for D-headings.

- A-head = A, B, C, and so forth
- B-head = 1, 2, 3, and so forth
- C-head = a, b, c, and so forth
- D-head = i, ii, iii, iv, and so forth

These heading designations help to visualize the sections and length of the document. The outline at this point looks like Fig. 9.1. This gives you an idea of how the document may come together. Each of the subheads in the body ends up as at least a paragraph and possibly as several paragraphs or several pages. If you go to small fonts, you can add a topic sentence for each subhead. This helps define each section and ensures that you discuss your work in a logical manner. When you get into the thick of writing, you may want to repeat this outlining procedure for each of the B-heads in the outline.

In summary, an outline is a plan or strategy for writing. An outline is needed on almost all documents. Even letters can benefit from an outline. A detailed outline will include topics for each paragraph. There is probably no technical writing situation that could not benefit from an outline. Use the suggested format or whatever feels right, but do not skip this step. An outline keeps a writer from straying, it collects ideas, and it helps assess an entire document for logical presentation. The more detailed the outline is, the easier it is to write the document.

RULE

Make an outline before you start to write any report.

OUTLINE

Title: Something about fretting corrosion

Readership: 1. Company machine designers
2. Public – paper in TLC proceedings

Purpose: 1. Summarize lab studies
2. Make recommendations on material selection

Objective: Prevent fretting corrosion damage of tool inserts in capping presses and reduce tool replacement costs 20%

A. **INTRODUCTION:**

1. Background – manifestation, cost, importance, literature
2. Purpose – (above)
3. Objective – (above)
4. Format of document

B. **LABORATORY STUDIES:**

1. Scope of work (#2 range of materials tested)
2. Type of test used (#1)
3. Damage measurement (#5)
4. Lubricants tested (#7)

C. **TEST RESULTS:**

1. SEM's of worst and best couples
2. Volume loss data (#3)
3. Lubricant effectiveness (#9)

D. **DISCUSSION:**

1. Why long stroke was acceptable
2. Comparison at results (#8)
3. How to apply results (#6)

E. **CONCLUSIONS:**

Enumerate

For in-plant only

Fig. 9.1 Formal report outline

9.2 Title

A title is an essential part of the outline. All reports must have a title, and the outline process must define the title, establishing the readership and type of report. Many letters do not have or need a title, but often letters do have subject lines for the benefit of the reader. Some computer mail software also prompts you to title a letter or note. It can help.

Titles must be carefully chosen. They are essentially abstracts of documents, and if you want somebody to read a document, you must create a title that encourages a person to read it. In addition to interesting a potential reader, a title must be accurate, concise, grammatically correct, and free of jargon and acronyms. The title must reflect what the document contains. If you wear tested six plastics, do not call the document "Wear of Plastics."

You did not test all the plastics that exist. Call it "Wear of Selected Olefin Thermoplastics." Make the title reflect the scope of the work. Do not make the title look like an abstract:

> Too Long a Title
>
> The Friction and Wear of Cellulose Triacetate under Conditions of Varying Humidity, Temperature, and Surface Temperature Using a Slow Speed Pin-On-Disk Apparatus

Try "Friction and Wear of Acetate Films in Controlled Environments." Do not make mistakes in the spelling or grammar of a title. This makes all of your work suspect. People expect at least the title to be checked and rechecked. If it still has an error, this may brand you as a lesser person.

> Title Containing a Typo
>
> The used of perfluoroethers to lubricate magnetic media

Often technical reports pertain to plant equipment with strange-sounding names. Try to convert the machine names for the layman; avoid jargon. Instead of writing a report on "D-Min Rating of CFMB-A on Wide Roll Trials," write it on "Sensitometric Properties of a New Cardiology Film." Sensitometric, which is in the dictionary, is more informative. In fact, you should try to limit the words in a title to words in the dictionary. If anyone wants to find out the meaning, they can easily look it up.

Finally, titles are the first thing that your prospective reader sees. They decide to read or not read your report based on the six or seven words in the title. [*Once at a conference, I ended up in a very boring talk. I passed the time critiquing some of the article titles at the conference. As shown in Fig. 9.2, they ranged from too long to too complex to simply not appealing in tone.*

Example A is too long. When your title exceeds one line, it is a candidate for trimming. Example B does not tell the reader what he or she will learn about chemical coatings. All of the articles were about coatings; it was a coatings conference. A title needs to be specific enough to allow a decision on whether the article is worthwhile to read. The edited title tells the reader that the article is about certain types of chemical coatings (chemical vapor deposited polycrystalline diamond) and the author will concentrate on tribological (friction and wear) properties.

When I read example D, I knew the talk was about ellipsometry—an optical technique for measuring the thickness of films that may be only 10 atoms thick, but the title suggested that it was a review article. A review article is one that reviews previous work and does not contain new information. However, the article was really about instrument improvements that make this measurement technique very easy and cost effective to use.

a. **Original title:**
Finite element solutions comparing the normal contact of an elastic-plastic layer medium under loading by (1) a rigid and (2) a deformable indenter.

Edited:
Effect of indenter stiffness on finite element modeling of coatings

b. **Original title:**
Diamond Coatings

Edited:
Tribological properties of vapor-deposited polycrystalline diamond on cemented carbide

c. **Original title:**
SADS technology and its implication for deep submicron devices

Edited:
Silicide diffusion sources for manufacture of semiconductors

d. **Original title:**
Modern ellipsometry: A new surface and film analysis method with an old name

Edited:
Ellipsometry improvements that make film thickness measurements easier

e. **Original title:**
Vanadium oxicarbide films prepared by CVD from vanadil acetylacetonate

Edited:
Vanadium coatings to solve wear problems

f. **Original title:**
Superhard tips for AFM applications

Edited:
Superhard atomic force microscope tips

Fig. 9.2 Titles (real) that may be improved by editing

The new title attempts to tell the reader that there is new information in the article that may help the reader solve a coating-thickness problem.

Example E may frighten readers away with big words. It was really a readable article with usable information. Unfortunately, many readers would pass this article by because of "title scare."

Example F needs little editing. It is concise and tells AFM people that it is about new tips. These tips are the key to the instrument. As much as I dislike acronyms, this particular one is known to all coatings people, but to a newcomer to the field, it may mean "Air Force Museum." Acronyms are never right in titles, but they are widely permitted in conferences and proceedings on specialized technical subjects.]

Summary. Your choice of words in a title is very important. It advertises your work, and everyone knows how important advertising is. The title must accurately describe what is in the document; it must be honest. It needs to be concise; potential readers may sense a wordy document if the title is too long. It must not be too foggy or intimidating. Titles must be just right. Give title selection the consideration it deserves.

RULE

Make your title "sell" the document.

[After due consideration, I decided that the appropriate title for the fretting corrosion report is "Fretting Corrosion Resistance of Tool Materials." This reflects the type of damage to be discussed (fretting corrosion), and it states the types of materials studied (tool materials).]

9.3 Front Matter

Most documents require some sort of front matter that identifies the document and contains background information. For example, letters have the traditional salutation with the title and the address of the recipient.

Salutation for a Business Letter

January 2, 1997

Susan Pross
Travel Director
Eastman Kodak Company
343 State Street
Rochester, NY 14614

Dear Ms. Pross:

In reply to your letter of 25 December concerning my account, please find enclosed…

I hope this satisfies your inquiry. Please contact me if you need additional details.

Very truly yours,

Kenneth G. Budinski
KP Materials Engineering Laboratory
5/23/Kp, MC 23423

Some computer mail systems allow the use of letter templates that automatically record information about the document and/or the recipient. This simplifies record keeping and letter generation in a consistent format.

Investigation reports in many organizations are required to use a salutation format. They are informal technical documents directed to an individual or group. Figure 9.3 shows a form that works very well for a laboratory that writes many reports on short-term projects and studies. The reports are assigned a number to aid access from department files. At the end of the year, they are reviewed for retention or disposal for the next year. These reports often are not retained for more than two years. If the information has

MATERIALS ENGINEERING LABORATORY

To:	Organization /Division:
Phone:	
Title:	
Written by:	**Location:**
	Phone:
Copies to:	
Contributors:	**Report #:**
	Date:

This entire document is the property of The AA-Company. Partial reproductions or omissions of one or more sections of this report are strictly prohibited to preserve data integrity.

INTRODUCTION (PROBLEM DESCRIPTION)
Click here and enter text

INVESTIGATION AND RESULTS
Click here and enter text

DISCUSSION
Click here and enter text

CONCLUSION
Click here and enter text

Fig. 9.3 Report form for small investigations

long-term value, it should be documented as a technical report or document that can be archived in the corporate library.

Formal technical documents and reports have long-term value and are not necessarily directed to an individual or group. Along with the outline, a formal report requires the appropriate front matter that describes its source and reason for being. The first page should contain front matter that identifies the author and contains other fields to identify the report, the organization, date, coauthors, contributors, key words, and other items that may be needed for tracking (Fig. 9.4). Some research organizations list the sponsor of the research in this front matter. Articles submitted to a journal also often require a formal cover sheet. The author instructions for the journal specify what information is needed on the cover sheet.

Front matter in a book includes:

- Copyright—legal requirement
- Dedication—to a person
- Foreword—usually written by another praising the book and author

- Preface—written by author to say why the book was written
- Table of contents—listing of chapters/sections and major headings

None of these are needed (or appropriate) in a formal report unless your particular organization wants them. For example, long project proposals sometimes have a table of contents. Title, author(s) organization, key words, preface, forward, and table of contents are only needed on extensive documents such as theses or books. A formal report will not need one unless it exceeds 20 pages or so [*then it may be a tome instead of a report*].

Another important aspect of the front matter is the distribution list. In some organizations, standard procedure is to have the distribution list on a separate sheet of paper so that it can be detached (Fig. 9.5). This can affect litigation. Every company employee who has any documents in their file pertaining to a product lawsuit may be asked to turn them over to the legal department. Formal reports containing a distribution list indict every person on the list. If 20 people receive a document relating to a problem with

TECHNICAL REPORT

REPORT ACCESION NUMBER	DATE 7/17/2000

SECURITY CLASSIFICATION	AUTHORS APPROVAL
☐ Unrestricted Internal Use ☐ Restricted Information. Reclassify ☐ Confidential Information. Reclassify Controlled Distribution	All new chemicals described herein are registered. All chemicals with assigned numbers referenced in this report are listed below. Report has been keyworded by author(s). Signature

ORGANIZATION / DIVISION: Worldwide Capital & Process Reliabily / Worldwide Engineering Division / Materials Engineering Laboratory

TITLE

AUTHORS (Last name, first initials):   CONTRIBUTORS (Last name, first initials):

Fig. 9.4 Cover sheet for a formal report

PLEASE DISCARD BEFORE FILING

TECHNICAL DOCUMENT NO: RCD 93016   AUTHOR: Kenneth G. Budinski

TITLE: Wear of Carbide Dies on Conventional Perforators

cc: J.J. Doe/K. Budinski
   T.G. Smith
   A.L. Jones (Abstract only)
   S.Q. Summers
   B.L. Springer
   J.V. Dale

Fig. 9.5 Example of distribution list

a company product, they can become a party to that problem if it leads to an injury lawsuit. [*My company encountered a patent suit that ended up unnecessarily involving many people through distribution lists. One billion dollars later (yes with a "B"), we now must detach the distribution sheet from formal reports as soon as we receive them.*]

Nowadays electronic distribution of reports is very common. Front matter is still needed because some people will want to print out a hard copy and libraries retain hard copies. There are no assurances that electronic copies will be in a retrievable form 10, 20, or 50 years hence. Some electronic distribution systems send reports to team suites or to databases. Team members download the report if they are interested (another reason for good titles).

Key words are becoming more important as computer databases attain more powerful searching capabilities. Most technical journals ask for key words or indexing numbers when a manuscript is submitted. Establish a hierarchy for the key words based on how you anticipate potential readers of your document may search. If the document is about a failure of a cast iron pipe hanger, the most important term would be pipe hanger, followed by failure, followed by cast iron, followed by less obvious terms like safety or brittle fracture.

If formal documents are coauthored, the traditional protocol is to have the principal investigator's name first in the front matter. Sometimes, research organizations have a protocol that the name of the laboratory head occurs first on all articles written by his or her staff. Hopefully, this protocol has passed away. If someone did not do the work, they do not deserve the credit. Do not list contributors as coauthors unless they wrote a portion of the article or made a substantial technical contribution. Technicians who performed planned work can be listed in an acknowledgment section at the end of the article, if this is the custom for the journal in which the article is published. Some articles do not have acknowledgments except to thank research sponsors for financial support. Do not make an "Oscar" acceptance speech where you thank every member of your department.

RULE

Do not list contributors as coauthors unless they did some of the writing.

Finally, many journals publish the mailing address of authors so that readers can write for reprints. Make sure on published papers that the listed address is adequate to get correspondence to you. [*Pretend that you will get a royalty check.*]

9.4 Writing the Introduction

The introduction in college textbooks is sometimes regarded as less important than the "inner Chapters." Some students (and teachers) feel that the "meat" does not come until Chapter two and beyond. This is not the case with most technical reports. In fact, many technical journal editors reject articles because the authors fail to say in the introduction why they are doing the work. The introduction is where you state who you are, what you did, what you hope to accomplish with the report, what you are going to write about, and a few words about your results.

The introduction is a sales pitch for the report; some executives, in fact, feel that they are too busy to read entire reports, so they only read the introduction and recommendations. This is why some reports include an executive summary, which is a more detailed abstract corresponding to the sections in a report (See Appendix 8). An executive summary is placed after the abstract, and it essentially contains abstracts for each of the major sections of a report. Busy people may only read selected sections, and the abstract is followed by the report with complete sections. Each section should stand alone.

The basic ingredients of an introduction are:

- *Background:* circumstances that prompted the report, including the importance of what you are writing about
- *Purpose of report:* why you are writing the report
- *Objective:* what you hope to achieve
- *Format:* brief list of report sections
- *Hint of conclusions/recommendations*

Background

The background information in the introduction of a report should include the following items:

- The problem addressed by the report
- Who you are and who asked you to work on the problem
- Why is the problem important—why should the reader bother to read this

It may also be appropriate to include the chronology of events that led to your work. If you were simply asked by someone to work on a problem, then state this. The following is an example of an introduction with inadequate information:

Introduction

The majority of ribbon yarn manufacturers use a razor blade slitter in their production line. Blade life is rarely more than one or two days; frequent shutdowns are, therefore, necessary to change blades. The purpose of this experiment was to design and build a slitter, which allows blade change without interrupting operations.

The background information is inadequate and confusing. What is the relationship between other manufacturers and the author? What is the purpose of the report? What is the objective of the report? What was the experiment? Adequate background in a report is shown below:

Rewrite

Razor blade slitters, which are used on yarn lines 9 to 17, require frequent blade changes. Blade change shutdowns are required every day or two, and monthly replacement costs average $2000/line. This is a common problem in the yarn industry; we can have a competitive advantage if we develop an automatic blade changer.

The Manufacturing Development Department was asked by the Staple Fiber business group to design and build such a device. This report describes the steps in the development of automatic blade change: concept development, laboratory experiments, finalized design, and prototype manufacture. This report is intended as the project summary management.

Figure 9.6 is an illustration of report situations and the required background in the introduction of a report. Conversely, the following are some guidelines for omitting items that should be left out of the report background:

- Never use personal names (in fact, never use personal names in a report except in referring to listed references or in an acknowledgment).
- Never whine (We have been asking for this repair for three years, etc.); state only facts.
- Never use trade names unless the purpose of the report is to evaluate trade name products. A copy of your report may end up in anybody's hands.
- Never blame a person for a problem.
- Never include anything that could in any way cause litigation against you or your company (write as if your report is to be published in a newspaper).
- Never express prejudice.

The use of names in reports is a common problem. A good rule that solves all such problems is to never use personal names in a report except as a reference or acknowledgment. Use of trade names should be avoided. Lawsuits have arisen from reports that evaluate, for example, several potential suppliers of a commodity. A supplier sees a copy of your study and takes issue with your testing plan. He believes that your testing was flawed or biased. Evaluation reports involving suppliers or products should be listed as confidential with controlled circulation. A rule of thumb to prevent problems with evaluation of suppliers and competing products is to write a document using generic names or internal codes (brand x, brand y, and so forth) for the competing items.

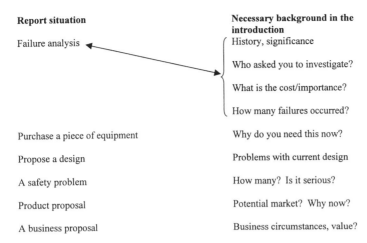

Report situation	Necessary background in the introduction
Failure analysis	History, significance
	Who asked you to investigate?
	What is the cost/importance?
	How many failures occurred?
Purchase a piece of equipment	Why do you need this now?
Propose a design	Problems with current design
A safety problem	How many? Is it serious?
Product proposal	Potential market? Why now?
A business proposal	Business circumstances, value?

Fig. 9.6 Necessary background for the introduction in different report situations. Necessary background for the introduction of a failure analysis report is indicated.

Scope

As mentioned in Chapter 4, "Writing Strategy," a writer needs to decide what will and will not be included in a document. This essentially defines the scope. In the following example on the subject of fretting corrosion, the report is not going to cover all instances where fretting can occur. To the contrary, the report is limited to just some aspects of this very large field.

Scope Statement

The laboratory fretting corrosion tests were conducted on a variety of surfaces that were candidates as materials of construction for metal-forming tools.

Information could be added on what the report does not contain, but this could be too extensive. For example, the paper does not address fretting of plastics, the fundamentals of fretting corrosion, nor the models that apply to fretting, and so forth. You only need to say what is not being covered when you anticipate that some readers may question why you did not discuss closely related subjects. For example, a medical report may report tests on males, and readers may question why the work does not apply to females. State this in the scope.

Stating the scope of the work and the scope of a document essentially tells the reader what is coming. Avoid confusing the scope statement with the statement of purpose and the objective of the work. If there is likelihood of reader misinterpretation, delete the scope. In fact, many times the scope statement can be blended with the purpose statement. Content of the scope statement is determined by the objective of the work.

> **RULE**
>
> **Make it clear what your report is about.**

Purpose and Objective

Continuing with "Writing Strategy," Chapter 4, the reason for a document must be conveyed in terms of purpose (the stated intention of the document itself) and objective (the stated outcome of the work reported in a document). As noted in Chapter 4, purpose and objective have similar meanings in common usage, but those distinct meanings should be reflected in technical reports, where you must deal with competing words:

- Purpose = intention = reason
- Objective = goal = deliverable = end result

These two essential elements are required in the introduction of any formal or informal report. There may be multiple purposes and objectives, but do not belabor these points. Be concise and direct. For example, the purpose of the fretting corrosion document is as follows:

> Purpose
>
> This paper summarizes the results and conclusions of the laboratory test program.

Usually, the purpose statement flows into the objective. What is the value of the work?

> Objective
>
> This work was commissioned by the packaging department to determine if there is a mating couple that will not be subject to fretting corrosion. The current couples are very prone to fretting corrosion, and fretting damage on production machines is a significant department cost.

> **RULE**
>
> **Introductions must describe the scope of coverage, the purpose of the document, and the objective of the reported work.**

Statement of Format

The introduction for a report should clearly describe its format or the overall organization and sequence of its major sections. If you have five sections in a report, state what is in these five sections in their order of appearance within the report.

Statement of Format

This report describes candidate materials for the fretting corrosion study, the laboratory tests, the test results, and the recommendations to solve the problem in production machines.

If there are additional sections, then include statements that list them. Do not just tabulate the section headings in the outline; put them into sentences. Describe the coverage in each of the major sections that are contained in the report outline. It is important because readers want (and deserve) to know what is coming.

The format statement can be deleted from an introduction if a short report has the required headings, such as:

- Problem
- Investigation
- Results
- Summary/Conclusions

For example, if a two-page report follows the format in Fig. 9.3, the reader, in a glance, sees the organization and format.

In an informal report, the format might even be described in a simple sentence, such as: "The following are the results obtained from the tests that you requested." Whatever the situation, try to give the reader an idea of what follows.

Finally, in big reports, it may also be appropriate to provide a hint of the results:

Hint of Results

These tests indicated that cemented carbide is a preferred counterface in fretting-prone systems.

This does not eliminate the need for conclusions and recommendations. It is just a one-sentence abstract of the conclusions and recommendations.

RULE

The introduction should describe the major topics (or sections) in a report.

9.5 Putting It Together

Combining all the elements of an introduction, Fig. 9.7 shows an introduction for the example of the fretting corrosion report. This section also describes a second example of developing an introduction for a formal report. This is for a formal report on a device called a "nibbler." It was developed to simulate notching and perforating of film, and this example also is used in subsequent Chapters on report writing.

[*Even though there are two more Chapters on reports with this example, completing an introduction with its key elements really gives you a "semi-finished" report. You have stated the problem, why it is important, why you are writing, and what you want from the readers. Then you tell the reader what is in the report and the outcome (hint of the outcome). The format, if complete, is an outline for the remainder of the document. You merely fill in the details. I should not tell you this because there are two more Chapters on reports, but completing an introduction is the major step. Forget that I mentioned this. It will be our secret.*]

Formal Report on "The Nibbler"

The first step in writing this formal report is to establish an outline and refine the title. "The Nibbler" sounds facetious, and the author may lose credibility with it. How about "Simulation of Perforating?" This is better,

---

Fretting Corrosion Resistance of Tool Materials

Fretting means oscillatory motion of small amplitude. When a solid surface "frets" in contact with another conforming solid, the contacting surfaces are subject to fretting corrosion or fretting wear. The former involves reaction of the rubbing surfaces and wear debris with their environment. If there are no environmental reactions, the damage that occurs is called fretting wear. This paper describes laboratory fretting corrosion tests conducted on a variety of surfaces that were candidates as materials of construction for metal-forming tools. This paper summarizes the learnings of this study. The work was commissioned to determine if there is a mating couple that could be used in metal forming dies that works and will not be subject to fretting corrosion. The couples currently used in a particular forming die are very prone to fretting corrosion. Repair of fretting damage on production machines is a significant department cost and needs to be reduced or eliminated.

This report describes candidate materials for the fretting corrosion study, the laboratory tests, the test results and the recommendations to solve the problems in production machines. These tests indicated that cemented carbide is a preferred counterface in fretting-prone systems.

Fig. 9.7 Example of an introduction for a formal report

but it still is not adequately descriptive. Remember that a title is an abstract of the paper. The final title is:

Final Title

A Laboratory Device to Simulate Notching and Perforating of Web Products

---

This title tells the reader that the report is about a laboratory or off-line device and that it is intended to simulate the notching and perforating done on many products made from thin plastic or paper (web products).

Now we need an outline for the introduction. It must contain background information, a purpose, objective, and format. Start the outline with the definition of headings:

Outline for the Introduction

Title

A Laboratory Device to Simulate Notching and Perforating of Web Products

Background

Why we need this device

Applicable products (scope)

How it will be used

What it will cost—what the savings are

Purpose

To demonstrate that new products can be tested for abrasivity and new tool materials can be tested for improved wear resistance

Objective

To reduce the amount of material required for new product tests and to lower tool costs

Format

Design

Laboratory tests

Results to date

Conclusions/recommendations

Hint: it works well

As can be seen in the preceding example, the outline for the introduction forces you to think about the body of the report. The completed introduction contains all of the basic elements, and it can stand alone in presenting the reader with the five "Ws" of writing: who, what, where, when, and why.

A Laboratory Device to Simulate
Notching and Perforating of Web Products

Introduction

Photographic film, packaging materials, polyester copier belts, and many other materials that are manufactured as continuous webs are finished to final product form by slitting, chopping, notching, perforating, and blanking the web. The tools used to finish web materials are the same types of tools used to shape and form sheet metal. They include slitters and punch presses, as well as specialized machines that cut "on the fly." Tools notch the web speed as the web is conveyed on rollers. The makeup of these flexible webs is continually changing.

Web coatings are upgraded for better durability; photographic films receive new emulsion and backside coatings to improve performance. Every time that the web composition changes, the manufacturing staff must determine if there will be a significant change in the life of the tools used to finish the web. This lesson was learned the hard way when an overcoat was improved to make the product scratch resistant and the tools used to finish the web started to wear out at one-fifth of their normal life.

The normal way to measure the effects of web changes on tool life is to make small lots (narrow widths) of the new product and run them in a pilot operation. The problem with this approach is that it may require finishing hundreds of thousands of square feet of product to obtain a signal on finishing tools. This greatly increases the cost and lead time for product changes. If we had a tool that could evaluate the abrasivity of a new web material using only a small amount of material, we could greatly reduce both the cost and lead time required to evaluate new products.

The problem was addressed in the Materials Engineering Laboratory by the development of a nibbling device that simulates notching and perforating of webs and uses very little material. It is the purpose of this report to present a description of the design, development, and use of this device so that manufacturing and development staff can be made aware that this device exists. The objective of this device is a significant reduction in the cost and time required to assess the finishability of webs after a composition change. A concurrent benefit of this device is the ability to evaluate new tool materials off-line.

The device that was developed to assess tool wear and web abrasivity is called a "nibbler," and it essentially nibbles a sheet of test material until there is none left. Each nibble cuts a small piece from a sheet. Tool wear is measured by assessing edge changes on the tools that produce the nibble. The smaller the nibble, the greater the number that can be made in a sheet. A one-quarter millimeter cut, for example, would allow about a million cuts from a letter size sheet of paper. Advanced edge measurement techniques allow tool life assessments after one million cuts by the nibbler.

This report will describe the design of the nibbler device, how it is used in the laboratory, and the types of results obtained to date with its use. The device has been very effective, and its use in product development is recommended.

End of Introduction

Summary

The basic elements of an introduction describe the background, purpose, content and organization of a report, and the objectives and conclusions of your work. All reports need these elements addressed in some appropriate fashion. Tailor the introduction (and the report) to the intended readers. If the report is about "hopper shims," some people on the distribution list may not know what a "hopper shim" is. In this case, you must define it in the introduction. Do not use acronyms at all if they can be avoided. If you must use them, define them at the first use. Make the introduction the most readable part of the report, because it (along with the conclusions and/or recommendations) may be the only parts that are read by some. It needs to make your readers eager to read on.

Additional points to keep in mind when writing the introduction are the following:

- Write to the reader with the least technical background, but do not state that you are writing "down." This insults the reader.
- Put in enough history, but do not sacrifice concision. Cite references when appropriate.
- An introduction should be written so it can stand by itself.
- State the scope of your work in the introduction.
- The purpose of an investigation report is simply to present its results. It is adequate to state this.
- The objective of a study is the same as the goal of the work. It is the final outcome and value of the work.
- The objective of a report may be different than the objective of the study. If it is, state this.
- The purpose of the report may be different from the purpose of the study. If it is, state it.
- Choose a title very carefully. It is the ultimate abstract of the report.
- Decide on distribution (readership) before writing anything.
- Informal reports need background, purpose, and objective. Sometimes these can be put in one sentence.
- Always tell the reader what is coming in the report. State the format at the end of the introduction.
- Purpose is the same as intention.
- The introduction and conclusions/recommendations may be the only part of a report that is read by managers. Write with this in mind.

Important Terms

- Introduction
- Format
- Scope
- Objective
- Salutation
- History

- Preface
- Purpose
- Title
- Foreword

- Table of Contents
- Front Matter
- Background

For Practice

1. What is the difference between the purpose of a report and the objective?
2. What is the scope of a project and where do you state it?
3. Write an appropriate title for a report that you are writing on the development of an expert system for selecting rolling element bearings.
4. What are these report elements and when do you use them?
   - Index
   - Table of Contents
   - Preface
   - Foreword
   - Title Page
   - Distribution List
   - Report Heading
5. Write an introduction for the project in question three.
6. List the basic elements of an introduction to a formal report.

To Dig Deeper

- B. Beer and D. McMurrey, *A Guide to Writing as an Engineer,* John Wiley & Sons, 1997
- G. Blake and R.W. Bly, *The Elements of Technical Writing,* MacMillan, Inc., 1993
- H.A. Pichett and A.A. Larter, *Technical English, 7th ed.,* Addisen Wesley Educational Publishers, Inc., New York, 1996
- M. Rose, *Technical Communications, the Practical Craft,* Prentice Hall, 1997

CHAPTER 10

Formal Reports:
Writing the Body

CHAPTER GOALS

1. *Know the basic parts of a report body*
2. *Understand how to write a reproducible procedure*
3. *Understand how to present results*
4. *Understand what should be included in the discussion*

WRITING THE INTRODUCTION, as described in Chapter 9, "Formal Report: The Outline and Introduction," may be the most difficult part of a technical report. You must set the stage for the reader and convince the recipients to read your report. You must let the reader know in unequivocal terms why you did the work and why it was important. You must also include the conditions that led to the work, the scope of the work, the purpose of both the work and the report, and the objective of the work and the report. All this must be done in a complete and concise fashion.

With the challenge of the introduction behind, the next step is to write the report body. The body is much easier to write because each section in the body only deals with one issue. You should have already done the difficult work in the introduction. For example, most literature survey results should be stated in the introduction, where you must show that you examined the work of others before venturing into the project. Your introduction must convince the reader that the work has not been done by others.

The basic elements of the report body are:

- *Procedure:* what was done, expressed in sufficient detail so the work can be repeated by others

- *Results:* outcome of the tests conducted in the procedure expressed in chronological order but with essentially no interpretation
- *Discussion:* analysis of results (including statistics); statement of why things happened the way they did and how these results compared with the results of others (from your references); description of models/equations/laws, and so forth that evolved from the work

These three sections apply to investigations, experiments, and feasibility studies. The concept applies also to reports that have other purposes. If you are proposing a machine design, the procedure section may be called "Design Options." Each option is reviewed in that section. The results section may be termed "Final Design." The discussion section can still be called "Discussion," but it may contain costs or advantages and disadvantages of the selected design. This methodology applies to most technical report situations. It uses the logic of:

- What was done
- What happened
- Explanation of what happened

This Chapter describes the basic content that should be included in these elements of the report body. There may be three, five, or more report subheadings under each major head. For example, a very big project may have involved tests at room temperature, 200 °C, and 400 °C. These may become subheads in the procedure results and discussion. This is proper, and this approach may apply to any section of the report body. You can even call them something else, but they need to contain the basic elements of the report body. These are mandatory elements. Technical journal reviewers will reject a paper if these elements are missing. However, the best reason for patterning a report to these guidelines is that they are a logical way to report work.

RULE

The body of a formal investigation report must contain a procedure, results, and a discussion of results.

10.1 Writing a Procedure

This section of a report body could be called "Investigation," "Laboratory Tests," "Field Tests," "Design Concepts," and so forth, as long as it contains sufficient detail that what was done can be repeated by others. This is the requirement and the key attribute of this part of a report. There are two main reasons for details that allow the repeatability of a procedure. First, if you need to repeat the tests sometime in the future, you will have the necessary information. Second, if your work is being published, others may want to repeat it to verify that your model, equation, and so forth are correct. Essentially, repeatable procedures are the basis of objective, scientific progress.

How do you make a procedure repeatable? Do what cooks do. A cooking recipe is a nice example of a repeatable procedure.

KGB Recipe for Cabbage Soup

1. Buy a big head of cabbage.
2. Cut in half.
3. Cut the halves in half (you now have quadrants of a sphere).
4. Boil the quarter heads in a large pot for 15 minutes at boil in 2 quarts of water which contains 3 tablespoons of salt.
5. Let the pot cool to hand warm, and mash the cabbage with a potato masher until it is disbursed in the water (the quarters are not recognizable).
6. Add the following to the pot:

    • 1 tablespoon paprika
    • 1 tablespoon garlic salt
    • 3 tablespoons olive oil
    • 4 tablespoons vinegar
    • 1 beef bouillon cube

7. Reheat the pot to 150 to 170 °F and maintain there for 7 hours.
8. Serve in shallow bowl and decorate with parsley (serves two people).

Needless to say, this soup may not be edible. [*I just made it up. However, it might be a good class project to make it and see what it tastes like.*] Most people could repeat this recipe, and it demonstrates what kind of information is needed for reproducibility. It details every step, gives specific sizes and quantities, and states conditions (temperature and times). A proper procedure should contain this level of detail. For example, the use of testing data sheets (Fig. 10.1) helps to ensure collection of appropriate data and facilitates sorting and other processing.

The imperative mood (where sentences start with verbs) is usually reserved for writing instructions or operating manuals like in the preceding example of a recipe. In contrast, descriptions of tests or investigations may be better expressed in the past tense.

EXAMPLE

Procedure in an Investigation

Four test webs were submitted for evaluation. They were identified as follows: XBC 2140, P10 overcoat; XO-2387, A7 overcoat; and XO-437, control. The test samples were in the form of 35 mm wide sensitized reels, 50 feet in length.

Ten samples, 18 in. long, were tested from each material using the capstan tester (Figure 1) and ASTM G 143 procedure.

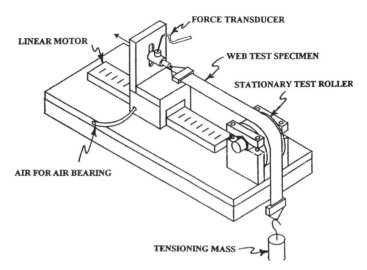

Figure 1 Schematic of capstan friction tester

The ambient test conditions were 70 °F and 50% RH. All samples were equilibrated in the testing environment for 24 hours prior to testing. The test counterface for both sides of the three films was an XP3 test roller, and the roller surface was cleaned with a spray and cloth wipe of toluene between tests. Sample testing was randomized per statistical procedure X23. The test speed was 0.5 cm/min. The length of sliding was 20 cm and the friction force was recorded for the duration of the test. The static and kinetic friction coefficients were recorded in accordance with ASTM G 143 procedure for both sides of each film. The test statistics were based upon the 10 replicates.

The preceding example is written in the passive voice. This is traditional in many technical documents, where the writer is communicating information as an observer. An active voice is optional, and the same example could have started with:

- "The laboratory received four test webs for evaluation..." (the preferred impersonal tone for formal reports)
- "I received four test webs for evaluation..." (first person suitable for some informal reports)

Using the active voice is a general rule. [*I prefer the passive voice for the test procedure because it was in the past. The tests are over, and the document describes what was done from the perspective of an objective observer. This is the tone and tense that relates well to the reader's perspective, too. In many cases, an active voice may be preferable. But if you try to maintain an objective tone in the active voice, do not use first-person singular pronouns (I, me, my) in a formal report or a paper for publication. Most technical journals do not tolerate it. I tried it once, and I had to remove them before they would approve publication. In other words, avoid*

Test ID and Type

Test Identification:	96
Date of Test:	8/1/98
Testing Organization:	Materials Engineering Laboratory
Standard Test Specification:	Loop Abrasion test
Laboratory or Field Test:	Laboratory
Nature of Sliding Test:	Non-Lubricated
Test Machine Description:	Loop Abrasion Test
Test Duration:	8 hours

Close   **Print**   Before printing, set this 'Page Setup' to Landscape Orientation.

Test Conditions

Load Conditions:	Steady
Load Value:	200 grams
Pressure:	
Velocity Conditions:	50 FPM
Velocity Value or Range:	
Total Sliding Distance:	
Total Sliding Distance per Cycle:	
Test Temperature:	70 F
Relative Humidity:	50%
Ambient Temperature:	
Type of Motion:	Continuous
Continuity of Motion:	
Contact Environment Description:	Line contact
Lubricated Contact Description:	
Abrasive Contact Description:	30 micron alumina oxide finshing tape
Other Test Information:	

Fig. 10.1 Test database form (for a complete file of test results)

them and write in the passive voice if it is necessary to maintain an objective tone when describing how a test or investigation was conducted.]

Chronological or sequential order is usually desirable in most discussions or procedures. The steps in an experimental procedure are usually written in chronological order. If a report describes a design, the discussion should start with the chassis or machine bed and build on them. Essentially, put the information in an order that would make sense to the readers. Illustrations, in the form of schematics, photos, tables, and figures are encouraged. It is very difficult to describe a special laboratory test rig without a schematic. On the other hand, there is no need to describe common tests (like tension or hardness testing) to an engineer. Most engineering schools expose engineering students to mechanical properties of materials. However, always identify the standard used when conducting even the most common test.

It is also appropriate to include references in a procedure, if you are reproducing a test developed by others. However, do not just replace the discussion of procedures by referring to procedures published elsewhere.

Inadequate Procedure

The test rig used in these experiments was previously described (7), and we modified it with a proximity sensor for these experiments.

This is rude. You are asking your reader to go to the library and get a copy of reference seven so that he or she can find out what kind of test rig you used. Readers are not going to do this [*at least not this one*]. It is an imposition. Readers are busy people. They need to be presented with all necessary facts in one document.

Finally, check the procedure for omissions. Scan it and see if the work could be repeated with the data presented. As in all aspects of technical writing, a review by a peer or trusted person will go a long way in preventing procedure omissions.

The Nibbler Report—Continued. The procedure section for the nibbler report, which was started in Chapter 9, "Formal Report: The Outline and Introduction," would be significantly different from a laboratory study because the evaluation probably will not be repeated. However, it is still necessary to describe what was done, as in the following example.

Report Body on the Nibbler Tester

Tester Design. The design concept is to "nibble" small rectangular "bites" from the edge of the test material. There is no space between nibbles, and this would maximize the amount of tool wear that could be obtained from a given piece of material.

A minimum of about 10^6 or 10^7 nibbles was anticipated to be the number required to obtain measurable tool wear. Other design constraints for this test device included:

- Make nibbles as small as 0.1 mm on a side.
- Cut materials ranging from 1006 steel to buna-N rubber.
- Cut a web thickness ranging from 25 to 250 μm.

The original punch and die concept is illustrated in Figure 1.

This design concept was tested by building the prototype tool shown in Figure 2. It nibbled web materials as we anticipated, but it was learned that a "clean" (no hairs) nibble required close punch/die clearances. The ability to precisely control these clearances was added as a design constraint.

The prototype rig established the cutting tool design, and the next phase was to design the punch, the holders, and the web transport mechanism. The former was quickly established. The Film Flow Department gave us a C-series perforating punch press complete with an experimental punch and die. The punch was modified to hold the nibbler punch, and the die was set up to hold the nibbler die in a set. Essentially, the tools were fastened to the side of the existing die set. The die set was outfitted with preloaded ball bushings, and the lateral punch motion was measured to be less than 1 μm at the bottom of its stroke with a 200 N force applied in any radial direction.

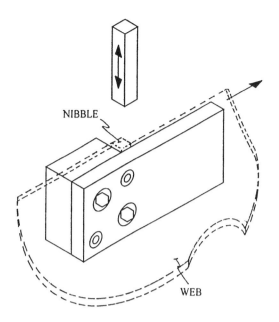

Figure 1 Original design concept of the nibbler

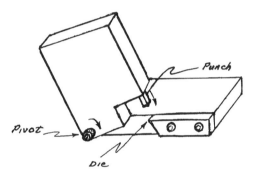

Figure 2 Schematic of prototype nibbler tool

The transport mechanism concept is illustrated in Figure 3. We would position the edge of the test sheet under the punch and make a nibble. Then the web is transported in the x direction the predetermined x dimension of the nibble. This would be repeated until we had traversed the x width of the test web. The manipulator would do a carriage return and feed the test web forward one nibble distance in the y direction and start nibbling.

A programmable commercial ball-screw positioner was also purchased to move the test web in the x-y direction. A z-direction servomotor was also obtained to provide punch motion.

Motion in the y-direction was obtained by feeding the web through the feed roll assembly from an FW-series monotone copier that was made available after some other laboratory tests on the copier. The roll assembly uses a silicone rubber roller 4 cm in diameter and 50 cm long. The maximum web width that could be tested was 45 cm. XYZ motions were programmed with C-base code on a standard 286 microprocessor PC.

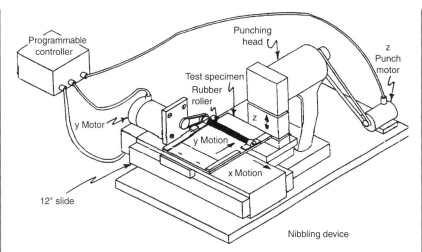

Figure 3 Nibbler web transport device

The test procedure became:

• Enter the x and y dimensions for a nibble, the nibble rate per minute, the web width, and the total number of nibbles into the PC control program.
• Insert the test web into a fixed stop.
• Enter the execute command on the software.

The system was debugged by making 10^6 nibbles on two different paper webs and three plastic webs. It worked as anticipated.

Tool Wear Measurement. The next phase of this project, assessment of tool wear turned out to be the most difficult part. When the nibbler design was formulated, the idea was to measure by profilometry the wear volume on the edges of the punch and die after x number of nibbles. For standard number of 10^6 nibbles in a test, the wear volume would be expressed in cubic μm per 10^6 as the test metric.

This concept is illustrated in Figure 4. Unfortunately, it took three months of measurements and three different profilometers to produce a procedure that produced sufficient repeatability and accuracy for a standard measurement technique. Essentially, 10 traces are made with a noncontact laser profilometer on each of the four wear surfaces, b, c, e, and f in Figure 4, and a software package was developed to integrate the wear areas from each trace into a wear volume. The spacing of the traces was one-tenth ($^1/_{10}$) the nibble width. The measurement software was written in C-base to operate on a Pentium type PC.

Test Procedure: Effect of Cobalt Concentration on WC/Co Wear. The first systematic wear tests with the nibbler were performed on three grades of cemented carbide with varying cobalt concentrations. Once punch was made from each of the following grades of cemented carbide: 3% cobalt (Co), 97% tungsten carbide (WC) (2 to 3 μm); 7% Co, 93% WC (0.8 μm); and 13% Co, 87% WC (1 μm)

The grain size of the carbide phase was in the range of 0.8 to 3 μm. The die was made from type 01 tool steel at 60 HRC and the test web was 30 μm aluminum oxide finishing tape.

The tape was 12 inches wide and it was nibbled with the abrasive facing the punch. The nibble rate was 100/min. The punch was removed every 10,000 nibbles. Its volume was measured, and another 10,000 nibbles were completed. This procedure was repeated on the punches made from the three different carbides.

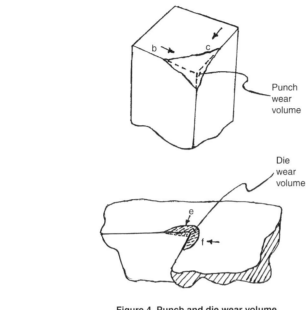

Figure 4 Punch and die wear volume

The procedure in this example contained three sections: one on the tester design, one on how to make measurements, and the final one on the procedure used to test three materials. The procedure part of the report can have as many sections as necessary to describe a test or study and how it is run.

Summary. Key requirements are that procedures be repeatable, but procedure sections should not contain test data. Only describe the test and what was tested. Describe the test material and the procedure details but omit trade names. Be concise. Do not report blind alleys that you ventured down, unless discussion of failures is key to the objective of the work. Keep focused on what was done. Do not describe unnecessary details. If you are discussing x-ray inspection of a pipeline concerning a failure, do not describe how x-ray machines work unless the report is about x-ray machines. Assess readership and decide on the level of detail necessary. Chefs do not describe how ovens and cook tops work in their recipes.

RULE

**Test procedures must be detailed enough so that
you or others can repeat the test in the future.**

10.2 Describing Machines/Processes

Describing a machine or process is a common occurrence in technical reports. There is no formula for making descriptions effective. Mostly, it requires practice. It helps to define all terms. If you are describing a new type of screwdriver, for example, define the blade, the part that contacts the screw, and the handle. Then describe the overall screwdriver.

Description of a New Screwdriver

The blade is selectively hardened steel with a flat rectangular tip that transitions to a 6 mm diameter round, 14 mm from the 0.5 × 4 mm tip. The blade is 200 mm long, and it is polished and plated for a pleasing appearance.

The handle is the unique part of this screwdriver. It has a trichoidal shape (see Figure 1), and it is made from natural material finish rosewood. The trichoidal shape in cross section is similar to a three-lobe cam. The shape is circumscribed in a circle with smooth blends to the legs (see figure).

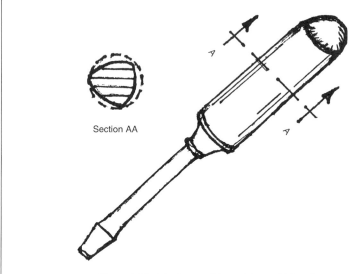

Section AA

Figure 1 Sketch of screwdriver with special handle

An illustration is necessary to complete the preceding description. It is apparent that a description by words alone would be very difficult, if not impossible. Do not waste a paragraph when a photo or sketch is more effective.

After adequately describing the mechanism, state how it works or what is unique about it. In the screwdriver example, the handle shape has been ergonomically designed to fit the average person's hand. The trichoidal shape prevents slipping, and field tests indicate that assembly line operators can use this screwdriver twice as long as a round-handle screwdriver before developing hand fatigue. The rosewood handle is also impregnated with raw linseed oil. This oil impregnation protects the wood and produces a coefficient of friction with human skin that is higher than that produced between skin and most painted wood and plastic handles.

You could now go on to describe how to use this screwdriver to perform some operation. You would use the same type of writing procedure in describing the process. Define terms; describe the overall process/machine. Then describe features that are unique. The recommended test of a proper description is the same as an effective test procedure. Ask a colleague or general acquaintance to read the description. Does the person know, understand, and appreciate the unique features of this screwdriver?

RULE

Always consider using a photo or sketch when describing special geometric shapes, assemblies, or other complicated interrelationships.

10.3 Writing Test Results

Test results mean just that. They are the outcome of what happened in each part of a study. If you measured the specific gravity of ten types of rocks, show the data in a table or, preferably, in a graph or a bar chart. Tables of numbers are not always easy to interpret whereas a graph or bar chart can provide a visual comparison of the test results. The reader is accommodating you by reading your report. In return, make reading the report as effortless as you can make it.

If a report concerns a design project, the section on results may contain a description of a device or some sketches or schematics that convey the design to the reader. In the nibbler example, the design was part of the procedure, because we also wanted to present test data on the use of the nibbler. We could have made the description of the nibbler design the test results. The author should make the decision on what to include in each section of the body. Make the sections chronological and logical. It is logical to describe the test procedure before the test results, and it is logical to clearly state the results (sometimes they can be enumerated) in concise statements. Wherever possible, use graphs for presenting results. Graphs were invented for this purpose.

[*One of the technicians in my laboratory loves spreadsheets. She always tries to give me data in this form, and I must negotiate a graph from her. She says that I have a brain disorder that prevents me from finding results in a page full of rows, columns, and numbers. She can make conclusions from them. I suspect that she is right. I have some kind of table phobia. I usually cannot draw conclusions from anything but graphs. However, there may be others like me out there, and we are the lowest denominator of your readers, so please use graphs for presenting results to us. You will be rewarded for your act of kindness to us.*]

In summary, the section on results presents test data along with any illustrations that help readers interpret the results. You should also state what the data show in words, but stop there. Explain the results in the discussion section of your report.

RULE

The section on results should focus on just that: results.

Present the test data and outcomes with good illustrations and state what the results show in a concise manner.

The Nibbler Report—Continued. The report on the nibbler continues as an example. The section on the results (along with the discussion section) for the nibbler report follows:

The Nibbler Report (Continued)

Test Results

The wear rates of the three different carbides are compared in Figure 5. The 13% cobalt material had the highest wear rate. The 7% and 3% had similar equilibrium wear rates, but the total wear on the 3% cobalt grade was higher than that of the 7% cobalt grade because the initial wear rate on the 3% grade was higher than the 7% grade.

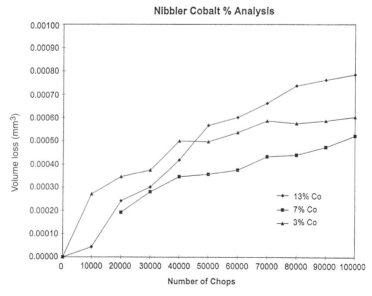

Figure 5 Wear results for the cemented carbide tools in the nibbler test

Discussion

The first four or five wear-measurement cycles indicated that the cutting edge on the 3% cobalt test material showed material removal by microscopic chipping. The 7% grade and the 13% grade only showed polishing wear of the type observed by Chalmers and his associates in their studies on the role of cobalt content on cemented carbide (9).

The fracture toughness of the 3% grade was significantly lower than that of the two other grades, 3 ksi-in.$^{-1/2}$, compared with 9.4 for the 7% grade and 14.3 for the 13% grade. It is felt that this explains the unanticipated results. The low toughness of the 3% grade as well as the large grain size combined to produce microchipping of the edge during the break-in part of the wear test. The remaining results followed the anticipated pattern of higher cobalt with lower wear resistance.

10.4 Writing the Discussion Section

The section on results describes what happened, and the discussion section explains why it happened. The discussion section of a report is where you explain the results to the reader and present closing arguments for your thesis. You describe the reasons why you think the results happened as they did. For example, the previous discussions on the test results of the nibbler explain the reason for some unanticipated results. Another example follows for the test results of a fretting test:

Fretting Corrosion Test

Discussion

The fretting test results suggest that fretting corrosion damage in hard-hard fretting systems can be mitigated a significant amount by making one member from cemented carbide. Similar results were obtained by Chalmers [18] in reciprocating metal-to-carbide systems and by Campbell [19] in rotary sliding systems. Our explanation of why cemented carbide is a favorable counterface for fretting systems is that its high compressive strength and modulus of elasticity prevent counterface asperity deflection in the real areas of contact, which in turn reduces the fatigue action tending to fracture asperities.

In addition to providing a high-strength, high-stiffness surface, the cemented carbides tested all contained cobalt binders and tungsten carbide particles. Both of these materials are resistant to oxidation in the test environment (air). This atmosphere resistance reduces the other component of fretting corrosion, reaction of fractured asperities with the ambient environment. Samuals [19] produced mitigation of fretting corrosion by removing the reactive environment of a fretting system.

Overall, these test results have a profound effect on the problem that prompted these tests. The production problem attributed to fretting corrosion, increased die deflection, can be solved by a thermal spray coating (HVOF) of WC/Co on one member of the fretting couple. This conversion will cost between $70,000 and $90,000, but the anticipated tool replacement savings will be in excess of $500,000.

Note that the use of a personal pronoun ("Our explanation") may be appropriate here, as the discussion section may be based on the opinion or perspective of the author(s). In fact, it may improve credibility, as the author is not proclaiming opinion as a proven fact. Critical readers appreciate this as a good indication of an objective tone.

Sometimes, statistics or math models are applied to the test results in the discussion section. If the results were graphed and this produced an exponential curve, it may be appropriate to have the computer-generated curve fitting. The author then would explain the equation.

You are also encouraged to compare your results with the work of others. Go back to the references in the introduction and relate them to your results and interpretations. Comparison to references helps to substantiate claims.

Of course, if your investigation results are not what you anticipated, the discussion section is the place to explain why your investigation did not work out. Avoid introducing new information in this section. You could report the results of a simple experiment conducted to explain a part of the results, but try to keep all the results in the results section.

As with all the sections of the body, the discussion should be tailored to the situation. Sometimes it can be omitted. If you are writing a formal report on extensive tests on blind samples, you cannot explain the results. This happens frequently in testing laboratories. A client may send thirty film samples for friction testing. He or she states the tests to be run but does not identify the samples in a way that can contribute to a test understanding. The samples come identified as lot A, B, and C. They are tested, and the data are graphed as lot A B C. It is appropriate to simply present the results and conclusions. The conclusion will be something like: "Lot A has higher frontside-to-backside friction than lot C and D, and lot D has the highest variability."

It is appropriate to omit the discussion in blind studies or service work. It is not possible to comment on why lot A is different from the other lots because nothing is known about their manufacture. On the other extreme, you may conduct only a few tests, but the results are so complicated that it takes much verbiage to produce a plausible explanation.

In summary, the discussion can be difficult, easy, or omitted, depending on the situation. It is normally the appropriate section to explain results, compare your results with those of others, and convince readers that your results are accurate, usable, and have value. If your project was a failure, say so and explain why here.

RULE

In the discussion section, explain why things happened the way they did.

Summary

The body of a formal report is the substance of your work. It does not have to be long. It could be only one page, but it must include procedure, results, and where possible, an explanation of results. A common mistake in writing technical reports is to intermix these sections. This confuses and sometimes loses readers. Keep the procedure separate from the results and the discussion separate from the results. Avoid presentation of new data after the results. Data is contained in the results section.

The following are some parting comments on this important part of a technical report:

- If you use standard tests in a study, reference them by number in the procedure section but still describe them.
- Make the procedure repeatable; ask a coworker or general acquaintance if they could repeat this test. Find out if you omitted key details.
- The body must include procedure, results, and discussion.
- A result is not a conclusion. It is a statement of the outcome of a test, study, design, or experiment.
- Never mix procedure with results or results with discussion.
- List all the parameters that could affect the results of your work. Check that you state test conditions for all these parameters in your procedure.
- Result sections contain only results—no interpretation. Just present results in a form that makes interpretation easy. Describe your results in words, too.
- Interpretation and explanation of results is done in the discussion section of a report.
- If you are reporting on an investigation, the purpose of the report is to present the results of the investigation and no more.
- The discussion is the place to compare your results with the work of others.
- Results usually require graphics to assist the reader.
- Try to avoid the presentation of data in tabular form only. In some cases, readers appreciate data presented in both tabular and graphical form.
- Keep the procedure and results sections free of opinions. These belong in the discussion.

Important Terms

- Procedure
- Tables
- Repeatable
- Discussion
- Standard tests
- Attributions
- Parameters
- Conclusion
- Results
- Statistics
- Graphics
- Thesis

For Practice

1. State the basic elements of the body and what is in them.
2. Write a repeatable procedure for changing a tire on a bicycle. Give it to a classmate and see if he or she can follow it.
3. Do a study on product variability; measure the length of 20 pretzel sticks, matches, toothpicks, diameter of pieces of popcorn—whatever. Write the procedure for the measurements and the results and then discuss the results.
4. Present the results of the experiment in 3 as a table, bar graph, line graph, and histogram. Which shows the variability best?
5. Present the results of your courses last semester and discuss your results. Why did you get all A's, and so forth?
6. State six report elements that belong in the discussion.
7. Write four reasons why the results need to be separate from the procedure.

To Dig Deeper

- A. Eisenberg, *Writing Well for Technical Professions,* Harper Row Publishers, New York, 1989
- M. Forbes, *Writing Technical Articles, Speeches and Manuals,* 2nd ed., Krieger Inc., New York, 1992
- *Form and Style for ASTM Standards,* 10th ed., ASTM, W. Conshohocken, PA, 1996
- J.E. Vinder and N.H. Vinder, *Engineering Your Writing Success,* Professional Publications, Inc., Belmont, CA, 1996

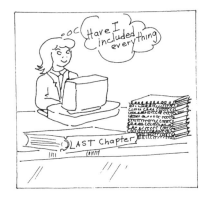

CHAPTER 11

Formal Reports: Closure

CHAPTER GOALS

1. *Understand what a conclusion is and how to write one*
2. *Understand what a recommendation is and how to write one*
3. *Know how to list references*
4. *Know how to write an abstract*
5. *Understand archiving and report distribution*

THE END OF A FORMAL REPORT requires special consideration, because in many cases it is more important than other parts of the report. As mentioned earlier, busy executives often read only the abstract; others read only the introduction and conclusions, and still others read only the title and conclusions. [*I must admit that I fall into the latter category when reviewing reports written by me or the technicians that work with me. Often, we get jobs that build on work done one or several years ago. We dig out the old reports and immediately go to the conclusions to recall what happened. I usually find what I need to know there.*] Good conclusions and end matter are crucial to a proper technical document. Formal reports need formal endings; less formal reports need less formal endings. All technical documents, however, need an ending.

This Chapter describes the details of a proper ending for a formal report. Specifically, the Chapter discusses the conclusions, recommendations, abstract, and other optional end matter, such as acknowledgments, appendices, and indexes.

11.1 Conclusions

A conclusion is a judgment based on results of a body of work. It is the final outcome of, for example, the investigation, the results obtained in the investigation, the analysis of results, and the comparison of the results with the work of others.

It is important to not present a result as a conclusion. If a survey conducted on preference between the colors red and gray yields a result that 75 out of 100 prefer red and 25 out of 100 prefer gray, this is the result of the study, and it is put in the results section. The analysis and discussion section explains why you think the color red had the higher percentage in the preference survey. A conclusion from the survey simply could be that most people prefer red over gray. Some additional examples of results and conclusions are listed:

Difference between Results and Conclusions

Result: The divorce rate in India is 5%; it is more than 50% in the U.S.

Conclusion: The divorce rate is much higher in the U.S. than in India.

Result: Three of seven lots of steel had yield strength below the specification limits in May.

Conclusion: It is possible we are receiving steel below specification.

Result: Three of twenty parts showed rusting after a 30 day humidity test. None of the 60 competitor samples showed any rust.

Conclusion: Our product's corrosion resistance may be inferior to that of the competitor's product.

Result: The design approved by the review team had a production capability of 200 parts per minute compared with 80 parts per minute with the old design.

Conclusion: The proposed design will increase productivity by more than 100 percent.

Results are tabulations of facts. Conclusions require inference by the person preparing the report. They prepare the reader for the upcoming recommendations. Do not introduce any new material in this section. If there is more than one conclusion, it is preferable to enumerate the conclusions in an indented list. It is preferred that conclusions be one-sentence statements.

The conclusions from the example report on the nibbler need to reflect the part of the project that involved design and building of the device and the conclusions related to tests performed to rank different grades of cemented carbide. What conclusions can be drawn from the design work? First it was determined that the nibbler tool cuts web materials like a punch and die and that the wear patterns look similar to those on production tools. This leads to the following conclusion: The nibbler device can simulate production punch and die wear.

A significant part of the project also involved the use of microscopy and profilometry to measure edge wear. There were pluses and minuses to the various techniques, but the report ended with integration of wear areas from ten noncontact profilometer traces. The second conclusion from this work is that this is the best of the techniques evaluated: Noncontact laser profilometry (10 traces at 0.1 mm spacing on 2 punch and 2 die edges) is a satisfactory technique for assessing wear of test punches and dies.

Investigators also learned that it takes at least 10^6 nibbles to get significant wear measurements on cemented carbide tools. This conclusion can be made: Cemented carbide tools need at least 10^6 nibbles in order to develop wear that is reasonably easy to measure.

The experiments performed on the three grades of cemented carbide indicated that the high-cobalt grade (13%) wore the most, while the 3% cobalt grade was expected to wear the least. It did not. The discussion section suggests that grain size (the 3% grade had large grain size) caused this unanticipated behavior. The conclusion is that abrasion resistance relates to cobalt content when grain diameters are comparable: The abrasion resistance of WC/Co cemented carbides increases as the cobalt content goes down if grain diameters are the same.

It is recommended that conclusions be numbered and presented with only a sentence or two introduction. Readers often read conclusions first, and numbering makes them concise and more visible. The nibbler project work led to the following conclusions:

Report on the Nibbler

Conclusions

1. The nibbler device can simulate production tool wear.
2. Noncontact laser profilometry is a satisfactory technique for assessing punch and die wear.
3. Cemented carbide tools need at least 10^6 cutting cycles to get wear deep enough to be measured by profilometry.
4. The abrasion resistance of WC/Co cemented carbides increases as the cobalt content decreases if grain diameter is the same.

An example of a poor conclusion is shown below:

Conclusion Mixed with Discussion

Grinding method, current versus double, is less critical than obtaining required clearance when rolls mate. In other words, precision in roll grinding has a greater influence than method of grinding chosen.

- Defect measurements indicated a significant difference between Shop A and Shop B *grinds* with both Shop A grinds producing less defects. Defect measurements showed the regular Shop A grind producing somewhat lower levels.

> No main effect differences were observed when grinds were on sidewalls, and no noticeable differences were seen between defect types.
> - Defects were significantly improved by increasing *relative clearance* with higher roughness. A small main effect difference was also seen where higher roughness reduced defect level from 30 to 20.
> - In this experiment, 750 rpm produced lower dirt with no difference between *speed* for the chip defect.
> - The thin product displayed the same behavior for dirt as many past experiments. This product typically produces long, thick slivers that are pulled away from the top of the product by the adhesive during defect testing. Apparently, these slivers are not easy to catch during edge evaluation. Dirt from the heavy product comes from the backside and also consists of surface coating but is thinner, less continuous, and more dispersed along the product. A small number of very fine hairs are seen distributed on the backside. Fortunately, large-roll roughness drastically reduces surface dirt and significantly improves edge dirt on thick products as seen in the individual condition plots.

This is paraphrased from a current report, and it is an example of mixing results and discussion and calling it conclusions. You can tell that it is not an appropriate conclusion simply by observation. The bulleted statements are almost paragraphs. Conclusions should be concise statements. The last bullet statement in particular directs the reader to plots in the results section. This is discussion, as is most of what was written. The author also italicized some terms. This is inappropriate, because it is not stated why these terms are italicized.

Conclusions should be like first principles. Many facts are scrutinized and distilled into a statement that is inclusive. The individual facts are results, and the interpretation of the facts is done in the discussion. The conclusion is for numbered or bulleted statements only. Never present new information or data or refer to graphs or tables in the conclusion. The above example is converted into an acceptable conclusion as follows:

Conclusions

1. Precision in roll grinding is more important in roller quality than grinding method.
2. Shop A produces less roll defects than Shop B.
3. Higher surface roughness reduces defects.
4. Higher roll speeds during grinding reduce defects.
5. Thick products produce the lowest dirt on rough rolls.

In summary, the conclusion section contains statements of fact inferred from test results and analysis of results. Results are not conclusions. Conclusions are more global and may be based on several results. They are the end product of a body of work. They should be in the form of an investigative report. Because they should be written to stand alone, the conclusions should be prefaced with a brief introduction. Almost all investigations require conclusions, but if for some reason you cannot draw conclusions from the work reported in the document, end the report with a summary of what was reported.

RULE

Never add new material or discussions in the conclusion.

11.2 Recommendations

A recommendation is a statement suggesting a particular course of action.

Recommendation

The spinning facility should be shut down by the end of the next fiscal quarter.

Recommendations come at the end of reports, and they often are the final product of a study in formal reports. Studies or investigations are done because there may be insufficient information available on which to base a decision. You, the technical person, proposed a study (with a proposal report); you received funding, did the work, and now are writing the report. The report should close with a statement to the funding organization on a course of action. It is the logical closure for reports with the purpose of analyzing a process, business, material, and so forth. Management people want the technical people to tell them what to do. Recommendations are normally written in the imperative mode:

Recommendation

Reduce production rate 50 percent.

You, the writer, are directing the action. Recommendations should be concise and never more than one sentence:

Wordy Recommendation

It would be well for the field division to consider limiting instrument surveys to daylight hours because of highway safety concerns and the possibility of error due to low light levels.

Preferred

Limit instrument surveys to daylight hours, sunrise to sunset.

Remember that the discussions supporting the recommendations should have already been presented in the discussion section. Recommendations may include a short lead-in but no discussion. The "lead-in" to conclusions and recommendations requires some creativity. The conclusion section usually needs no transition. It can be introduced with a statement as simple as:

> Simple Lead-In
>
> Based on the laboratory tests and the corroborating field trials, the following conclusions are made:

If this statement is used as a lead-in phrase for the conclusions, the recommendation should have a similar type of statement:

> Recommendation
>
> DEC should apply a herbicide (Floxym) to the plants in Sector A.
>
> Recommendation with Lead-In Sentence
>
> We will conclude our laboratory tests on purple loosestrife, but we will continue to monitor infestation in the Braddocks Bay Wildlife Management area. Our recommendation is that the DEC apply a herbicide (Floxym) to the plants in Sector A by September of this year.

When there is only one recommendation, try to make it part of a short paragraph. However, do not add new information or discussion; only add transition statements like the one in the preceding example. It is never appropriate to enumerate conclusions or recommendations with only one item. This rule applies to any type of list. There cannot be an item "one" if there is no "two."

Recommendations must contain an action verb. You are requesting somebody to do something. You should also identify who that "somebody" is:

> Recommendation
>
> It is recommended that the wildlife division dissolve the quail-counting program by year end.

The following are examples of action verbs to use in recommendations:

- improve
- expand
- create
- develop
- close
- move
- build
- enforce

- study
- continue
- reduce
- purchase
- design
- install
- delay
- adopt

A very common problem in recommendation sections is indecision.

Weak Recommendation

You could close the sheet metal department, move it to another location and reduce staff, or outsource the operation.

Which is recommended? When a report requires a management decision, people usually want succinct statements to consider. There are often three answers to the same problem. Maybe all will work, but the reader only wants to know about the best one—the author's opinion of which is best. [*This is a daily problem in materials selection. Machine designers come to my desk with drawings for a part and say "What should I make this out of?" After getting answers to a litany of questions, I offer a short list of two or three materials that I have confidence will meet the customers' (the designer and his customer's) needs. The designer cannot list three materials on the drawing. This is not done. A detail drawing can only list one material of construction. This is accepted traditional engineering practice. From the more pragmatic standpoint, all drawings are checked by the design supervisor for accuracy and completeness. He or she would never approve a drawing with more than one material of construction specified. I must choose one of the three materials that will work and give that recommendation to the designer.*] This is what must be done with recommendations. They must not be nebulous or indecisive. Each recommendation should be singular in nature.

Enumerated Recommendations

1. The compounders must reduce silicone levels from 2 to 0.5% in the September test lot.
2. The mold shop must reduce the surface roughness on the X-mold from 1 to 0.5 μm before the September melting.
3. The mold shop must install a larger gate (2.5 mm) in the mold before the September melting.

Recommendation for the Nibbler Report

The development of the nibbler device has produced an off-line way of assessing the abrasivity of web products and tool materials. It is the recommendation of this report to initiate the use of nibbler tests on all new web products.

If the purpose of a technical report is to present the results of, for example, an analytical service, it may be inappropriate to have recommendations. Enough information to use as the basis for the recommendations may not have been provided. As an example, if you work in a laboratory that performs analytical services, a client submits a sample of an unknown substance and wants it identified. The purpose of the report will end with conclusions. This is acceptable. If you lack enough information to solve a problem, state that in concluding remarks.

In some instances it may be appropriate to substitute a summary for recommendations. For short reports, the summary can contain the conclusions and recommendations.

Summary

The knives with the short service life were determined to be significantly higher in hardness than the knives with a service life of 17.1 million chops. This suggests that reduced hardness is beneficial and future knives should be made from A11 tool steel with a hardness in the range of 54/56 HRC.

For very short reports, the summary may be only one sentence. In all cases, end a report in words. In fact, end each report section in words. Each section of a formal report should also have a summary and transition to the next section.

Summary. All projects need closure. The report body is closed by a summary or by conclusions and recommendations. Make recommendations concise, directed, unambiguous, useful, and with a timeline—who is to do what by when. The same is required of summary statements. Make this section stand on its own. It should be written so that it could be handed to a person and that person would have a well-defined action item, the objective of the report.

11.3 References

The reference section in a technical report lists the works of others that you referred to in a work. It includes books, journals, and articles, and under some circumstances, unpublished works such as company/organization reports or university theses. References are often cited in the text with square brackets [10], but this may depend on the style for a given publication. Brackets are good, because they are distinct and can be easily found when you are editing the text electronically.

Citation of References in Text

There are other ways to produce veneers [12, 13, 14], but the most common way is to use lathe skiving.

Some journals prefer a superscript (Smith[17]) for reference citations. Check the style and form manuals for a journal before you write a paper intended for publication.

It is common to name the author of a referenced work, but it can become distracting to the reader (as mentioned earlier) if there are many references.

Citations with Author Names

Waskow [45] applied blankholder pressure on the strip. The strip is drawn and blanked and ejected from the die.

Dodge and Kramer [46] separated the removal force. Simonds [47] used a modified champing technique that…

It may also be appropriate to cite numbered references several times in a work. References are numbered in the order of their appearance; they are given in numerical order if combined a second time with other references.

Referring to References a Second Time

The results that were obtained in the creep tests agreed with some other investigators [2, 6, 9] but were in disagreement with Cavenaugh [15], who performed similar studies.

Some journals use a reference style without a numerical reference. References are cited by author and date only.

Reference Citation by Author and Date

Archard in 1979 performed experiments that led to a widely used wear model.

This system is useful when authors or editors do not want to waste time renumbering references when articles are revised or adapted. It also allows the reference list to be organized in chronological order or by author. This has advantages for an ongoing body of work that is continually being updated or enhanced by many investigators. However, this system can be distracting when there are multiple references to a statement:

Distracting Attributions

Archard in 1979, Wolfgang in 1983, Petcher in 1984, and Spicer in 1993 worked on friction models that employed shear stress calculations.

Similarly, it can be "messy" looking up references by name rather than number. It also adds unnecessary words to the paper. This system is not recommended unless required by an organization or a particular journal.

The purpose of a list of references is to give the reader information on related work. The objective is to add credibility to your work. A reference list shows that you have examined the literature as background to see if others have done similar work or if the work has never been done previously by others. This is especially true with published articles. Archival journals only publish original work unless it is a review article.

Bibliographic Information in a References List. Now that we know how references are used and cited in text, let us address the correct way to list them. Each journal and organization has different guidelines on paper listing of attributions. The basic information required in a reference citation is:

- Author
- Publication title (title of book, title of article plus journal title)
- Publisher (and sometimes place of publication if the reference has limited distribution and access by your readers)
- Year of publication
- Page numbers

This is the minimum information needed by a reviewer to get a copy of the reference. Some journals want author names listed first, followed by the title of the paper in quotations followed by the name of the journal or book in italics, followed by the volume, the number, the date, and the page. Commas separate each and "pp" is commonly used as the abbreviation for "pages."

Citation of an Article in a Journal

R.J. Archard "The Role of Friction in Friction Models," *Journal of Testing,* Vol 2, No. 23, 1997, pp 207–209

Some journals prefer the listing of author names with names spelled out; others require the surname placed first as in: Archard, R.J, "The Role of Friction..."

One system for listing references is shown in Fig. 11.1. Each publisher is different, but this is a good default style if you do not have instructions for some other style. The various styles usually contain the same information. The presentation is just different. One exception that should be avoided is to omit the title of the article in a journal or multiauthored book:

Title Not Listed

18. D. Summers, *Res. Correspond.,* 8 (1955) 575
19. M. Graham, *Proc. Roy. Soc., A,* 212 (1952) 491
20. H.G. Howell, *Proc. Swed. Inst. Text. Res.,* 23 (1953) 589
21. J. Huffer, *J. Res.,* 12 (1959) 10
22. J. Graflex, Private Communication (1994)

This listing style is undesirable, because it does not give the reader enough information to make a decision about obtaining a copy of the cited work. It could be that J. Huffer only made a glancing reference that pertained to the work at hand, or it may be very closely related to the work at hand. The title of an article helps the reader decide if the reference is appropriate. Readers working in the topic area of a paper often get copies of references cited by others to add to the comprehension of their work.

The preceding examples point out another potential problem—using acronyms for journal titles. This practice is discouraged because it may not be clear to readers. The preceding citations, which were taken from a book with only the author names changed, are not clear. What does *"Res. Correspond."* mean? Or what does *"Proc. Swed. Inst. Text. Res."* mean? It may mean "Proceedings of the Swedish Institute of Textile Research," but this is only a guess. If you must abbreviate journal titles to save space or time, use the same rule for any acronym or abbreviation. Define on first appearance.

In the above example, it may also be unclear as to what the numbers mean. Usually subscribers for a periodical (such as a trade magazine or archival journal) receive issues on a periodic basis (monthly, quarterly, and so forth). Each issue has a number, and all the issues for one year are often bound into one volume for long-term shelving or archiving in libraries. Generally, the numbers in journal citations refer to the volume number for a given year. This is the number for the volume to be pulled off the library shelf. However, even when volume numbers are given, it is wise to include the year of publication.

It is also helpful to include the issue number (or month of publication) for many periodicals. This is not required for many peer-reviewed archival

---

Journal article:

18.   W. Schwartz, M. Maskow, "Carbonitride Case Depth Measurement," *Journal of Heat Treating*, 67 (1994) 3627-30.

Unpublished work:

12.   F. Crantz, "Effect of Solute Concentration on the Yield Strength of Alpha Iron" (unpublished Masters Degree Thesis, Alpha University, 1990) p 11.

Book:

7   Paul D. Branfdon, *Surface Science* (New York: McGraw Hill Co., 1992) pp 114-15.

Article in a composite book:

13. K.G. Brown, "Solid Friction," in *Polymer Friction*; K.R. Ludens, Ed. (London: Alpha Press Ltd., 1991) pp 21-27.

Book with no author:

2.   *A Manual on Heat Transfer Fluids*, 12 ed. (Boston: Brandon University Press, 1989) – 80-84.

Internet note:

      <http://www.kettering.edu/~princel>

Internet database:

      Encyclopedia online. Version 98.4. Americana Encyclopedia, 12 June 1998 <http:www.ae.com/>

Fig. 11.1   One system for listing references. Some journals have variations, but this can serve as your default system.

journals, because the issues are all paginated so that each article can be found in the bound volume directly from its page numbers. Nonetheless, it may still be useful to include issue numbers or the month of publication, unless you know the readers can find the article from page numbers. In some cases, the issues of a periodical are not paginated in the order that they appear in the bound volume for a given year. This is especially true for trade magazines (that is, periodicals with advertising). In this case, the issue number (or month of publication) must be included in the reference:

Citation by Volume and Issue Number

Wear News, Vol. 1, No. 12, 1988, pp 10–12

Lastly, some technical writers use a reference of a person's name followed by a "private communication" reference. The motive is noble. It recognizes discussions with knowledgeable colleagues, but it does not help the reader or add credibility when the reader does not know the person listed.

RULE

Never list a reference that cannot be obtained by the reader.

This rule means that you should not list private discussions or internal company reports in a paper for publication. Readers of the journal cannot get copies of internal reports to check what was said in them.

In summary, all formal reports and papers need references. They should be cited in text as numbered references with complete attributions listed after the last section of the report or paper. Remember references are listed to help the reader assess a work and ultimately to show credibility. Make the attribution complete enough to allow any librarian or individual to get a copy.

11.4 Writing an Abstract

An abstract is usually only used on formal technical reports or articles for publication. The abstract is usually placed on a separate page that includes the report title and the author name and affiliation. The location and format for an abstract for a technical journal is dictated by the journal. The format for articles differs with journals and technical societies. All have style and form guides, which are included in the author instructions for a journal. These directions should be obtained before writing anything.

Regardless of the specific style requirements for an abstract, the content of all abstracts must contain certain "ingredients":

- Objective of the work
- Scope of the work and what was done

- What was accomplished (results)
- Conclusions and recommendations

The first sentence is a likely place to state why the work was done—what you hoped to achieve. The investigation may be the hardest part of an abstract to keep brief. You must be ruthless in eliminating details yet comprehensive in conveying what was done. For example, there may be an urge to describe unique software used in interpretation of data. If the tests conducted consisted primarily of tensile tests, state that "tensile tests were conducted and indicated that ...." Do not mention how they are done if the reader will know what was done without mentioning it. The detail can be left to the body of the report. Do not include references or information that is not in the body of the report. The scope shows the boundaries of the work. What types of materials did you work with? Do not abbreviate unless it is absolutely necessary. Define acronyms at first use.

The results of the work are stated in words. Do not use data in the abstract unless it is very brief. It is acceptable to say that "the feature material had only 50% of the reflectivity of the control material," but it is not appropriate to state that "the feature had average reflectivity of 22.47, s = 7.6, compared to 155.90 s = 4.3, on the control." If you obtained ten different results in your tests, group them so that they can be stated in a sentence or two. The same type of consolidation must be done with conclusions and recommendations. If you have ten conclusions, convert them to one general conclusion and one general or overarching recommendation.

One of the most common mistakes with formal reports is too long an abstract (Fig. 11.2). Try to make the abstract no longer than 250 words (count-

Hydrolysis and Condensation-Coupling of Phenyltrimethoxysily1-Terminated Polystyrene Macromonomers

Anionically grown monofunctional polystyrene macromonomers are coupled through the hydrolysis and condensation of trimethoxysily1 end groups. Hydrolysis is initiated by adding acidified water to a tetrahydrofuran solution of the macromonomer. Condensation is facilitated by evaporating the solvent and heating the polymer under vacuum above the glass transition temperature. The macromonomers couple to form high-molecular-weight polymers and reach a finite size, beyond which further growth is inhibited. The final products are completely soluble star-shaped polymers, which are characterized by size-exclusion chromatography with molecular-weight-sensitive detectors. The molecular weight distribution of the stars are surprisingly narrow, although there are definitely mixtures of stars with different numbers of arms. The average number of arms in a star decreases as the molecular weight of the macromonomer increases. The final, limiting structures of the stars can be explained by the free energy changes associated with the number and length of arms. The results strongly suggest that the prevalent mode of growth at later stages of condensation becomes addition of macromonomer to stars, rather than addition of stars to stars. Studying the condensation-coupling of the macromonomers provides understanding for more complicated network-forming systems and also provides a unique method for synthesizing start-shaped polymers that has several advantages over other synthetic methods.

Fig. 11.2 Abstract with some of the basic elements missing (purpose of work, scope, what was done, recommendations)

ing all words). Some journals impose word limits like this. Sometimes extended abstracts of 300 to 400 words are used in lieu of published papers, but these are special cases. The next most common mistake in writing an abstract is that it contains so much technical jargon that nonworkers in the field cannot understand what was done. Remember that some readers may not have your technical background.

Figure 11.2 is not only very long and highly technical, but it also misses the mark on content. Why was this work being done? What is the scope of the work? Was it part of a study to invent a new plastic? What does the author want readers to do with the data in the report? Is he or she recommending the use of condensation coupling of macromonomers? This abstract was widely circulated, and the author knew that people other than polymer chemists would read it. It should have been changed in level accordingly.

Abstracts must contain the basic elements, written for the perceived readers. Abstracts with about 250 words are about the right length. Figure 11.3 presents an example of an abstract acceptable in length, content, and general format. Keep in mind that an abstract is not a summary of everything done; it is the essence of what was accomplished. It is like an advertisement of a report; it is your statement to readers of the value of your work.

Use of Wear Tests for Plastics

By
Kenneth G. Budinski
Eastman Kodak Company

For presentation at
The Annual Meeting of the American Chemical Society
August 1998, Boston, MA

Abstract:

It is estimated that over 50 percent of all new products in the United States are made from plastics. Many of these products employ plastics in some type of tribosystem. The plastic can wear or it may produce unanticipated wear on another surface. This paper describes the various wear tests that are frequently used to assess the tribological characteristics of plastics. Tests are described that measure how plastics wear in sliding contact with other solid surfaces as bushings, guides, and the like. Another section discusses tests that can be used to assess how abrasive plastics are to other solid surfaces, plastic abrasivity. Special tests are described that are applied to plastics in web form; these tests assess how they wear mating surfaces. Finally, erosion testing of plastics is described. These tests apply to the use of plastics in applications such as pumps and piping. The goal of this paper is to familiarize designers with the tests that can and should be used to determine if a plastic part will provide anticipated life in service. Guidelines are presented to assist in test selection and use of these tests.

Fig. 11.3 Typical abstract on a formal report (paper)

Make it brief, complete, and informative. Also be aware that it may be published by itself. It must stand on its own. Write it to ensure that it covers all of the work including final recommendations.

RULE

Write the abstract last and make it a concise summary of the body of work.

Another example is shown in Fig. 11.4. This is the abstract for the report on the nibbler. This particular abstract is on standard form with a variety of other administrative information and front matter.

TECHNICAL

REPORT/

DOCUMENT

Maintenance Organization	

Technical Report X

TIS Accession # _____

Technical Document # _____

Type of Document _____

DIVISION	Engineering & Project Management Division
UNIT/LAB	Materials Engineering Laboratory

Date Written	Remove From File	Review Before Destruction		No Yes	New Date for Removal
96 12 29	99 12 29		xx		

TITLE: A laboratory device to simulate notching and perforating of web products

KEY WORDS:	tools dies	carbide cutting	EWO, SER, BO, BP#

AUTHOR(S) (LAST, FIRST, MI) Budinski, Kenneth, G.	CONTRIBUTOR(S) (LAST, FIRST, MI) Kohler, M.

ABSTRACT:

Plastic and paper web products are often notched and perforated in finishing operations. The tools that perform these operations erode from a combination abrasion from hard particles in web coatings, from adhesive wear from rubbing on the web, and from corrosion from chemical coatings on the webs. Some webs erode tools at a much higher rate than others. A project was conducted in the Materials Engineering Laboratory to develop an off-line test device that can simulate web finishing and assess the erosive effects of various webs on tool materials. The objective of this work was a test that could be used to develop web coatings with low erosivity to tools and tools with maximum erosion resistance to a spectrum of existing web products.

The test device was designed and built that makes small "nibbles" from webs with a punch and die that could be removed from the device for periodic wear measurement. The normal test was 10^6 nibbles and wear is assessed with laser profilometry of cutting edges.

The design of the nibbler is described and its use on the evaluation of the relative erosion resistance of grades of cemented carbide tools is discussed. It was determined that the best resistance is obtained with a carbide with 6% cobalt binder and the worst resistance of the three grades was obtained with the sample with 13% cobalt binder. The nibbler is now a standard tool for evaluation of new web products and for evaluation tool materials and coatings.

Fig. 11.4   Abstract for nibbler report

11.5 Back Matter

You are not finished yet. Formal reports and papers can contain back matter that may be needed to conform to an organization or the style required by a journal. These sections can include:

- Acknowledgments
- Appendices
- References
- Illustration captions
- Illustrations
- Index

Acknowledgments. The acknowledgement section can be optional or not, depending on the organization. It is the proper place to list the names of people and organizations that contributed to the work described in the document or helped in writing the document. Almost all U.S. universities and U.S. government research organizations require acknowledgment of the source of funding:

Acknowledgment of Funding

This work was funded by the U.S. Department of the Interior Erosion Control Center, Contract 21742368.

Individuals or organizations and their contributions can be named:

Acknowledgment of Individuals/Organizations

Appreciation is expressed to ASM International for the use of their metric conversion charts in Appendix 2, to Yolanda Principle for her help in microtoming tissue sections, and to Jacques O'Brien for his help in conducting laboratory corrosion tests.

As mentioned previously, this is the only place in a technical document for mentioning people's names. Acknowledgments are not normally used on informal reports (for concision), but they are allowed on papers and formal reports. If they are commonly used in your organization, then use them. If not, skip it unless there is some individual that you would like to thank for his or her contribution to your work.

Appendices. An appendix is the "attic" of a document It is added to the end of a document, because there is no room within the document. You may have a three-page table that is essential to a message. It would be very disruptive to the reader to put such a long illustration in the middle of text.

You also could have a long list of special definitions used throughout the paper. An appendix can solve this problem.

This book has several appendices. Some appendices explain examples that would be disruptive in text. Another appendix contains grammar details. Grammar rules could have been a separate Chapter, but it would be boring for the reader who was well-versed in these rules. Putting these rules in an appendix makes them available to those who need them, while not distracting readers from the main message of an effective and practical writing methodology. An appendix can provide supplemental information that is not critical to the understanding of the text material but is useful reference information.

An appendix should stand by itself. It should be numbered and referred to in the text, like a reference. It should have a title that describes the content. This title should be as complete as the title or caption used on an illustration. It should state what is in the appendix as well as any other information that the reader may need to know to use the appendix.

In summary, appendixes are used to include information that would be too long in the text or for supplemental information. Appendixes add length to a document, so do not append anything that is not useful and pertinent.

RULE

Only include useful and essential information in appendices.

References. This Chapter already discusses the use and content of references, but they are part of the back matter of a formal technical report or paper. The usual format for a reference section is to list references by number on a separate sheet with the title "References" at the top of the list.

A bibliography is a list of books, articles, or other archived material consulted in preparing a document or relating to a document subject. It is not customary in industry to use a bibliography in investigation reports, only references. Bibliographies are more appropriate in essays and or general reference publications where readers may be seeking reference in formation from a variety of perspectives. For example, the "To Dig Deeper" listing of technical writing books at the end of each Chapter is essentially a bibliography. The books listed offer opposing or complementary points of view to the Chapter subject, but they are not references for specific statements in the Chapter text. They offer additional information for readers who feel the need to "dig deeper" into the Chapter subject matter. Often the bibliographic list is in alphabetical order of the lead-author surname.

In summary, use references with papers and formal reports, but not bibliographies, unless they are dictated by the style of the publication.

List of Figure Captions and Table Titles. Articles submitted to journals for publication usually need a list of figure captions and table titles captions that is separate from the actual illustrations. This is not always true of tables, which generally always have the title above the body of the table. This is not generally true of figures, because original figures may not include caption text. Often illustrations are numbered (usually in pencil) as "Fig. 1" with the complete caption listed on a separate sheet with the other captions.

List of Illustrations

Figure Captions

Figure 1 Schematic of integral quench vacuum hardening furnace

Figure 2 Effect of dew points on carbon potential of an endothermic gas atmosphere

Table Titles

If table titles must be listed, they go in a separate list:

Table 1 Conversion of Centigrade to Fahrenheit temperature

Table 2 Conversion of HRC hardness to HR15N

This approach to captioning is useful when a publisher (or another production person) is reformatting and typesetting the article into a specific type of page design. The need for this is changing now with electronic documents, because authors can compose pages and submit them electronically. This type of captioning may also not be required for internal formal reports. In most cases, it is sufficient if complete captions are placed on the page with associated illustrations. Each illustration (with its complete caption) should be placed on separate pages. This helps handling if the article is being reformatted. It may also help readers or reviewers, who may refer back to a figure while reading text. Whichever way figures are presented, keep it simple, consistent, and easy for yourself and the readers. The main objective is to provide good-quality illustrations that are useful.

Index. The index of a formal document is a list of subjects contained in the document with the associated pages where these subjects are addressed. Teaching and reference books must have an index (publishers require them), and the completeness depends on the diligence and indexing skill of the author or indexer. Some word processing programs offer indexing as an option. There are also stand-alone computer programs to do indexing. The laborious part of indexing is alphabetizing, which computers can do with the right program. Indexing of a technical document is generally not necessary, unless it is a very lengthy work with many topics covered in several places. Most often publishers or book production departments handle indexing. It is not something recommended for authors.

Summary. The content of back matter in formal reports and papers depends on the writing situation. Under normal circumstances, reports only include acknowledgments, a list of references, and illustration captions, or only the list of references and illustration captions. The other back-matter elements are used only as dictated by the report situation.

11.6 Report Distribution

Electronic or Print Distribution. In large organizations with local-area networks or intranet capabilities, reports and documents can be distributed by e-mail. The report is attached to an e-mail message, which contains a brief note about the attachment. This note should describe briefly the subject and any conclusions/recommendations contained in the report:

E-mail Note with Attached Report

Attached is the formal report on the cemented carbide seals that failed to meet expectations because of cracking in finishing. We propose continuation of the project using a different fabrication approach.

Attachment Icon, Report A

The readers can open the report by clicking on the attachment icon to read it or save it as an electronic file on their computer. This system works well, but some readers still prefer to review a hard copy [*as do I*]. Also, if the report has graphics and photos, the electronic file may be too large or unwieldy for distribution as an e-mail attachment.

Your choice of distribution method depends on the type of document, its length, the size of the distribution list, and the system used for archiving. If you are sending a document to many people, then electronic distribution has obvious benefits. However, if a large document is sent to just a few people, it may be best to route an original or send copies. This is a more cost-conscious alternative, as it costs more to have each recipient print out separate copies on printers. In addition, if formal reports are often archived in print form, reviewers may want to look at them in the final print form.

For distributing a large document to many people within an organization, another alternative is to announce completion of the report by e-mail and to advise recipients of its location on a local-area network (LAN) drive:

E-mail Notice of Report File

A formal report on the cemented carbide seals that failed to meet expectations because of cracking in finishing has been completed. The document is available as a read-only file on the network at:

file:///K/REPORTS/REPORT-A.doc

We propose continuation of the project using a different fabrication approach.

In this e-mail notice, readers can open the network document directly from the underlined name for the file path. Clicking the mouse arrow on the file name (or hitting the enter key with the scroll bar in the file name) starts the appropriate application and opens up the document. This approach is more efficient for large files, because the large file is not copied and sent to each recipient as an e-mail attachment. However, ensure that the document is posted as a "Read Only" file so that readers cannot change the work. This approach may work in the distribution of some reports, although it depends on the system for distributing and archiving of reports. Be aware of how documents are stored and archived so that you distribute the document in the most efficient manner.

Finally, if you are unsure about the best means of distribution, ask your recipients when completion is announced with the electronic file path of the read-only report. Ask if they prefer to receive hard copies. It never hurts to give them options.

Distribution Lists. As noted in Chapter 4, "Writing Strategy," identify readers before any writing commences. How do you decide to whom to send a report? This question is answered in many organizations by management. Some supervisors give engineers and scientists lists of people who are to receive copies of every report. If your work is a support organization, the distribution list should contain the customers who requested the work, as well as other customers who may be interested in the results.

It is usually a good idea to include one's immediate supervisor to any distribution list. This lets your supervisor know what you are doing (as any boss wants to know). Also include technicians and/or coworkers who assisted in the work. If they did substantial work on the project, they should be listed as contributors with your name or in an acknowledgment at the end of the report. In addition to your supervisor and helpers, the distribution list on a comprehensive project should include a report library (if you have one). The others on a distribution list should be carefully selected as potential users of report recommendations, or they may be potential new customers.

Distribution lists are normally attached on the front of a report. In some cases, formal reports may require a special form like the one shown in Fig. 11.5. This form helps identify some key factors in the distribution process. For example, the distribution cover sheet in Fig. 11.5 stipulates its destruction upon receipt. In this case, the distribution list can be detached and destroyed by the reader, because (as noted in a previous chapter) this may be required for legal reasons. Remember reports are often the only product of technical people. They should be distributed in the sense that any product is marketed.

RULE

Distribute your report with a strategy.

TECHNICAL REPORT DISTRIBUTION LIST

The Classification of the Attached Report is:
☐ Unrestricted Internal Use
☐ Restricted Information
☐ Confidential Information Controlled Distribution
✓ Must check one

_____
Date

_____
Report Accession No.

Title: _____

Author(s): _____

Organization: _____

Issuing Secretary: _____ Phone: _____

| Proprietary Information 2/83/R 02217 (ORIGINAL + 1 copy) |||||
| *Asterisk – by name indicates individual – received abstract only |||||
Name	Address	Name	Address

DESTROY THIS PAGE UPON RECEIPT

Fig. 11.5 Distribution list form

11.7 Saving Reports

As in the case of a distribution list, some technical organizations have a formal procedure for saving a technical report. The usual procedure is to put a hard copy in a personal file, the department file, and possibly the corporate or organization library. Needless-to-say, it is a good idea to save technical reports on diskette or some other computer database. A system that has proven successful for reports issued by a single department is to number the reports with a unique number and file them in a central file for at least two years. The documents may be reviewed after the first two-year interval to determine if they should be retained for another two-year cycle. The reports should be indexed, and this index lets all department report writers see what kinds of reports are being written and by whom.

Report No.	Title	Date	Author
95002	Tensile properties of Milliken Lightlock	1/6/98	L.B.J.
94003	Failure of A 10 tool steel cavities	1/10/98	K.G.B.
95004	Microstructure of staked dies	1/21/98	R.D.F.
95005	Sliding counterface for CD transport belts	1/24/98	F.P.D.
95006	Film can drawing problems	1/28/98	D.D.T.
95007	Interpretation of ball-on-plane abrasivity tests on film	2/3/98	R.A.F.
95008	Cavity failures around Lee Plugs	2/13/98	S.a.m.
95009	Service failures of Kodel splice tape	2/18/98	J.F.K.
95016	Friction and surface texture of intensifying screens	2/20/98	R.I.P.
95011	Failure of aluminum exhaust fans	2/24/98	D.O.M.

Reports on significant work must be archived.

Reports that become published papers are always available if they were published in an archival journal. Articles are usually available indefinitely even if the journal no longer exists. This is the ultimate archiving. Most archival library systems have the ability to complete computer searches. Key words are used to conduct searches. Give considerable thought to appropriate key words for reports or articles if you want them to be located by computer searches. The best key words are single nouns or verbs. If you must modify a noun or verb to make it clear, select precise modifying words (for example, tool steel, tool steel wear, wear failures, and so forth). Just use key words that you would use to search for the report in a database. Try to only use one word, not a phrase.

In summary, give significant attention to who should get copies of your work and how it will be saved for future use. Good reports are valuable for many years.

RULE

Save your technical reports and use a filing system that allows retrieval.

Summary

All reports, like any project, need closure as described in this Chapter. Formal reports should have a conclusion section, a recommendation section (if appropriate), references, and an abstract. Sending a report to an appropriate list of people is just as important as writing the document. This is especially true for reports that request action on the part of readers. Authors of technical reports should also ensure proper saving and archiving of reports that may have value five years or more in the future. Many reports are distributed in electronic form, but saving both hard and electronic copies is recommended for formal reports with long-term value. The vicissitudes of computers are known by all. [*I have two obsolete computers in my attic in case I need to get into old electronic files. My new hardware cannot deal with my old disks.*]

A final admonition on report closure is to be meticulous about conclusions, recommendations, and the abstract. Many times these are the only part of a report that management reads. Each needs to stand by itself. The reader should not have to refer to any part of the text in order to understand anything in these report elements. Acknowledge contributions from others in a way that is consistent with the situation and organization that will receive the report.

Important Terms

- Abstract
- Archiving
- Private communication
- Conclusion
- Closure
- Credibility
- Recommendation
- Enumeration
- Scope
- Attribution
- Imperative mode
- E-mail attachment
- References
- Timeline
- Index
- Acknowledgments
- Action verb
- Bibliography
- Distribution list
- Unambiguous
- Appendix

For Practice

1. You evaluated five different breakfast cereals for use by yourself or children. Write a conclusion section for the study.
2. Write a recommendation section for the situation in question 1.
3. State five differences between conclusions and material in a discussion section.
4. When is it appropriate to omit conclusions and recommendations from a report?
5. Describe the information that should be included in citing a journal article as a reference.
6. List a reference taken from a handbook on tool steels. The editor was Shuly Fess, and the section referenced was entitled "High Speed Steels" on pages 28 to 131.
7. List the basic elements of an abstract.
8. Write an abstract on the cereal study in question 1.
9. You are writing a formal report on a big study. Two technicians, J. Wallace and R. Smithers, performed the laboratory work. Acknowledge their contributions.

To Dig Deeper

- G. Blake and R.W. Bly, *The Elements of Technical Writing,* Macmillan and Company, New York, 1993
- J. Gibaldi, *MLA Handbook for Writers of Research Papers,* 5th ed., Modern Language Association of America, New York, 1999
- J.M. McCrimmon, *Writing with Purpose,* 5th ed., Houghton Mifflin Co., Boston, 1974
- T.A. Sherman and S.S. Johnson, *Modern Technical Writing,* 5th ed., Prentice Hall, Englewood Cliffs, NJ, 1990

CHAPTER 12

Informal Reports

CHAPTER GOALS

1. *Understand when to choose an informal report over a formal report*
2. *Know the elements of an informal report*
3. *Understand how to write informal reports*

INFORMAL REPORTS include various types of documents (Fig. 12.1) that are considered distinct from a formal report. The main difference between a formal and informal report is that informal reports usually do not need an abstract or references. They are usually too short to require them. Informal reports usually contain only one or two pages of text. A one-half

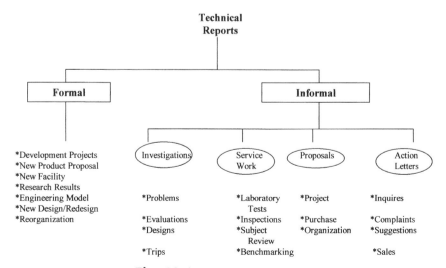

Fig. 12.1 Types of informal reports

page abstract for a two-page lab report is neither concise nor necessary. Similarly, references are usually not needed because the introduction and discussion sections are too brief.

This Chapter is about informal reports with a focus on four application areas:

- Investigations
- Service work
- Action letters
- Proposals

Figure 12.2 is an example of a typical informal report. There are many report templates available in word processing software. Filling in the form completes the document, and supporting graphs and tables are made into attachments. The report is sent to the individual who requested the work and to others that the customer suggests. Management personnel should also receive a copy so that they are aware of what you are doing. Depending on the type of document, some informal reports may be filed by number and be kept on file for a few years. At the end of the retention cycle, the report may be reviewed for another retention cycle or more permanent archiving.

12.1 Elements of an Informal Report

The informal report in Fig. 12.2 has three sections: problem, investigation, and summary. These three sections contain the basic information required in any report:

Introduction (problem)

- Nature of problem/situation
- Why the report is written (purpose, objective)
- What is in the report (format)

Body (investigation)

- Procedure
- Results
- Explanation of results

Closure (summary)

- Conclusions
- Recommendations

This content reflects the basic content elements of a formal report but with the omission of an abstract and references. The example could have

Materials Engineering Laboratory Report

To: Michael Smith	Location/Phone: PSET-MSTD/61250	Date: 6/18/98
Subject: Silicone Elastomer Abrasion Test		Report No.: 98095
Written/Phone: John R. Duford/72019		Approved:

PROBLEM

The Materials Engineering Laboratory was asked to evaluate the abrasion resistance of various elastomers for improved wear properties compared with thermoplastic elastomers (TPE) for processor rollers. It is the purpose of this report to present the results of laboratory abrasion tests on feature and control elastomers.

INVESTIGATION

We applied the ASTM D3389 abrasion test to six rubbers: TPE 55A, TPE 64A, transparent urethane 50A, Gray 50A, APR 50A, and urethane plus antistat 80A. The test conditions were CS17 wheels, 500 g load on each wheel and 1000 revolutions. Samples were weighed to measure the mass loss in 1000 revolution tests. Three replicate tests were conducted on each rubber. The wheels were resurfaced before each test with 150 grit aluminum oxide bonded abrasive. The mass change data are presented in Table 1. The test results are compared with historical data (from MEL Report 96040) in Figure 1. All of the polyurethane elastomers had lower abrasion than the TPE's and the 50A polyurethane.

SUMMARY

These data suggest that the CF1-135 silicone elastomer tested in our prior study is more abrasion resistant that the TPE's, but the 80A polyurethane rubber containing the antistat is the most abrasion resistant of the group. It should be the prime candidate for more abrasion-resistant rollers.

Table 1
Weight Changes in Tabor Abraser Test

Material	Average Weight Loss(s)
CF1 135 Primer	0.0146
TPE 55A Durometer	0.1027
TPE 64A Durometer	0.1096
Transparent 50A Polyurethane	0.2740
Grey Cast Urethane 50 A Durometer	0.0158
APR Urethane 50A Durometer	0.0126
Urethane + Antistat 80A Durometer	0.0086

Figure 1

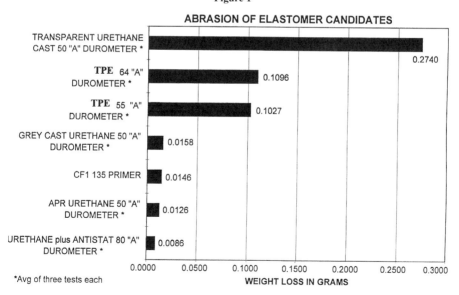

*Avg of three tests each

Fig. 12.2 Informal investigation report

had more details on why the tests were conducted, but what was said was certainly adequate for the customer (addressee). The customer brought the sample in and requested the test. If the addressee is the only recipient, the introductory information is adequate. However, most reports are distributed to more than one person, and so the readers may need enough introductory information to know why the work is being done, what the scope of the report is, and what the scope of the work is. It is usually not necessary to state the format for a report like in Fig. 12.2. There are only three section headings, and the readers can clearly see all of the headings on this one-page report.

In summary, an informal report should contain the basic content elements of a formal report with the exception of references and an abstract. Try to use a single-page format similar to Fig. 12.2 for investigation reports. It is very important to include all test details in the investigation section. As in the example, the previous tests were conducted three years earlier. Without adequate test details, the test procedure would not be repeatable. Finally, provide the answer to the investigation—what was found and what should the customer know?

RULE

Informal reports must contain an introduction, body, and conclusion.

12.2 Investigation Reports

Investigation reports present the results of studies, problems, failures, safety concerns, and environmental concerns—anything that requires methodical probing or testing. These reports usually need all the elements described here for the three basic sections, problem, investigation, and summary.

If an investigation involves a design concept, the introduction would state the problem—why a new design is needed or why the work is being done:

Problem Statement for a Design Concept

Improve the wear properties of rubber rollers

The description of the problem in the introduction may include the history of the situation—what was tried in the past. There may be no history in the preceding example if the report is directed to individuals who are well aware of the history of the device. This is proper on reports where the author knows the background of all intended readers.

The scope of the work is implied; it is to evaluate several rubber rollers. The objective is to improve roller abrasion resistance; the purpose of the

Materials Engineering Laboratory

Report Number 98018   Date: 9/7/98

Subject: Trip to Tecmet Inc.

Author: K.G. Budinski, KP Materials Engineering Laboratory

Contributors: T. Skevar, Film Flow, 4/58

Purpose of Trip:

There had been accelerated tool wear in the ECT department in cutting a new web product (8823). It appears that a new grade of Tecment cermet, A8, will solve this problem. The purpose of this trip was to establish a quality control procedure and specification for this new tool material. The group visiting the Tecmet plant included myself, the production supervisor, R. Tooth, the tooling supervisor, T. Skevar and D. Muscal, the quality statistician.

Trip Details:

The plant is in Alpena, Michigan. We arrived in the afternoon of September 21 and we spent the remainder of the afternoon touring the manufacturing facilities, the metallurgical laboratory and the quality control department. The morning of September 22 was spent reviewing Tecmet's quality procedures and a strawman specification that was presented to Tecmet for their review two weeks prior to the trip. We returned to the plant on the afternoon of September 22.

Summary:

The trip reassured us that Tecmet has superior manufacturing capabilities and their quality tests on the new grade of cermet are quite adequate. They disagreed with some of the composition limits in our specification, but through negotiation, we arrived at a specification that was mutually agreeable. We will purchase to this specification and Tecmet will supply monthly process control data to D. Muscal so that we do not need in-house quality testing.

Fig. 12.3 Typical trip report

report is to present test results. These basic elements of the introduction are then followed by the report body, which describes the test. Depending on the nature of the design investigation, the body of the report could be called "Improved Design" instead of "Investigation." Obviously, complicated designs or tests may require substantial text. Schematic drawings could be added to show test equipment. Sections can be merged in informal reports as long as the essential report elements are retained.

Trip Reports. A common form of an informal investigation is a trip report. Trip reports document observations, and they can be chattier in tone than technical reports because they often involve the use of names and personal statements. Figure 12.3 is an example of a trip report to a company supplier. These types of reports need to state why the trip was needed, the purpose and objective of the trip, what transpired, and the outcome of the trip (introduction, body, and conclusion). In this example, the historical incident that prompted the trip was:

Problem Statement for a Trip Report

Investigate accelerated tool wear in the ECT department

The purpose of the trip was stated to be:

> Purpose of Trip
>
> Establish quality control procedure and specification for the new tool material.

The objective of the new tool material is not stated. It could be, but because all the readers know about the problem, it is left out. The objective of the trip was to solve the tool wear problem. The purpose of the report is not stated in the text. Most trip reports have the same purpose and objective. The purpose, implied in the very nature of a trip report, is to inform interested parties and management. Your purpose in writing is to show that you are doing your job properly.

The body of a trip report describes the important details of what transpired regarding the problem or objective of the trip—who you met with, what you observed, and what decisions were made. Any information that may be of value to others is put here. Avoid unnecessary details regarding travel or accommodations during the trip. Nobody cares where you stayed or dined. The same is true of travel details. Only present information that could have value to others.

Finally, the conclusion of a trip report should state the outcome or value of the trip.

> Conclusion of a Trip Report
>
> We arrived at a specification that was mutually agreeable, and we do not need in-house quality testing.

These kinds of statements show the value of the trip and how you are working to resolve the problem that prompted the trip.

In summary, trip reports are an excellent tool that a technical person can use to show value derived from company travel. They show professionalism on the part of the author and they provide needed communication with others. Try to use them.

Problem Report. Problems occur in all organizations, and technical people are often assigned the job of solving them. A report is usually issued to lay the problem to rest. Figure 12.4 is an example of a report written to resolve a problem with material safety datasheets (MSD) for laboratory chemicals. In the United States, manufacturers are required by law to have an MSD on file for every chemical used in the lab or production. This includes even purchased commodity items such as eyeglass cleaner, household cleaners for the coffee area, even the paper correction fluid used by secretaries to fix typing mistakes. Environmental auditors monitor conformance to MSD regulations with periodic unannounced laboratory inspections.

The problem section gives necessary background; an audit citation was the incident that triggered the report. The last information illustrates the importance of the problem. The purpose of the report is stated to remind lab members about their responsibility and to introduce a new MSD procedure. The objective of the report is stated to be to avoid fines, and the implied objective is to not get another citation from company or state auditors. The boss gets performance demerits, which are passed onto responsible department members.

The body of the report describes the remedial action. More details could be added, like the other problem-solving options that were considered, but the concise statement in the example does the job. This three or four sentence paragraph replaces the procedure, results, and discussion of a formal report.

Closure is again obtained by the summary section, which replaces conclusions and recommendations. The imperative statement telling lab members to adopt the new procedure and to be meticulous in safety record keeping is a recommendation with a hint of threat. The summary section is also an abstract of sorts. The reader can get the essence of the report from this statement. Everybody loves a bottom line. The bottom line of a report is the closing section, which tells the reader what to do. A suitable transition

---

Materials Engineering Laboratory

Report Number: 98076   Date: 7/17/98

To: Laboratory Staff

Subject: Lab safety

Author: Diane Summer, Materials Engineering Lab, 2/35/KP, MC 34267

Problem:

On June 27, the Laboratory was given a safety citation by the division auditor for having two chemicals in the laboratory without appropriate MSD safety sheets. In 1997, the company paid $180,000 in fines for lapses in chemical record keeping. It is the purpose of this report to remind all laboratory members that each individual has the responsibility to ensure that these sheets are available in the central file on chemicals and safety procedures. These sheets convey potential hazard information to chemical users and we need to have these sheets available for our own safety as well as to avoid fines from state inspectors. We changed the procedure for logging-in chemicals and this report will outline the new procedure.

New Procedure:

The Laboratory Safety Awareness Committee was convened to decide on the resolution of the recent citation. Many possible approaches to preventing reoccurrence were discussed and the consensus procedure changes are summarized below.

1. Individual laboratory members will still have the responsibility for obtaining MSD sheets from the supplier of all chemicals.
2. No chemical will be allowed on laboratory shelves or workbenches unless they contain a fluorescent red dot in a conspicuous location.
3. The dots will be issued by the laboratory safety officer upon submission of a valid MSD sheet.
4. All chemicals awaiting MSD sheets must be stored in the hazardous waste management area.

Summary:

This recent citation did not result in a fine because it was discovered by in-house auditors. However, this safety violation could have been discovered by state inspectors. The Safety Awareness Committee is requesting all lab members to actively support our new procedure and to make an extra effort to have all chemicals arrive in the lab with their MSD sheets.

Fig. 12.4 Typical problem report

to the action statement is a one or two sentence summary of the problem and the study.

Evaluation Reports. Some evaluations warrant a formal report; some can be handled by an informal report. The deciding factor is usually the long-term value of the report. If the evaluation is likely to be of interest five years hence, a formal report is appropriate. If not, the informal report can be used.

Figure 12.5 is an example of an evaluation of knives from two different suppliers. The evaluation consisted of measuring the metallurgical properties of knives from both suppliers to determine why one wears at twice the rate of the other. This report has all of the elements of a formal report, except for sections with discussion, abstract, and references. The introduction describes the

---

Materials Engineering Laboratory

Report No.: 98163   Date: Sept. 12, 1998

To: A. Wang, PPD Dept. 12/4/KT

Subject: Comparison separator paper chop knives

Author: K. G. Budinski, KP Materials Engineering Laboratory, 5/322/KT

Problem:
The PPD Department requested the Materials Engineering Laboratory to investigate why separator chop knives purchased from Arun Corp. last twice as long as knives purchased from Arkansas Knife Co. The knives allegedly are made from the same steel, AISI type M2, and they are made to the same drawing. The company has a preferred supplier agreement with Arkansas Knife Company and they are twenty percent lower in cost than knives from Arun Corp. The purpose of the study reported in this report is to determine why the Arkansas knives do not perform as well in service as the Arun Knives. It may be that Arkansas can modify their procedure to improve wear characteristics. This report will present the results of laboratory studies on the metallurgical characteristics of both knives and make recommendations to improve wear life of Arkansas knives.

Investigation:
Three knives from each supplier were submitted for laboratory testing. The following tests were performed on each knife.

Test	Description/procedure
1. Hardness	ASTM E8 – Rockwell C and 15N
2. Microstructure analysis	ASTM E22 – Optical microscopy
3. Chemical analysis	ASTM T15 – X-ray fluorescence
4. Retained austenite	ASTM E45 – X-ray diffraction
5. Reciprocating wear	ASTM G132 – WC/Co rider

Three replicate tests were performed on each material for each test. The test averages were compared for significant differences.

Results:
Both knife materials had the same hardness, 58 to 60 HRC. The chemical analysis confirmed that both knives were made from steels that conformed to M2 tool steel in chemical composition and carbide phase microstructure. The retained austenite was 8.3 percent (s = 0.4) on the Arkansas knives and 0.6 (s = 0.5) on the Arun knives. Three Arkansas knives were re-hardened with a triple temper and deep freeze to reduce the amount of retained austenite. The G 132 wear test was repeated and it was determined that both knives had the same wear rate.

Conclusions:
The Arkansas knives had more retained austenite (soft structure) than the Arun knives and this appears to be the reason for their wear rate difference.

Recommendations:
The study determined that the Arkansas Knives were probably heat treated differently than the Arun knives. We recommend adoption of heat treat procedure HT 9110 for future knives from Arkansas Knife Co. This procedure should ensure retained austenite levels below one percent and the Arkansas knife life should improve to match that of Arun knives.

Fig. 12.5 Typical evaluation report

problem and why the evaluation was needed. The objective is to get one type of knives on par with another type of knives and thus save money.

The procedure section is very abbreviated. The use of standard tests allows this. The author does not have to describe the test and how it was run. This test information is described in ASTM standards, which are available on the Internet or from most technical libraries. Most industrialized countries use ASTM or similar standards for many types of engineering tests.

Note that results are separated from the procedure and that procedure was renamed "investigation." This latter term is useful for informal reports because it is more inclusive than "procedure." This heading can include procedure, results, and discussion—the entire report body. The investigation heading is often preferred for very short (one-page) reports. [*This is the default heading for the informal report form on our lab Intranet.*]

12.3 Service Work

In U.S. industry, most companies have two types of employees: production and support. Production employees make saleable product; support workers do all the things necessary to make products.

- *Production personnel:* manufacturing managers, assemblers, machinists, forming press operators, and so forth
- *Support personnel:* research scientists, engineers, and testing, sales/marketing, purchasing, production planning, tool making personnel, and so forth

Support people often write service reports for production departments as well as for other support units. They are often asked to test something, to fix something, to improve something.

Laboratory Tests. Figure 12.6 is an example of the simplest type of service report. It differs significantly from the formal and informal reports in previous discussions because it is only a statement of results for a blind test. The person submitting the work to a service department may not divulge anything about the material submitted for testing. A laboratory may get requests to conduct an ASTM A 3240 abrasion test on samples A and B. They may have no more information than that, so the report will probably look like Fig. 12.6. These kinds of tests go with the business, but a testing lab may provide more value (at the same cost) if they know what the problem is and the objective of the test. This kind of testing is not very fulfilling to the person who runs the tests. [*The attempt of our laboratory to make testing more valuable to both customer and tester is to make customers fill out a job write-up sheet that requires input on the nature of a problem and more information on the objective of the work.*]

Inspections. Testing or service departments often make inspections or measurements on submitted parts or materials. These situations are often misreported. The inspection department may simply scribble measurements on a scrap of paper and consider the job completed. [*Our laboratory has a group of four people who spend their day performing measurements of surface finish on samples that are submitted by many customers. Test results are given to customers on reports similar to the one illustrated in Fig. 12.7.*]

```
From: 618902  --KP26                  Date and time    09/04/98 11:25:08
To: 55861    --KP26      BUDINSKI K G


From:            Analytical Technology Division, IRAD
       Atomic Spectroscopy Group of ESCC
       Building 34, Floor 1,
Subject: Samples for Trace Elements

Inductively coupled plasma atomic emission spectroscopy (ICP-AES) was used to
measure the samples for trace elements.  The samples were prepared by digesting
them with a mixture of sulfuric and nitric acids.

Inductively coupled plasma atomic emission spectroscopy (ICP-AES) was used to
measure the samples for trace elements.  The samples were prepared by diluting
them with deionized, distilled water.


All values are in ug/mL.
```

Sample ID	ug Co/mL	ug Cr/mL	ug Fe/mL	ug Ni/mL	ug Ti/mL	ug W/mL
3A-32	---	< 0.3	5.2	---	---	---
	---	< 0.3	5.3	---	---	---
3B-32	---	< 0.3	< 0.3	---	---	---
	---	< 0.3	< 0.3	---	---	---
3C-32	---	< 0.3	< 0.3	---	---	---
	---	< 0.3	< 0.3	---	---	---
3D-32	---	< 0.3	9.1	---	---	---
	---	< 0.3	10	---	---	---
3E-32	---	< 0.3	0.3	---	---	---
	---	< 0.3	< 0.3	---	---	---
3F-32	---	< 0.3	< 0.3	---	---	---
	---	< 0.3	< 0.3	---	---	---
3G-32	---	< 0.3	< 0.3	---	---	---
	---	< 0.3	< 0.3	---	---	---
3H-20	---	34	569	---	---	---
	---	36	588	---	---	---
3I-20	---	< 0.3	0.6	---	---	---
	---	< 0.3	0.6	---	---	---
3J-20	---	< 0.3	< 0.3	---	---	---
	---	< 0.3	< 0.3	---	---	---
2H-20	34	---	---	---	---	4
	35	---	---	---	---	4
2I-20	7.2	---	---	---	---	< 3
	7.6	---	---	---	---	< 3
2J-20	29	---	---	---	---	< 3

Fig. 12.6 Service report that only presents results

Needless to say, inspection reports depend on the type of inspection. The main point is that some type of written report should be offered. A spreadsheet or graph is not a report. A report contains the requested information and any observations from the inspector that may be of value. When replicates are tested, at least the numerical average and standard deviation should be calculated. In fact, the desired statistical treatment of the data should be determined at the time of the job write-up. Writing inspection reports that add value increases the satisfaction of both customer and inspector.

Subject Reviews. Review articles can be found in many journals. They are usually commissioned by an organization, and they consist of digesting the literature for some period of time on a particular subject. Reviews provide a service to people who work in that particular field and are usually written by a senior researcher in a particular field.

The significance of a review is often the listed references. As shown in Fig. 12.8, short statements are supported by literature citations. The citations that are applicable to a particular problem, for example, with web friction can be obtained and studied in depth. As mentioned previously, some books are written in review style, but the average technical writer may have occasion to write a review report on a subject of interest to a particular organization.

Metrology Laboratory Inspection Report

Report Number: 2176   Date 9/9/98

Subject Ball-on-plane wear test

Customer: S. D'Amato, PCT MC 21443

Inspector: S. Callister

Request Description:

Measure wear volume of 10 ball-on-plane wear scars, using the average of 10 transverse traces taken at 2mm increments along the 25mm scar length. Take the "down" area from the 10 traces and multiply it by 25mm to arrive at wear volume in mm^3.

Test results:

The profilometer traces for all ten samples are appended. The individual averages are below.

Samples	Avg. wear volume	standard deviation
1	22 mm^3	0.07
2	48	0.03
3	10	1.4
4	220	1.6
5	123	4.6
6	110	1.3
7	43	0.9
8	12	1.3
9	37	0.4
10	38	3.6

The sample results are compared graphically in Figure 1.

Comments:

The surface texture within the wear scars appeared to vary significantly with the various test materials. You may want to quantify these differences.

Fig. 12.7 Inspection report

[I wrote a review article once on the friction of photographic film using all of the formal reports in the company's library. It was very interesting reading fifty-year-old reports that dealt with problems that I was currently working on.]

Well-researched review articles are often of great benefit to the author and the organization. They are written in the form of Fig. 12.8 with subheadings that reflect key aspects of the subject. Review reports are closed with a summary section that presents an overview of past work and some predictions on work that needs to be done. The objective of reviews is often to identify areas for research or improvement.

Benchmarking. In business vernacular, the term benchmarking refers to a comparison of competing methods or alternative operations. It may involve the study of other companies and compare what you do in a particular area to what they do. If they do that thing better, you lose. You must take some action. In the past, the term benchmarking has been jargon among some business consultants for popularizing business experiments where benchmarking reports served as the basis for many personnel reductions, divestitures, and other company changes.

Companies with experience in corporate intelligence often conduct industrial benchmarking studies. If you want to know what your competition is doing in a particular area, you can hire one of these companies. They give you a benchmarking report with details and analysis from information acquired in a variety of legal ways. They attend public speeches given by corporate officials; they scour the local newspapers for pertinent information. They go to the bars and restaurants surrounding factories and devour annual reports, inspect public patent documents, and review building permits and plans in local zoning departments.

The result is a report with business information on production rates, cost, and capacity. Figure 12.9 is an example of the type of information in a benchmarking report. These reports examine a number of candidate companies, and the closure consists of a summary of the collected data and a financial analysis of these data. Taxes, environmental concerns, even cur-

---

Web Friction – A Review

Friction Fundamentals

Web friction research was documented by Leonardo DaVinci in 1495 [1]. He made detailed sketches of ropes on capstans supporting loads and other friction devices. Amontons [2] in 1675 developed the classic $F = \mu N$ equation that relates friction force (F) to normal force (N). Woltec and others [3] researched the roll for surface textures on friction, but Coulomb in 1795 [4], performed classic experiments on the role of individual asperities in friction and Bikerman [5] used modern mathematics to model friction force as a function of surface texture.

The work of these early investigators was merged into friction fundamentals for flexible webs in a classic text "Roller Traction" by Altman [6]. This test documented important tests such as roll spindown and roll dynamometer test which have been successfully used in paper web transport [7] ....etc.

Fig. 12.8 Typical writing in a review article

Roll Electroplating-KT compared with other suppliers

Problem:

The KT Electroplating department occupies 200,000 square feet of space. The department has full A & B shifts and a limited C shift. There are four grinders/polishers on the first shift, one laboratory technician, one department head and 0.6 engineers (from B2). The annual labor costs are 1.6 million dollars; the annual consumables, rent and utility costs are 2.3 million dollars. The work load for the past five years is shown in Figure 1. The average plating costs are tabulated below:

> Large rolls = $1.05 per square inch (Cr or Ni)
> Small rolls = $.35 per square inch (Cr or Ni)
> Electroless nickel = $.56 per square inch
> Anodizing = $.87 per square inch
> Hardcoating = $.92 per square inch

Compared with Tecmet Plating:

Tecmet Plating of Oakville, Illinois was toured on September 3 and it was determined that they could do all of the plating that is done in the KT facility with the exception of large rolls. Their charge for various plating services are:

> Chromium of nickel on small rolls = $.23 per square inch
> Electroless nickel = $.27 per square inch
> Anodizing = $.32 per square inch
> Hardcoating = $.34 per square inch

Fig. 12.9 Benchmarking report

rent interest rates are considered. The last section of the report is a recommendation to management on a path forward.

As with many other types of informal reports, the sections on a benchmarking report vary with the situation. However, these reports still require an introduction, body, discussion, conclusion, and recommendations. Oftentimes they are written as formal reports. The example cited would probably be a formal report because it involved ten competitors. Some studies are much smaller and may involve only benchmarking two labs who do the very specialized work that your lab does. It is difficult to write these reports if you are personally involved in the subject operations, but objectivity should be sought in the presentation and discussion of data. [*Yes, they shut down our plating department after 83 years in business.*]

12.4 Action Letters

[*I have computer disks full to the brim with letters to politicians and company presidents. My success rate on getting satisfaction from these groups is about 60% for company presidents and zero% for politicians (including U.S. senators, congressmen, cabinet members, the Vice Presidents and the President). Obviously, I am not in a position to recommend procedures for writing letters to elected and appointed government officials. I suspect that the success in writing may require closure with a campaign contribution check.*]

Letters in business, industry, and government often fall into two major categories: inquiries and complaints. Sales is a third important category, but one that usually only applies to people in marketing.

The purpose of a letter is usually to evoke action from an individual or organization. The objective varies with the situation. Inquiry letters usually ask for information. Complaint letters seek redress of the complaint. The basic parts of a letter are the salutation, the introduction, the body, and closure—much the same as with informal reports. The salutation format may vary depending on personal or organizational preferences, but the following style is probably the most widely used salutation format:

> June 23, 1998
>
> Dr. J. M. Miller
> Director of Research
> White Engineering
> Box 1006
> Evansville, TX 14617
>
> Dear Dr. Miller:
>
> XXXXXXXXXXXXXXXXXXXXXXXXXXXXXXXXXXXXXXXXX
>
> XXXXXXXXXXXXXXXXXXXXXXXXXXXXXXXXXXXXXXXXX

Letters to elected officials often require statements to acknowledge that these people are superior to others, such as "honorable," and so forth. [*I do not use these titles. (Maybe this is why I never get answers.)*] Some organizations may have another salutation format.

Internet e-mail letters have salutations that vary with the software. Usually there is no choice other than to type in the addressee's name. Letters to "whom it may concern" or to "the sales department" or "order department" usually are directed to the trash. It usually takes a name to get attention. Names can be obtained from web sites or by a phone call.

Inquiries. Letters are often required in business and industry to accomplish such items as:

- Check on status of work
- Obtain vendor information
- Request quotations
- Consult with technical colleagues
- Consult with universities
- Recruit employees
- Make employment offers
- Resolve technical issues with suppliers
- Purchasing information

There are probably a hundred other reasons for writing a letter, but the format and style for each is essentially the same (Fig. 12.10). It is desirable to begin with some pleasantries or some kind of statement to make the inquiry friendly rather than confrontational or challenging. Next, state the purpose of the letter. Make it clear what you want.

Purpose of an Inquiry

I am writing to request additional information.

Give details and tell the addressee what you want done and by when. Inquiry letters are closed with another pleasantry. These letters are more likely to receive a reply if they are addressed to a specific person.

Complaints/Adjustments. [*As mentioned in our introductory remarks, my success on complaint letters is far better to businesses than to politicians. My most recent success was a letter to the president of a major airline. My overhead light was out on a seven hour flight from Washington to Paris. I had planned to use that flight time to prepare a presentation that I had to give on arrival. I could not write for most of the trip because it was dark. In fact, all seats near me also had no overhead light. I was upset and on return, I wrote to the president of the airline. I received an apologetic letter and a certificate for $100 off my next flight with that airline.*]

Figure 12.10 is a general guideline for writing complaint/adjustment letters. This is the procedure that netted me favorable replies over the years. Sometimes I write to make product suggestions. The philosophy of this guide is to pretend that you are not mad and that you are a loyal customer who wants to continue to do business with them. If you say that you will never do business with them again, there is no reason for them to reply. A key part of a letter asking for a refund or some form of monetary return is to use euphemisms for money.

Euphemism in Asking for Money

I feel that an adjustment is in order.

Writing a letter requesting an adjustment or submitting a complaint:

1. Start by saying something nice about the company/product.
2. Introduce the problem/complaint as if it was the only one that ever happened to the company.
3. State what you want them to do. If you want your money back, say it nicely (an adjustment is in order).
4. End the letter reiterating how you like the company and want to continue to do business with them.

An angry letter will end up in the trash. A letter from a loyal (repeat) customer stands a much better chance of a favorable reply.

Fig. 12.10 General guidelines for writing a complaint letter

People in complaint departments like to decide the value of monetary awards. The term "adjustment" is the most popular euphemism for refund. [*However, when my wife writes to her delinquent tenants for overdue rent, she does not use euphemisms but backs up her demands with facts.*]

My rent records show no payment for March, April, and May. The amount in arrears is $1500, and by June 3 your total in arrears will be $2,000. Please contact me by May 30 with your payment plan.

[*She still makes the letter as cordial as possible by following the guidelines in Fig. 12.10. She has even paid a cash bonus to get a down-on-their luck couple to leave. She was pleasant even though they were essentially being evicted. The cash was offered to help with their moving. It was cheaper than the cost of a lawyer to get a court ordered eviction.*]

Maintaining composure and civility is key in writing complaint/ adjustment letters. Figure 12.11 is one with less pleasantries while direct and to the point.

Sales. Everybody in the United States gets sales letters on a daily basis. Many recipients [*moi*] do not even open letters that appear to be sales-related by the appearance of the envelope. Advertisers, of course, realize this and resort to clever techniques such as "handwritten" (by a printing machine) envelopes.

How do you write a sales letter that will get read? Sale letters require a special type of writing that is probably outside the scope of this book, as

---

October 1, 1998

Starbutts Coffee Company
PO Box 3717
Seatech, WA 98124-8878

Dear Starbutts:

I am writing to bring to your attention a defective product that is being sold at a Starbutts store. I recently purchased a stainless steel Starbutts coffee mug at a store in Boston Transportation Center. The lid is constructed from a black plastic and the first time that I used the mug (back in Peoria) I learned that it was impossible to drink anything out of the mug with the lid on. The plastic that the lid is molded from contains some foul-smelling chemical that leaches out into the coffee making it taste and smell terrible. You can examine the enclosed lid and see that the four coats of enamel that I applied did not stop the chemical smell.

I would appreciate it if you could send me a replacement lid made from a non-smelly material and I suggest that you change materials of construction for these particular lids. Thank you for your attention in this matter.

Very truly yours,

Kenneth G. Budinski
3110 Pearl Road, Suite 146
Peoria, IL 14412

Results:
I received a non-smelly lid and two coupons for free coffee.

Fig. 12.11 Complaint letter

there are many marketing techniques discussed in numerous textbooks or popularized how-to books. However, a simple formula for writing some types of sales copy is the "you-what-so" methods. Begin your sales message with a direct you statement and quickly say what the readers need to know and why it is important. For example:

> Sales Copy with a "You-What-So" Lead
>
> You will be happy to know that an upgrade for the "AlGorithm Polling Software" is available at reduced price for previous voters who want to express their opinions more effectively.

This simple formula is often useful when communicating by direct mail (e-mail) to a large, general audience. It is just one example of a technique used in sales copy.

In a way, the you-what-so formula has the elements of an informal report that might serve as a default format without more advanced marketing techniques gleaned from books dedicated to sales techniques. In a sales letter, try to address a person and make a personal remark if you know them. State the scope of the sales proposition and make it clear what you are offering. The body of the letter should contain details of what is offered. Include dates, costs, extras, and anything else that pertains to what you are selling. Finally, state why your product/service is so important and useful. Close with stating appreciation for consideration of the plan and state dates and terms that may relate to the purchase.

In summary, a sales letter is a special type of writing covered in other books on marketing techniques. However, the simple guidelines presented in this section may be useful in the occasional instance of writing for marketing purposes. For example, Fig. 12.12 is a letter for selling exhibit space at a technical conference. [*This was a marketing duty that I acquired as a member of a steering committee for a technical society conference.*]

12.5 Proposals

Even though you may be trying to sell something in a proposal, proposals are different than sales letters. A proposal is often directed to a small group or specific set of individuals. It is not like a sales letter that is distributed to a very wide audience.

Large sales proposals (e.g., a new heating and ventilating system for the local school) should be written like formal reports. It should include a letter of transmittal, executive summary, introduction with past experience, testimonials, and other information that would help a sale. The body should consist of project details, costs, labor details, insurance, terms, and comparisons to competitors. The conclusion should contain the best argument for the reader's acceptance of a proposal.

Dear Tribology Colleague:

Enclosed is a brochure on the International Wear of Material Conference (WOM), which will be held from April 25 to 29, 1999 in Atlanta, Georgia, U.S.A. at the Grand Hyatt Hotel. I am writing to ask your company to consider advertising in the commercial exhibit at this event. This is the premier conference in the world on the subject of wear and friction and it is the ideal opportunity for companies that make wear testing equipment or wear prevention materials/treatments/devices to advertise their products or services. The exhibit is intentionally limited to only ten tables and 20 posters in the common break area of the conference (see prefunction area in the enclosed hotel map).

Exhibit hours are from 13:00 to 19:00 on Monday, April 26, 09:30 to 19:00 on Tuesday, April 29 through Thursday April 28. Exhibitors can set up displays on Monday, April 26 starting at 09:00 and exhibits must be removed before 21:00 on Thursday, April 28.

This conference will provide your firm or product exposure to several hundred delegates from all over the world. These are the people who determine what happens in the field of wear and friction and they are interested in your products or services that deal with friction, wear, lubrication and treatments that extend wear life. The deadline for reservations is October 30, 1998, but we would like to hear of your intentions before that time. Please use the enclosed form for your response as soon as possible. Thank you for considering advertising at The International Wear of Materials Conference.

Very truly yours,

K. G. Budinski
Commercial Exhibit Chairperson, WOM '99

Fig. 12.12 A marketing/sales letter

This section deals with the informal types of proposals—the ones that are short, do not contain all of the elements of a formal report, and are relatively short lived. Three types of proposals are discussed: project proposals, purchase proposals, and organization proposals.

Project Proposal. Many engineers and scientists spend much of their careers working on projects that started as written proposals. Chapter 3, "Performing Technical Studies," presents some examples illustrating a form for a government-funded project. In industry, research and development (R and D) projects are funded from previous profits, although sometimes the cost is tax deductible if they meet certain criteria. If you did not do the work, the government would take the money for taxes or you would give it back to stockholders.

Research and development must be directed toward something new, not an incremental improvement, and there must be a risk. Buying and installing a new lathe is not R and D. There is no risk. It will run. Buying a new lathe to try to hard-turn brake drums rather then grinding them is R and D and would probably be allowable. However, if the process worked and was used in production, the lathe would have to be capitalized and depreciated, which has different tax consequences than R and D deductions. If the lathe did not work and the project was cancelled, the lathe would have to be destroyed if it was purchased with R and D dollars. Thus, there are tax consequences that pertain to R and D projects, and project proposals must meet government guidelines for R and D projects.

Project Proposal

Title: Clearance slitting of magnetic tape

Purpose:

This project is intended to investigate the feasibility of slitting magnetic tape with rotating knives that do not contact. Current knives have an upper knife that is spring-loaded against the lower knife. This rubbing contact produces metal-to-metal wear that is additive to abrasive wear from the product. We propose a knife system where the upper knife never touches the lower knife and this will eliminate one component of knife wear.

Scope:

An experimental knife set will be fabricated and tested for slitting quality.

Objective:

Eliminating the metal-to-metal component of knife wear will increase overall knife life by a factor of five. The savings, if this development is successful, will be $200,000 annually.

Plan:

Experimental knives of cemented carbide (for upper and lowers) will be installed on an off-line slitter and tests will be conducted on three tapes to determine if knife geometry and running conditions will produce a slit edge with quality comparable to product slit with contacting knives. Quality assessment will be conducted in the Research Laboratory's Physical Performance Center.

Timeline:

The clearance and edge geometry experiments will be completed by mid-year. Implementation on #27 slitter will take place by year end.

Cost:

Engineering = $30,000, Technician = $20,000, Equipment = $10,000; Total = $60,000.

Fig. 12.13 Project proposal

The essential elements of an R and D project proposal are:

- Purpose—what is to be investigated
- Scope—boundaries of work
- Objective—what will be achieved (savings)
- Plan—steps/milestones/procedure
- Timeline—when will steps be completed
- Cost—budget (staff, purchases, etc.)

Figure 12.13 is a simple proposal for funding of an R and D project in industry. This proposal report addresses all of the above elements, and it is sufficient to be considered for funding in some organizations. Sometimes, like the federal government, they may want more details; one page is not enough for approval. Proposals must be patterned to comply with organization guidelines. Regardless of the length, the basic elements are similar.

Purchase Proposals. A document similar to a project proposal is required by most organizations to get approval to purchase capital equipment. Capital equipment may be given complicated definition by government taxation entities, but in some organizations, anything that costs more than $2500 is

Capital Purchase Proposal

Requested purchase description:

Table-top microscope with polarized and transmitted light capability.

Cost:

Basic unit = $25,000

Polarizing Accessory = $1,100

Video Adapter = $2,500

Total = $28,500

Justification:

The current metallograph is 12 years old and is in need of repair. Two lenses are scratched and need to be replaced. The light meter is inoperable and the specimen stage creeps under load. Our plan is to purchase this new microscope to be used during the repair of the central metallograph and it will offer additional user capacity and other benefits not presently available. We do not currently have polarized light or transmitted light, both of which are needed to do effective failure analysis on plastics.

For these reasons, we recommend approval of $30,000 for the purchase and installation of this microscope.

Approved:

Fig. 12.14 Purchase proposal

a capital purchase. Capital equipment purchases, depending on the amount, may require approval by more than one layer of management. Often approvals can be obtained only by writing a capital purchase proposal.

Figure 12.14 is an example of a purchase proposal. This example is as simple as they come. Many want additional savings calculations or calculations pertaining to return on net assets. A proposal for a capital purchase should describe the purchase so that management understands what you want. Then provide the calculations necessary to demonstrate the need. Be advised that there are usually numerical criteria such as "cash flow ratio of return" or some other metric established by the financial staff. If your calculated number does not meet their criterion, you may want to rewrite or rethink the proposal. In other words, show that your expenditure will help the company save money or make more money than putting the money in the bank would earn.

Organization Proposal. Improvement projects sometimes follow a trend of being in vogue, but one way to improve a division or department may be a change in its basic organization. Proposed changes to an organization can be announced with an informal document called an organizational proposal. If you think that your department would run more effectively organized into four teams rather than two workgroups, then you might write a proposal for reorganization. If you are in a management position, you

Reorganization Proposal

Plan:

The Materials Engineering Laboratory is currently divided into two work groups, each with a group leader who spends up to eighty percent of his time on administrative matters. It is proposed that the laboratory reorganize into five technology teams and group leaders will be eliminated. The functional teams will be:

1. Tribology
2. Plastics
3. Engineering Materials
4. Surface Metrology
5. Testing

The functional teams will have from three to six members and they will have a common development goal.

Justification:

Almost two management staff will be eliminated by the proposed reorganization. In addition, teams can be closer aligned with customer requests. Customer service will improve.

Fig. 12.15 Reorganization proposal

may be asked to reduce the size of your department. This will require moving personnel from one position to another.

A reorganization proposal needs to be written to get approval of a plan and to show how the plan can be implemented. The elements of an organizational proposal are a proposed change, its benefits/savings, and its implementation costs. Figure 12.15 is an example of a reorganization proposal. It is similar to the capital purchase proposal—just state what you want and justify it.

Summary

After reading this chapter, it is probably apparent that informal reports can address many types of writing tasks for investigations, service reports, letters, and proposals. In these types of informal documents, technical issues can often be resolved with one or two page informal reports. You do not need all of the formal report sections, an abstract, or a list of references. The typical informal report follows a simple format:

- Statement of the problem/issue (introduction)
- The investigation and the results (report body)
- Conclusion and recommendations (closure)

This format should adequately address a very large percentage of your technical writing responsibilities. These essential elements need to be there. Some summary thoughts are also listed here:

- Informal reports typically have a useful life of about two years. They generally are not archived.

- Trip reports are almost always informal reports.
- Procedure needs to be kept separate from results whenever possible.
- Informal reports need closure with conclusions, recommendations, or a summary.
- Service reports may not need a problem statement or closure if project details are not known.
- Benchmarking projects may be suited to informal reports, because they are contemporary (not long lived).
- Correspondence should have a salutation that meets the guidelines of your organization.
- All letters should contain an opening, body, and closure.
- Complaint letters need to be written as if you are a long-time valued customer and this undeserved product/service is their first mistake.
- Project proposals need a purpose, objective, plan, justification, and cost.

Key Words

- Informal reports
- Reviews
- Reorganization
- Investigations
- Inspections
- Purchase request
- Evaluations
- Benchmarking
- Capital equipment
- Trip reports
- Inquiries
- R and D
- Laboratory tests
- Complaints
- Status reports

For Practice

1. Write an informal report requesting and justifying funds to purchase a new desk chair for your office.
2. Write a trip report on your last visit to a technical facility.
3. State the basic elements of an informal report.
4. Write an inspection report on a house that you are considering buying.
5. Benchmark your job with a comparable job at another local company.
6. Write a complaint letter to the last restaurant that gave you a disappointing meal.
7. What are the criteria for writing an informal report as opposed to a formal report?
8. List five occasions where a formal report is preferred and five where an informal report is preferred.
9. Write a report to present the results of tests where you measured the static friction coefficient of five different plastics identified as A, B, C, D, and E.
10. What are the basic elements of a project proposal?

To Dig Deeper

- C.T. Brushev, G.J. Alred, and W.E. Oliu, *Handbook of Technical Writing,* 4th ed., St. Martins Press, New York, 1993
- H.E. Chandler, *Technical Writers Handbook,* American Society for Metals, 1983
- J.M. McCrimmon, *Writing With a Purpose,* Houghton Mifflin Co., Boston, 1973
- R. Meador, *Guidelines for Preparing Proposals,* 2nd ed., Lewis Inc., Boca Raton, 1991
- R.D. Stewart and A.L. Stewart, *Proposal Preparation,* John Wiley & Sons, 1992

CHAPTER **13**

Review and Editing

CHAPTER GOALS

1. *Recognize the importance of a critical review*
2. *Understand the methodology of review and editing*
3. *Know how others review your writing*

YOU FINISHED WRITING a technical document. You spent a lot of time on it and checked it twice for mistakes. It is now time to present it to the readers that you selected. Wrong! A document should never be sent to the final readers until someone other than the author(s) reviews it. A critical reading by someone is always needed to uncover those ever-present mental slips that seem invisible even after a double check.

Some writing errors remain invisible, because as you read, your brain is telling you which words to write next. You know what you want to say, and you may read what you want to say without some of the necessary words in place. For example, you may delete the "in" in the "in place" statement in the previous sentence. You know that you wanted to say "in place," but you only wrote "place." Your brain remembers that you wanted to say "in place," so it reads the "in" when you reread the document. A critical reading by someone else will quickly uncover the missing "in." All documents can benefit from a review.

The purpose of this Chapter is to discuss the review and editing process. The goals are to convince you to seek reviews of your work and to show how review and editing are important parts of technical writing. The Chapter describes the types of review and editing and discusses the methodology and the psychological aspects of reviewing and editing. The Chapter ends with examples of edits on different types of documents.

13.1 Types of Review and Edit

The ultimate review of a document is the one by its readership. These are the reviews that you want to be favorable. Intentionally or not, every time that you read anything, you assimilate the facts presented, agree or disagree, and form an opinion on what the information conveyed and how it was done. You reviewed the piece and will make decisions on the piece. You may decide to instantly forget it. [*Your brain presses "delete" the piece.*] You may find a useable nugget in the piece [*your brain presses "save"*], or you may form a favorable or unfavorable opinion of the author [*your brain also presses "save"*].

If you want positive results from your intended readership, you need to make sure that the document is correct and does what you want it to do. A review will uncover "defects" in the product [*your document*] and editing addresses correction of these defects. There are different types of reviews; self, peer, and authoritative. They differ in intensity. You may also have reviewers edit your document. Like reviews, there are also different levels of editing. Some edits may focus on the proper and clear use of language (copy editing), while substantive editing may focus on the overall content, organization, and clarity of presentation.

Overall, reviews are necessary to identify a document's shortfalls. Edits fix them. Both need to be practiced. All documents need at least one review, which is then followed by an edit to correct the problems. Reviews and edits are part of writing, and they are essentially the quality control steps for any type of writing.

Reviews

Reviews can be as simple as rereading your own work, or the document can be reviewed by experts. The approach recommended here is to do a self-review and a peer review on all reports. Formal papers or papers for publication will get an authoritative review from your organization, the journal reviewers, and the publication editors. The more important the document, the more it should be reviewed. Glaring errors or omissions can undermine your reputation and the credibility of your document.

Self-Review. As previously noted, one of the main reasons for a review by others is that the author will miss defects because his or her brain can skip over its own mistakes. You know what you wanted to say, and this is what you read, even though some words may be missing.

Self-review, however, still should be performed on every document that you write. You and your computer can always find something that is not quite right. [*I have noticed that many of my daily e-mails contain abnormal amounts of grammar and punctuation errors. The better I know the person, the worse the spelling and grammar. I believe that this is because people that you know just want to convey a message; they write fast and they know that it does not matter that they hit some wrong keys. You will still be their friend. I used to do the same thing. I would send an e-mail to a close associate without reading it for errors. That is until a coworker made a transparency of one of my error-laden e-mails and discussed it in a meeting room filled with 20 not-close coworkers. Now I check spelling and proofread even two-sentence e-mail messages—even ones to my sons.]*

RULE

Proofread everything that you write.

In summary, self-review all documents, but be aware that you will skip over some of those mental mistakes. If you want a document to be perfect, go on to a more extensive review.

Peer Review. Everyone needs somebody that they can trust. Your first "other person" review could be by a trusted coworker, secretary, or friend—with whomever you feel comfortable as a candid reviewer with reasonable writing skills. [*In my case, it is the technician who has worked with me for the last 15 years or so. He is a good report writer; he knows the subject material. I trust him to tell me if what I wrote stinks.*]

Try to establish a trusted person to review your work. It does not have to be detailed. Ask "How does it read?" Are there any misspellings, mechanics problems, or grammatical and typographical errors that stand out? If you fix these, you should have a document that is relatively acceptable to your readership. This type of review does not take a significant amount of time, and you can repay your reviewer by offering to review his or her writing in similar fashion. [*That is the arrangement that I have.*]

If you work in a small firm and there are no people that you trust for this kind of review, then look within your family or hire a teacher or other person with writing skills. It is really important to have another set of eyes and another brain or another operating system read your words.

Authoritative Review. A critical reading by an expert is an authoritative review. If a person is an expert in writing, he or she will provide suggestions on format, style, grammar, flow, logic, sentence structure, and report mechanics. If a technical expert reviews a document, he or she will offer less of the former and more suggestions on technical content and correctness. Technical papers often receive both. In large organizations, a formal paper or report may require a review by a professional editor. The editor looks for errors in language usage and report mechanics.

Archival journals require reviews by three or four technical peers. This is sometimes referred to as peer review, although it should not be confused with your own peer review. The peer reviewers for a journal community are mostly concerned about the thesis, investigation, results, and analysis of results—that is, the technical substance of your work. Of course they will flag typos and glaring English mistakes, but usually they do not advise on how to rewrite a poorly written paper. They simply recommend rejection. This is another reason why it is important to establish a personal peer review system that you use on a continuing basis.

Edits

Editing means correcting mistakes and making the document more readable, while honoring the author's original intentions. The input for an edit comes from the reviews described in the previous section. There is no reason to restate the importance of making a document as good as it can be when it is distributed to readers. [*I suspect that I have done this, already to the point of nagging.*]

Similar to reviews, there are different types of edits. The three general types of edits described here are copy edits, substantive edits, and corporate edits. There may be different editing criteria in the latter two types of editing in terms of policy statements, legal confidentiality, and political correctness (such as using inclusive, nonsexist language). Others may suggest additional editing criteria, but this list reflects technical writing in industry [*my range of experience*].

Copy Editing. One definition of copy editing is the fixing of problems in language usage after review. A self-review usually does not result in any major changes. You just wrote the document, and you may not notice any major faults, unless you wrote it quickly or in some diminished state [*exhaustion*]. Typically, self-review results in fixing typos, misspelling, missing punctuation, bad grammar, and the occasional repeat of a word or sentence. A copy edit by a peer or other trusted person may produce similar results. It is unlikely that a peer reviewer has the time or inclination to suggest ways of fixing problems in style or language usage. It may take a significant amount of time to get in-depth copy editing to improve the language in a document written by a technical person with average writing skills.

Formal reports or papers for publication are usually copy edited by someone trained in the writing skills. Most organizations require a review and edit by a particular person. [*In my department, all formal reports were reviewed and edited by two levels of bosses, a group leader, and a supervisor. After downsizing, it was reduced to the department secretary and supervisor.*] These edits may still be rather superficial fixes of language use, but if one of the reviewers is knowledgeable in the field, he or she may suggest technical content changes.

A copy editor in a publishing house also focuses on the style, use, and clarity of language. Many copy editors have degrees in journalism or the language arts or a similar background, and they are very proficient in finding problems in grammar, punctuation, and sentence structure. They seldom have expertise in the technical subject, but be aware that they want to comprehend what is written. When you are writing for publication, the copy editors must comprehend the work. They are the first line of readers. If a paragraph is not understandable to them, it will be flagged for revision by the author.

RULE

A published document must be understandable to any reader.

You may know absolutely nothing about endocrinology, but you should be able to read an article on the subject and comprehend the message even with many words that are not in your vocabulary. [*Insurance policies, IRS instructions, and legal documents may be exceptions to this statement; they seem intentionally confusing. Sorry, my neutral tone is slipping.*]

Copy editing of a book like this may start with a once-over by the production editor, the person in charge of printing the book, designing the format, and the cover, and getting the finished product put into stock. Then there is a separate reading by one or two copy editors. The copy editors mark up manuscript pages (like in Fig. 13.1), or they edit the manuscript electronically. When editing is done on paper, the editors use standard proofreader marks (Fig. 13.2). If several editors are marking up pages, they use different color pens.

The result may be a multicolored manuscript with numerous changes and perhaps a series of questions directed to the author. The author may receive this version for review, but more often the technicolor manuscript is next sent to some poor soul known as a compositor. He or she must incorporate all the changes and decipher what the typesetting should look like. Of course, typesetting now means final computer input. In the past (before computers when type was physically set), the next step was to produce a set of "galley proofs," an intermediate form of typeset text where text is placed in a column width as intended for the final book. Galleys are just for proofing text and typesetting prior to final composition of the pages with text, tables, and figures all in place.

With electronic typesetting and desktop publishing, the step of sending galley proofs to authors and editors may be skipped or done in a different way. If the editing involves minor changes, the author may receive page proofs for final review. Page proofs are the result of all typesetting and page composition, where all text, tables, and figures are placed in the intended format for publication. Minor changes in text or illustrations can easily be done without upsetting the basic composition. However, if substantial editing and proofing of text may be required, then the author may receive rough pages without final placement of figures and tables. With electronic typesetting, documents can be easily structured in various forms. Thus the idea of galley proofs is really an anachronism from the past days of hot type.

Whatever the production steps might be, the author always gets a production draft for final review and approval. This is your final chance to fix mistakes. Depending on contract terms or other arrangements, authors may have to pay for any changes other than typos, misspellings, or errors introduced during the editing and production processes. It is still not over. The author's first task after publication is to read the book again to find any mistakes that made it through the gamut. There are usually some.

Substantive Editing. Copy editing may be superficial or significant, but usually it does not involve significant rewriting. Substantive editing does. Moving sections, deleting paragraphs, rejecting conclusions, and asking for section rewrites are all aspects of substantive editing. Substantive editing and author review of substantive edits are performed prior to any copy editing.

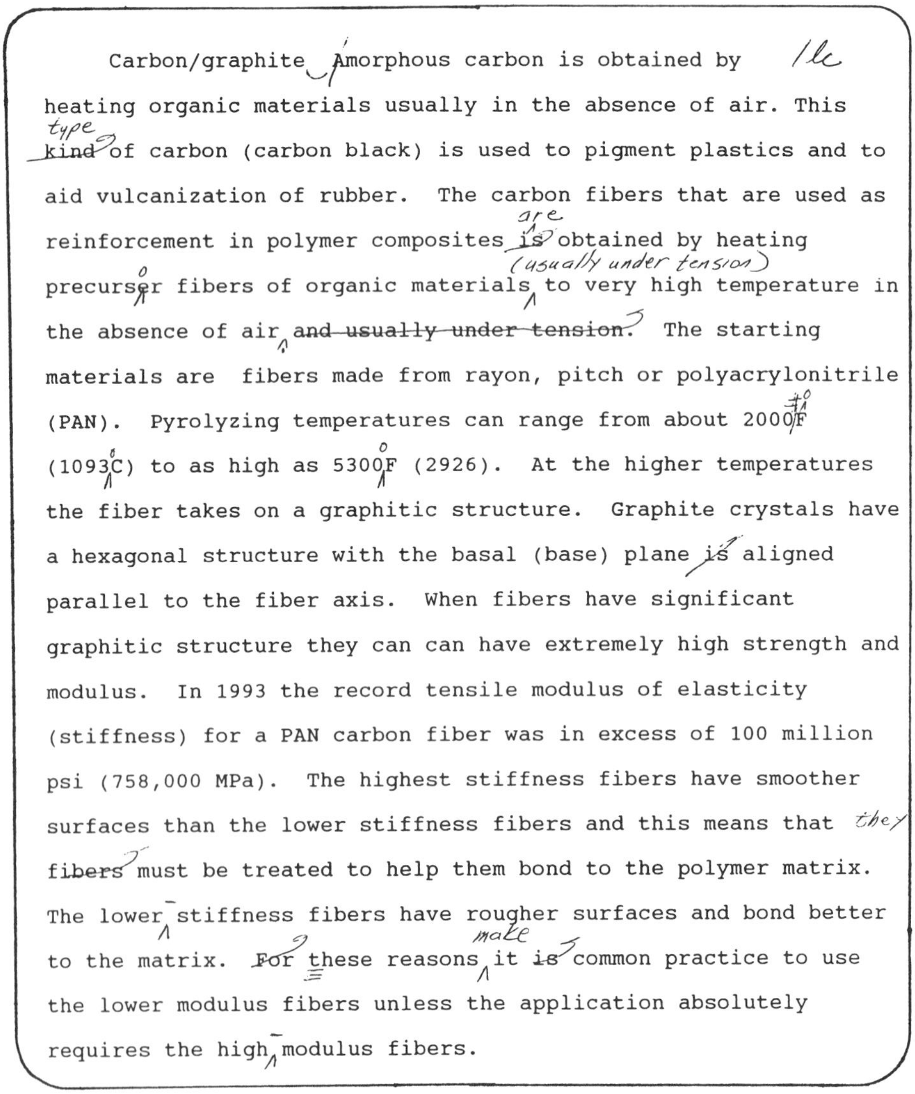

Carbon/graphite. Amorphous carbon is obtained by *lc* heating organic materials usually in the absence of air. This *type* kind of carbon (carbon black) is used to pigment plastics and to aid vulcanization of rubber. The carbon fibers that are used as reinforcement in polymer composites *are* is obtained by heating *(usually under tension)* precursor fibers of organic materials, to very high temperature in the absence of air, and usually under tension. The starting materials are fibers made from rayon, pitch or polyacrylonitrile (PAN). Pyrolyzing temperatures can range from about 2000°F (1093°C) to as high as 5300°F (2926). At the higher temperatures the fiber takes on a graphitic structure. Graphite crystals have a hexagonal structure with the basal (base) plane is aligned parallel to the fiber axis. When fibers have significant graphitic structure they can can have extremely high strength and modulus. In 1993 the record tensile modulus of elasticity (stiffness) for a PAN carbon fiber was in excess of 100 million psi (758,000 MPa). The highest stiffness fibers have smoother surfaces than the lower stiffness fibers and this means that *they* fibers must be treated to help them bond to the polymer matrix. The lower stiffness fibers have rougher surfaces and bond better to the matrix. For these reasons, it is *make* common practice to use the lower modulus fibers unless the application absolutely requires the high modulus fibers.

Fig. 13.1 Example of professional copy editing

Peer reviewers or copy editors seldom do this kind of editing. It may take a lot of time to go into the detail required for a major edit. English teachers do this kind of editing, but it seldom occurs in industry unless a person is a very bad writer or the occasion is of great importance, like a corporate merger proposal. On the other hand, substantive reviews are common in articles submitted for journal publication. Sometimes authors write an article to advertise a new company product or service. Such articles often require major surgery to return objectivity to them.

Substantive edits should not discourage a technical writer. If the reviewer's reasons for wanting changes are valid, just make them and move on. If you disagree with suggested revisions, you may want to get a second opinion. If two or more reviewers concur that changes are needed, just make them and move on. Rewriting is part of writing; there are few technical documents of importance that are published in the way that they were originally written.

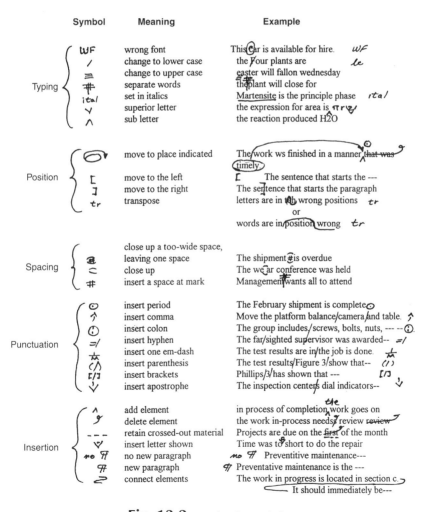

Fig. 13.2 Proofreader symbols

Corporate Editing. Many organizations review all documents that leave company premises to insure that they do not contain information that will have a negative effect on the business. The following is a typical corporate checklist:

- Does the document present a useful, significant finding?
- Can the information divulged help a competitor?
- Is the process/device described patentable?
- Does the document contain proprietary information?
- How will the company benefit from publication?
- Is there anything in the document that could produce a liability concern?
- Does the document contain any sexist statements?
- Does the document have inappropriate overtones (sexist, elitist, etc.)?

These kinds of questions are intended to make sure the document does not divulge proprietary information or contain any statements that could compromise the integrity of the company. This type of editing is essentially mandatory; it is part of company policy.

13.2 Review and Editing Methodology

How do you review or edit the work of someone else? How do you get a document reviewed? Addressing the latter question first, it has been suggested a number of times that a review is an essential part of your document strategy. Identify a trusted person as a reviewer of your work. Determine if he or she is qualified by reviewing his or her work, too. They need reasonable writing skills. Mandatory reviews such as those requested by your supervision and corporate editors should be done after your peer review. It may not be good for your performance appraisal to have your supervision uncover significant errors. Try to eliminate big problems through your own personal peer review.

The methodology for reviewing the work of others is to essentially go through the checklist that was presented in Chapter 6, "Criteria for Good Technical Writing." Read the work and determine if it contains all the elements that should be there. All documents need a logical format. Do the section heads make sense? Is there a start, body and finish? If it is a formal report, the document must have a good title, a proper abstract, a complete introduction, a body with a procedure, results and discussion, and an appropriate closure giving conclusions and recommendations. Does the document look right? Does it have sufficient white space and spacing of text, paragraphs, and illustrations? Is the grammar reasonable? Is the document free of misspelling? Are the sentences and paragraphs readable and concise? Is the document technically correct? Is the information useful? Is the document free of plagiarism?

The last question was briefly mentioned in Chapter 6, "Criteria for Good Technical Writing." A review is one check for plagiarism. By definition, plagiarism is the use of another's work without their permission. Most of what we know comes from reading the work of others. How does an author keep from repeating something that he or she read elsewhere, possibly years ago? It is still in the memory database. Using the work of others verbatim without previous permission, of course, is plagiarism. Paraphrasing can be fair use, depending on the level of detail used in the paraphrasing.

Direct Quotation—Needs Permission

From E. Rabinowicz, *Friction and Wear of Materials,* New York: John Wiley and Sons, 1965, Pg. 67:

"It is known that for most materials, the shear strength S is about $^1/_2$ of σ_y, the plastic yield strength in tension, and the penetration hardness P is about 3 σ_y. Hence, the ratio S/P has a value of about $^1/_6$, whereas actual friction values, which according to eq. 9.6 should equal S/P are about 0.4 or about two or three times as great as S/P."

Acceptable Use—Needs Reference

According to Rabinowicz [Ref xx], in metal-to-metal sliding systems friction, the coefficient does not equal the ratio of the shear strength and penetration hardness as proposed by most models.

Acceptable Use—No Reference Required

Friction models for metal-to-metal systems are not developed to the point where they can eliminate friction testing.

The paraphrased version needs a reference because it uses details from Professor Rabinowicz's statement. This usage, however, does not require written permission because it is essentially the author's conclusion obtained from reading Professor Rabinowicz's book as well as other factors.

A practical way to prevent unintentional plagiarism when researching a subject is to write a day or so after reading pertinent references. What you remember of the concepts probably constitutes fair use unless you have a photographic memory. Use an acknowledgment anytime you use even ideas and concepts of others. The use of quoted material requires permission from the publisher of the material, unless you are quoting a phrase or sentence in the context of an overall review of the published work or subject area. The main thing is to present properly the work of others by using quotation marks for direct quotes and proper attribution of thoughts and ideas.

Reviews by nontechnical editors may miss plagiarism. They may not be aware of related works. If your document is being reviewed by a nontechnical person, only you need to make a special effort in your self-review to ensure that the works or ideas of others have not been used unintentionally.

[The top five "reasons for revision" in my reviewing experience are use of trade names, not a clear purpose and objective for work, details omitted in the procedure, too many/unnecessary illustrations, and weak conclusions. The corrections for these are obvious. Do not use trade names; make sure the reader knows why you did the work. Make sure the procedure is repeatable, and do not use an illustration unless it is needed. Finally, enumerate conclusions and make them meaningful.]

In summary, the methodology of reviewing and editing involves using our checklist of "Criteria for Good Technical Writing," Chapter 6, and then reading for understanding. If you do not understand what is written, it probably needs editing. However, try to avoid being trite. Do not ask for a rewrite just because the author uses different words than you would use. Review with the intention of eliminating report defects; edit with positive suggestions. Only ask for changes that are absolutely necessary for correctness and reader understanding.

13.3 Examples of Reviews

This section is intended to demonstrate review and editing techniques for various types of documents. The objective is to encourage an effective review and editing style. Many times, reviewers slash and burn their way through a document. They feel obliged to find something wrong; they are not doing their job unless every page has a red hue.

One insidious part of review and editing is the risk of imparting psychological damage to the author. The wording of comments and corrections may have a debilitating effect on the author of a technical document. The author may be deliberate in the use of words and the construction of sentences and paragraphs. Authors may even make unwieldy sentences when a subject is complex, or when ideas may be contingent on several factors. Prudent authors often must resort to complex phrasing (especially in technical topics) so that readers do not jump to unwarranted conclusions.

Be a careful and respectful reader when reviewing or editing. If you come along and write "Wrong" next to a paragraph, you are being negative and capricious. The author probably does not know what is meant. Is the whole paragraph written in a manner that will not be understood by readers? Is there a misspelled word in the paragraph? What is the specific problem with what is written?

Negative editing can discourage people from writing anything. In addition, it does not help the author correct faults if what is lacking or in error is not identified and explained.

Proper Correction

This statement is too broad. Limit your conclusions to the specific materials tested, namely cast irons with combined carbon between 1.0 and 2.3.

A review should always make specific suggestions for corrections. Comments should never be directed at the author.

> Improper Review Comment
>
> At last a voice appears!

[*The above comment was written by an English professor in a review of an early edition of this book. The particular review (by a reviewer who was anonymous to me) contained enough cutting remarks to end the writing career of Ernest Hemingway. His or her comments were acidic and demeaning. This type of behavior should be avoided in any edit that you are asked to perform.*]

The emphasis in the following reviews is on fixing problems without destroying the self-esteem, feelings, and reputation of the author. A proper review is specific; it suggests, is objective, is positive, and should never include comments that would demean the author.

Reviewing a Letter. Figure 13.3 is an example of a letter that is in need of help. As stated in Chapter 12, "Informal Reports," the basic elements of a letter are:

- Format
- Salutation
- Opening
- Body
- Closure

Start the review of a letter with a check to see if the letter has the necessary elements. The format of the example is almost acceptable, but the date, the title of the addressee, and the title of the author are missing. Insert the date.

Frank Smith
Houghton Corp.
126 Clay Avenue
Port Huron, MI 09426

Dear Frank:

I have reviewed the drawing that you sent on the lamping collars, and I don't understand why the thickness has to be so thick. We can only tolerate a maximum collar width of 35mm using our present tool setup. Please see what you can do about it. How about you contact the factory and get a quotation of a special order collar that is 40mm wide or less. If they can make one, we might buy some.

Thanks,

Jim Smith

Fig. 13.3 Example of a letter in need of editing

The salutation, "Dear Frank," is only acceptable if the person is known. If you are writing to a salesperson that you never met, it is more appropriate to use a salutation of Mr., Dr., or Ms.

The opening of this letter is also too abrupt. It needs to start with a small pleasantry such as:

Open With a Friendly Statement

I received your drawings on September 2. Thank you for expediting them.

The body should start with stating the purpose of the letter. The example omits this, and it has a bad tone. It sounds confrontational. An improvement would be:

Purpose Statement for a Letter

I am writing to bring a dimension question to your attention. Our present tooling can only accommodate a collar width of 35 mm. Your current design specifies a 40 mm width. Would it be possible to special order collars with this "A" dimension?

The body also contains slang and some misspellings and ends on a bad tone. A more genteel closure is:

Closure

We would like an answer to the question for a September 20 design review. Please try to supply your answer to our query by that date. Thank you for your help.

Respectfully,

KGB

Structure a letter to read like one that you would like to receive. If you are asking for something, do not try to achieve it with a threat. Present reasons why the reader should give you what you want. Review letters for the presence of the basic elements and the proper tone.

Reviewing an Informal Report. Figure 13.4 is an example of an informal report that needs a constructive review. The title and header appear acceptable, but the "Problem" section appears inadequate. It is missing the objective of the work and the format of the report. The reviewer should suggest the addition of these mandatory elements as in the following example:

Problem

The Film Flow Department requested the Materials Engineering Laboratory to conduct corrosion tests on T-20 web backside coatings to determine if they are corrosive to cutting tools. This study is being conducted to assist in department projects aimed at a 10× improvement in tool life. This report will present the results of lab tests and recommendations on a path forward.

Materials Engineering Laboratory

Report 21371 – Corrosion of T-20 Web Coating   Date: 9/9/98

To: Frank Reynold, PCME, 5/23/KP, MC 24312
Author: T. Fritz, PCU, 3/22/KT, MC 20130
Contributors: R. Wald, F. Putte

Problem:

Corrosion tests were conducted on T-20 web to determine if the coatings on the backside were contributing to premature tool wear. The tests should answer this question.

Investigation:

4 materials were tested in the five chemicals that are mixed to make a low friction coating on the backside of T-20 web. The solutions were put in bottles with samples of each material. The for materials were types of carbide and 440C stainless steel. The samples were weighed before and after tasting. Attack was quantified by measuring the elements that dissolved in the solution with ICP. These samples of each material were tested in each solution for 30 days. The results are presented in Figure 1. These results tell us that two of the carbides corroded in the test solutions. The 440C and the 805 carbide had negligible attack.

Conclusion:

We will test a few other materials to check ourselves. The 805 seems like the best candidate at this time.

Fig. 13.4 Informal report in need of editing

The report body also contains some grammar and spelling errors, and the procedure description is not complete. The reviewer should annotate the grammar and spelling errors and ask for the missing procedure steps.

Proper Editing Remarks

Line 6—*4 materials*—Never start a sentence with a number; write it out, "four"

Line 8—the *for*—four

Line 9—testing

Line 11—*ICP*—define acronym

Line 12—state exposure conditions (temperature, state if were samples fully immersed, state amount of solution)

Line 13—put results in separate section

Line 17—rewrite in conclusion form, e.g., "The 805 material had negligible corrosion rate in all backside chemicals."

Informal reports need to be checked for format and basic elements just like the letter.

Reviewing a Formal Report. Formal reports and papers are reviewed with the same procedure as letters and informal reports; check for presence of the basic report elements. Then cite errors in word and report mechanics. Articles are usually reviewed to specific standards. Each reviewer is given a form that cites the criteria to be used in evaluating a paper.

Figure 13.5 is an example of a reviewer's form for articles at an international conference. The articles were published in a prestigious international journal. A review of the headings on the checklist shows the items considered important. Also note that the entire article is given a rating. The editors may decide not to publish an article with a fair or poor overall rating. It would be sent back to the author for rewriting or completely rejected. One report review item missing in both these checklists is listing of key words. Key words are important for searching literature, and some journals require them. Computers can now search electronic documents for any

Technical Paper Evaluation Form

1. Originality – Based upon what is known about a particular field, does this work present a new concept, process or analytical technique?.. □*

2. Significance – What is the probable importance of this work in this Area? Will it become a classic reference?.. □

3. Completeness – Does the paper suggest thorough, well planned, and executed scientific study building on the work of prior investigators?.. □

4. Organization of the manuscript – Are ideas, concepts, and procedures presented in a logical order?.. □

    a. Title – Is it brief and descriptive of the work?... □

    b. Abstract – Does it clearly present the purpose, scope, objective and results?.. □

    c. Body – Does it have a logical organization, purpose, description of the problem, a proposed solution, results, discussion and conclusion?.......... □

    d. Symbols – Does the paper use SI units and define acronyms?................. □

    e. References – Are they listed in proper form at the end of the paper and properly referred to in the text?.. □

    f. Illustrations – Are they properly executed and clear? Do they have a thoroughly descriptive caption?.. □

    g. Length – Is the paper less than 6000 words? (20 pages of double spaced typing)?... □

    h. Style – Is the paper written to accepted literary standards and understandable to workers in the field of interest?..................................... □

    i. Language – Is the use of the English language acceptable?...................... □

    j. Acknowledgements – Are they properly placed and to prevailing standards?... □

5. Freedom from commercialism – Is the paper free of trade names and implications of commercialism? Is it trying to sell a machine, material, or process?.................. □

Overall rating.. □

Should this work be published □ as is □ with revision □ not?
*Rate each block 1 to 10; 10 is best.

Fig. 13.5 Peer review form for a technical paper

word in the text. Many times, however, the article may only use a word in an offhand manner; the article is about medical malpractice even though you were doing a search on forging of steels. Archives should be set up to search on key words submitted by authors.

A more extensive checklist for review is presented in Appendix 12, "Document Review Checklist." It applies to formal reports and papers based on the good attributes described in this book. These examples illustrate what to look for in reviewing a report. Figure 13.5 and Appendix 12 illustrate what a reviewer looks for in a formal report or article. It is proper to annotate the text, but the best review lists errors and problems by page and line as in the review example of the informal report.

The last item to be covered is the reviewing/editing of a problem report. What does a reviewer do with a really bad report like the one in Fig. 13.6? Editors of journals and conference proceedings invariably review an occasional "stinker." In most cases, the work has technical merit, and the authors

---

New Elastomer Composite with Superior Performance and Cost Savings

Abstract:

Property tests are presented and discussed. The raw composite called "Purock" combines the toughness of rubber and hardness of ceramic. Purock can be used where the shock resistance of polyurethane is not adequate by itself. This material has been successfully used in a number of factory applications. Potential now exists to move into wider application.

Introduction:

This new composite offers a unique opportunity to solve wear problems in many industries. Polyurethane (PUR) has been used by others for wear parts, but it was always too rubbery to use for machine components. Other people [1,2,3] have tried to make polyurethane into an engineering material but to no avail. We have tested Purock thoroughly, and we are now ready to widely market it. The wear life of suitcase wheels made from plain polyurethane is only about 2 km. Our Purock shows no wear even after 10 km of rubbing. We will describe the development.

Background:

Purock was developed after much experimentation from castible polyurethane resin and aluminum oxide spheres. Polyurethane resins can be liquid-cast into any mode that will allow its release. The resins can be made to various durometers with different formulations [9]. We develop Purock with a formulation that yielded 90 Shore A hardness. Purock is a composite of 90A polyurethane and aluminum oxide spheres one mm in diameter. We tried many techniques for making the spheres uniformly distributed in the part. The problem was solved by a coating on the alumina spheres. The coating provides the optimum combination of wetting and buoyancy to make the spheres uniformly distributed in the cast shape. We received a patent on the process [5].

Project Objective:

We want to implement our new composite in a wide variety of parts. Our aim at this point is to use Purock for suitcase wheels. These suitcases are very popular these days. Everybody has them. The wheels are always wearing out so we want suitcase manufacturers to use Purock wheels because we know that nothing will wear them out.

Testing Program:

The test equipment used in the project was critical to the success of the study. They had to simulate characteristics of real luggage. Two wear machines were determined to be relevant and useful. One used abrasive wheels as the abrasive, and it has been used for decades to rate the abrasion resistance of floor tile (Sand Abrasive Test [SAT]). The other was an impact wear (IW) test used to qualify luggage wheels. This test impacts the test sample with the sharp edge of a steel cylinder on a rotating platen.

Fig. 13.6 Problem paper

Results:

Wear tests results are shown in Tables 1, 2, 3, and Figures 3 and 4. Table 1 lists the SAT rates for 10 samples of Purock compared with Hylon and Delron. Purock is three times better. The IW tests (Table 2) shows that Purock was five times better than Mylan and six times better than Delron. The other graphs show wear rates with time.

Discussion:

Wheel wear is an important luggage consideration because it affects the useful life of the device. Polyurethane was chosen as the binder for Purock composite because of its low wear and high impact toughness. It has been used as a replacement for metals in various applications. Polyurethane has a good combination of properties. It has a very high value in impact tests, very low compressive creep, excellent moldability, and excellent abrasion resistance. It has moderate temperature resistance, being limited to a maximum operating temperature 125 C. It is used as a gear and bearing material.

The had phase in the composite, alumina is dense, very hard and it has optimum wear resistance. The coating on the aluminum does not after its properties. Its role is simple as an "emulsifying" agent.

Production Processes

Initially, attempts were made to make the composite using air purging of the liquid resin to disperse the hard phase. High intensity sonic vibration was also tried. All processes failed to produce the desired uniformity in hard phase dispersion. Wear testing confirmed that these processes did not produce the desired degree of dispersion. Sone other processes were tried, but they were slow and expensive. Jerry Kramer used ultrasonic agitation in his work [6], but we thought that this would be too expensive to do for our part.

After months of experimenting we came on the coating that finally worked. The coating can be applied by a variety of processes, and this makes the aluminum spheres dispersion in coating. It only requires mixing of the spheres into the casting resin. We visualize a robot-controlled automatic casting machine for making suitcase wheels at high production speeds.

Conclusions:

A successful composite has been developed. It possesses very high abrasion resistance, and it should be accepted for roughness. Because its wear resistance and toughness depend on the alumina content, it can be tailored to specific applications. Purock can be easily manufactured into luggage wheels and other parts requiring very long life.

Fig. 13.6 (continued)   Problem paper

have done a lot of work in preparing the document. The work may lack organization and proper format, and it may be done in an amateurish style. If you are reviewing an article from a conference, you can simply reject the article as "poorly written," but this is not easy to do when the report was written by the person who sits next to you. Your peer is requesting specific suggestions and positive input.

Our recommendation on problem documents is to suggest a rewrite and make helpful comments on each section. Our comments for revision of the formal report in Fig. 13.6 are:

Review Comments for Rewrite

Abstract

- Why was this work done?
- Delete "Purock" name—use a generic name. State the scope of the work and what was done.

- What was the outcome of the work.

Introduction

- Give background information—what led to the development of the composite?
- State the purpose and objectives of the work.
- State the format of the report.
- Combine the introduction, background and project objective sections into one introduction section.

Body

The body should contain information on the nature of the new composite. How was it made? Then should come a reproducible description of the composite tests. The next section should state the results of the tests and the discussion should explain the results. Merge the production processes into the discussion. Show them as proposed ways that composite can be commercialized.

Closure

Enumerate conclusions

Overall, this article reads like a sales brochure. It should be revised to objectively illustrate the technical aspects of this new composite. Dealing with a "stinker" is the hardest part of reviewing. Try to do it with compassion. If you are the author of one, do not be discouraged. Take the reviewer's comments and act on them.

Summary

Rewriting is part of writing. This book was started over from a blank sheet twice because of reviewer comments. As an author, you really do not want to have any document, letter, informal report, formal report, or paper distributed in your organization or outside your organization if it contains errors or is not properly written. Your reputation could be at risk. A critical review by others prevents this.

Always get your work reviewed, and when you review the work of others, think about the guidelines presented in this Chapter. Avoid being picayune when you review the work of others, and do whatever you can to help writers who are writing in a language that is not their native language. Mark the text appropriately to correct language faults. Above all, do not be offended when reviewers find problems with your writing. Be constructive and not overly critical when you review.

Important Terms

- Review
- Report elements
- Completeness
- Proofreader

- Punctuation
- Edit
- Salutation
- Originality
- Concision
- Grammar

- Paragraphing
- Significance
- Proofreader marks
- Sentence structure
- Organization

For Practice

1. State the basic elements, style, and format of a business letter.
2. List five phrases that you use and show how they can be replaced by a single word.
3. State the basic elements of an informal report.
4. State the basic elements of an abstract.
5. Write a paragraph about your hobby and get it reviewed by another student. Comment on the items cited; were they right?
6. What is the proofreader's mark for: (1) add a comma, (2) delete a word, (3) capitalize, (4) insert a space, (5) change to lower case?
7. Explain originality in a technical paper.
8. What are the basic elements of an introduction?
9. What is the proper way of making acknowledgements in an informal report, in a formal report, in a paper?
10. List the ten most important factors to check in a formal report.

To Dig Deeper

- R.A. Day, *How to Write and Publish a Scientific Paper,* 3rd ed., Oryx Press, Phoenix, 1988
- *MLA Style Manual and Guide to Scholarly Publishing,* The Modern Language Association of America, New York, 1998
- L. Rozakis, *The Random House Guide to Grammar, Usage and Punctuation,* Random House Inc., New York, 1991
- M.E. Skillen and R.M. Gay, *The Chicago Manual of Style,* University of Chicago Press

CHAPTER 14

Oral Presentations

CHAPTER GOALS

1. Know how to convert a report into an oral presentation
2. Know how to prepare effective visual aids
3. Know successful presentation techniques

ORAL PRESENTATIONS complement written documents. Although 14 Chapters of this book are devoted to technical writing, the need for an oral presentation must also be considered when you are reporting your work.

Why make oral presentations? Shouldn't a document suffice? There are many reasons to make oral presentations in engineering and the sciences. In fact, the preferred way to complete a project in many organizations is first to make an oral presentation and then distribute the written report to those who want to refer to information and details in the report. An oral presentation can reach a larger audience than a distributed report, and it ensures that the information is transferred. You are not always certain that an addressee will read a document.

In addition, oral presentations are required to train others, to share information, to make proposals, to direct people, to review progress, to plan; there are countless occasions when an oral presentation should be used to complement technical writing. It can be said that oral presentations are part of technical writing. As will be pointed out, a technical document may serve as the outline of an oral presentation.

The purpose of this Chapter is to illustrate how to prepare and present a technical talk. The objective is to gain a working knowledge of how to make a paper and presentation for any occasion that may arise in one's technical career. The Chapter starts by describing the types of presentations that you may be called on to make as part of your job. Following this is a discussion on the preparation of a talk; that is, how to prepare effective visual aids and how to make a presentation.

14.1 Types of Oral Presentations

Oral presentations of technical work can be given to a group of peers, managers, a broad cross section of people within an organization, or to the public at a technical conference or public-affairs gathering. The audience may vary from teammates to hundreds of people at a big conference. [*I recently presented a colloquium to a group of graduate students at a university; this morning, I made a presentation on testing results to fifteen peers at a project team meeting. Later, I was also called on to be master of ceremonies at a coworker's retirement party. This is the spectrum of oral presentations that have typified my career in engineering, and I suspect that similar occasions arise in the careers of most technical professionals. Making oral presentations is part of the business.*]

Technical professionals are likely to encounter two basic types of oral presentations: informal and formal. Informal oral presentations are like the team meeting. Teamwork has always been important, but with the many buzzwords of total quality management, many industries in the United States in the 1990s began to assign teams to significant projects. These teams are usually diverse in nature and in larger companies may involve as many organizations as there are team members. For many teams, the universal method for communication is to get in front of the team with a few transparencies and an overhead projector and proceed to tell the team what you have accomplished toward the team goal since the last meeting two weeks ago. Needless to say, these types of presentations do not warrant the preparation and detail of a formal presentation.

Informal presentations require preparation, but when they occur weekly you cannot spend a significant amount of time on them. In addition, the audience (your teammates) does not want a full blown presentation. They just want to know what you accomplished since the last meeting. Teammates may each get five or ten minutes to present their work. [*In most cases, teammates follow up with written presentations. These are sent to a computer "team suite" that can be accessed by all team members and managers. Those who want to study results have the written document; those who only want to know what is happening were informed at the oral presentation.*] Even though a presentation is informal, it still needs to be done right, as discussed in the next section.

Formal presentations are like the example of a colloquium or a conference talk. You are expected to talk for a given length of time; you know the audience. The topic has been announced; you will be expected to thoroughly cover the topic. You must field questions from the audience, and often your presentation will be evaluated by the audience. This is the other extreme in oral presentations. Formal presentations are usually structured like a technical document. They have a beginning, body, and closure. They must be prepared like a technical document. More often than not, a technical document is the basis of the presentation. In these instances, the purpose of the oral presentation is to share the outcome of your work with a larger audience than the document readership. A technical paper given at a conference is a way of sharing work with the world (if it is an international conference).

Teaching a course is someplace between an informal and formal presentation. You usually do not have to structure a class that you are teaching as rigorously as a formal presentation. You need to be prepared to cover a certain amount of material in an allotted amount of time, but you can interact more with the audience (the class). It is acceptable to spend time during the class making visual aids on the blackboard or projector screen. You can sketch and write out equations in real time. You do not have to come prepared with premade visual aids.

There may be other occasions that do not fit into informal, formal, and teaching categories, but the preparation and presentation techniques that follow can be useful even for those special occasions like a retirement gathering or a presentation to management.

14.2 Preparation

Informal Presentation. A written report can often serve as the basis for an oral presentation. This is true for both formal and informal presentations. Prepare your talk based on your written work and using the headings of the written work as the headings for the body of the talk.

Figure 14.1 is a laboratory report on shaft failure. It was a relatively routine mechanical failure, but because it could have resulted in a personal injury, the department called a meeting of approximately ten people who wanted an explanation of what happened, why, and how a repeat failure could be prevented. The report served as the basis of the presentation.

All presentations require consideration of the following factors:

- An opening
- Purpose of presentation
- Objective of presentation
- Audience (content, size)
- Time allotted for presentation (also location and time)
- Visual aids/speaking aids available in room
- Expected audience response
- Written documentation/handouts

As in the case of report writing, the speaker must decide on the purpose of the presentation and the objective of the work/presentation. In this instance,

Materials Engineering Laboratory Report

To: J. Doe BI-FF, 3/326/KP MC2511   Date: 8/21/99

Subject: Shaft failure on 502 film slitter

From: K.G. Budinski, AB Materials Engineering Laboratory, 5/23/KP, MC2434

Problem:

The Materials Engineering Laboratory was asked to examine a shaft that broke in a lifting device on 502 slitter in B326. The shaft was integral with a worm in a purchased commercial screw jack that converted horizontal rotary motion to vertical rotary motion. The roll elevator is driven by a ball nut that rides on the output shaft of the jacking device. The weight of the roll of film and its ancillary cradle can be as high as 3300 lbs. This load is supported by two of these jacking devices; one on each side of the machine. When the jack input shaft failed, one side of the roll dropped and the roll could have hit an operator. The unit operates in the dark with an operator nearby.

It is the purpose of this report to present the results of laboratory tests on the failed shaft and to present our opinion on the root cause of the failure. The objective os this study is to prevent additional failures of these devices.

Investigation:

The broken shaft is shown in Figure 1. The shaft broke in two pieces at a grinding relief. The nature of the fracture surfaces indicated that the shaft failed from bending fatigue and the crack had been propagating from the corner of the sharp grind relief for some time. The screw jack was coupled to a 4 HP motor with a series of line shafts. A chain coupling connected the jack to the line shaft. The failed shaft showed severe fretting corrosion and wear on the end of the shaft that fits into the chain coupling (Figure 2). The fretting-damaged jack shaft end had as much as 0.008 inch worn from its diameter. The key lost 0.050 inch of dimension due to fretting damage (Figure 3). This fretting damage is an incontrovertible indicator of oscillatory relative motion between the jack shaft and the chain coupling. According to the manufacturer's catalog, the chain-coupling catalog can only accommodate a parallel offset of 0.012 inch. It is our understanding that the measured offset between the jack shaft and the line shaft was 0.125 inch. It is our opinion that this misalignment is the root cause of the jack shaft failure.

Radioisotope stimulated x-ray fluorescence on the failed shaft indicated that the shaft was made from a carbon steel. The laboratory results (Appendix 1) indicated that the shaft was made from a 1040 carbon steel. The worm OD had a hardness of about 50 HRC. The shaft had a hardness of 95 HRB. These hardness measurements suggest that the shaft was annealed or normalized and the worm teeth were selective hardened. The shear strength of the failed shaft is estimated to be about 30,000 psi, and the applied stress was 13,000 psi.

Discussion:

It appears that the jack shaft was adequate in size to carry the torsional stress of jacking, but it was not made from a premium steel and the grind relief designed into the shaft produced unnecessarily high stress concentrations. Radiused grind reliefs are preferred (the larger the radius, the lower the stress concentration). The shaft probably would not have failed if it did not have the bending fatigue load applied by shaft offset (misalignment).

The use of a coupling with that would allow 0.12 inch offset could also have prevented this failure. There are couplings that can accommodate more than 0.012 inch offset allowed by the chain coupling, but very few could accommodate a 0.12 inch offset. The Material Lab's Tribology Group felt that the chain coupling is acceptable for this application if aligned within specifications, but there are other couplings they prefer (more robust, in stock, etc.), but some may require even closer alignment than chain couplings.

Summary:

This failure occurred because of a significant misalignment (parallel offset) between a line shaft and the shaft on the elevator jacking device. It is our understanding that there was a recent jam of the roll elevator due to interference with a 2 inch square by 12 inch long wood roll spacing gage. However, the fretting damage on the jack shaft suggests that the misalignment had existed for months, not just a few days. The recent jam may have contributed to the fracture, but it was not the root cause.

To prevent future incidents of this nature, we recommend checking the alignment on the shafts wherever these devices are used. If the misalignment is outside of the vendor's specification, the shaft connections should be inspected for fretting damage. If none is present, the unit should be aligned and it can be used. Any shafts that show fretting damage should be inspected for cracks or replaced. We recommend that the shaft stress concentrations be brought to the attention of the manufacturer of the jacking units. They can be improved. Finally, we recommend consideration of alteration of the jacking system such that the elevator will not drop if there is a failure in a power transmission device (rack and pawl, etc.).

Fig. 14.1 Report that was made into an oral presentation

the safety coordinator of the department with the failure requested the presentation. The purpose of the presentation was to present findings on the cause of the failure. The objective of the work was to prevent recurrence. The audience was determined by the meeting announcement. About ten people were invited, and they included machine operators, maintenance mechanics, department engineers and department supervisors. The safety coordinator called the meeting, and he allotted 15 minutes for the presentation of laboratory results on the nature of the failure.

You are familiar with the scheduled conference room; you know that it always has a transparency projector in it and there are blackboards on two walls. You expect that the audience will be pleased that you pinpointed the origin and that they will accept your recommendations to prevent additional shaft failures. You will come out of the presentation with a completed job for this audience.

You decide to prepare transparencies for projection at the presentation. The transparencies come from the highlights of the report. A time-honored technique for preparation of presentation visual aids is to use 3 by 5 inch index cards to show each illustration. You do not have to make transparencies, but all presentations are easier when visual aids serve as guides for what you are going to say. The transparency or slide contains your thoughts. They guide your discussion. As you read the report, make appropriate visual aids on the index cards. The small size of the cards forces concision. Number the cards and use them as a guide in your talk in the event you cannot use the projector.

Figure 14.2 illustrates the transparencies you will prepare for the presentation. These are nine visual aids. After preparing the index cards, you make slides on a word processing program or other computer program. You make the letters with a 24 point font minimum. The final step in preparation is to put the transparencies in plastic jackets so they do not stick together. Put them in a notebook and go to the meeting with the notebook and the confidence of knowing you are prepared.

Formal Presentation. One significant difference between an informal and formal presentation, besides content, is that most informal presentations are made to coworkers and peers. This often means you can eliminate descriptions of machines, processes, and similar background information. Your peers know these things. If the formal presentation at hand is a paper at an international conference, assume that the audience knows nothing about your machine, process, or situation. You must dedicate a portion of your talk to the presentation of background information. Depending on the audience you may also want to add some tutorial information to bring the audience up to speed on the topic of your presentation.

RULE

Research the audience and tailor the presentation accordingly.

All talks, like written documents, must have certain elements. A formal technical presentation should contain the same sections as a formal report:

- Introduction
- Body
    Procedure
    Results
    Discussion
- Closure
    Conclusions
    Recommendations

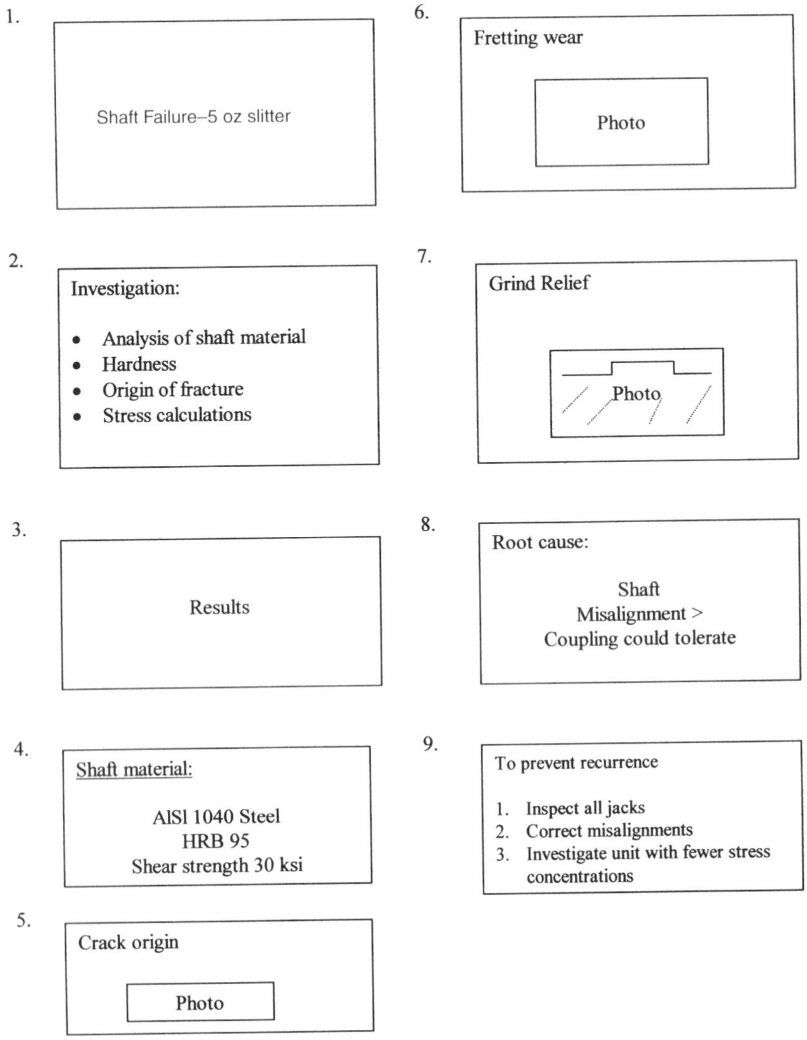

1.

Shaft Failure–5 oz slitter

6.

Fretting wear

Photo

2.

Investigation:

- Analysis of shaft material
- Hardness
- Origin of fracture
- Stress calculations

7.

Grind Relief

Photo

3.

Results

8.

Root cause:

Shaft
Misalignment >
Coupling could tolerate

4.

Shaft material:

AlSl 1040 Steel
HRB 95
Shear strength 30 ksi

9.

To prevent recurrence

1. Inspect all jacks
2. Correct misalignments
3. Investigate unit with fewer stress concentrations

5.

Crack origin

Photo

Fig. 14.2 Visual aids for informal presentation

A proper opening is important in any oral presentation. Informal talks can begin with some pleasantries to the audience, but a formal talk requires a proper lead such as a photo of a worn or broken part related to the topic to get audience attention. Assess the audience and decide on a tone and style for the presentation. It is not advisable to open a formal technical talk with a joke. This is especially true with an international audience. Each country has a different set of standards regarding what is funny or not. [*My wife and I have become close friends with a couple that emigrated to the United States about five years ago. They frequently tell us Russian jokes, and they never seem funny to us. They often have to explain to us why they are funny to Russians.*] A failed opening joke can diminish your entire talk. The use of jokes is entirely discouraged.

RULE

Jokes are inappropriate in technical presentations.

The preparation of a formal presentation is done in much the same way as preparation for an informal presentation. Use an index card for each point you wish to make during the talk, and the words, photo, sketch, and so forth on the index card can be visual aids. What is on the index cards can be made into slides or transparencies, or the cards can be used as an aid in a talk without visual overheads. However, if the presentation location allows the use of visual aids, it is always recommended. The file cards used for outlining a presentation should contain notes to show special effects that you want to use (Fig. 14.3). If you write in a normal size on the index cards, this effectively limits the number of words and lines on your slides, so they do not become cluttered or busy.

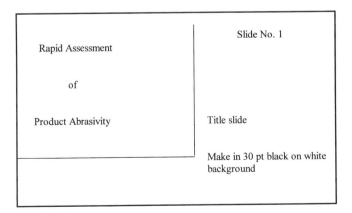

Fig. 14.3 Use of index cards for talk topics and visual aids

RULE

**Word visual aids shall contain a maximum of five
to six words per line and five to six lines per slide.**

This may be the single most important rule about visual aids for oral presentations. So often people take a page of text (like 25 lines of typing or 25 rows of a spreadsheet), make a transparency, and show it to an audience. Nobody can read the words, and the presentation and presenter fail. An oral presentation is an abstract of your work – not the entire work. The purpose of an oral presentation is to ensure that an audience gets a message. Cluttered or wordy visuals convey the wrong message. [*Before reorganization, we had an art department who made slides for presentations. We would submit a stack of index cards with our proposed material. The presentation specialists would edit our visuals to prevent wordy ones, and they would use colors to emphasize headers and important points. They made extremely effective visual aids. I miss them. Now we must rely on a computer program with no experience on what works with people.*]

Figure 14.4(a) illustrates the use of index cards to prepare a formal presentation on the nibbler project (used as an example in Chapters 9, 10, and 11 on formal technical reports). As was done with an informal presentation report, the formal report serves as the outline for the presentation. Begin with the introduction and make a card containing the title, the purpose, the objective, the importance of the problem, and the format. The same procedure is repeated for the body and closure (Fig. 14.4b).

There may be other ways to prepare an oral presentation on a technical subject, but this simple technique is the one recommended by the presentation professionals. [*We have used this technique in the past when professionals prepared our visual aids. It always served us well in practice.*] It is fast and orderly, and it works.

Teaching. It is outside the scope of this text to teach pedagogy, however, there are some simple techniques that may be useful to the technical person who occasionally teaches a technical subject in a school or within his or her organization. The following approach is recommended when teaching:

- Be prepared
- Interest the students
- Involve the students
- Have fun
- Test the students

If you are going to give an hour lecture, come prepared with an outline and plan that covers the hour. It helps to announce your plan to the students at the start of class. Let them know what you intend to cover and what you expect them to learn. Your presentation should be interesting. If you are teaching a dry subject, interject some hands-on demonstrations or other techniques to keep the attention of the class. Involving the students is another effective way to keep the subject interesting and promote learning. [*I have taken many classes in my career, and my most memorable teacher was one that I had for a graduate course in polymer chemistry. It was an evening course, and the other students were mostly graduate engineers in technical positions in industry. The professor would lecture for about 15 minutes on a subject such as olefins. Then he would ask students, "Who works with polyethylene or polypropylene?" Students who raised their hands were asked to come in front of the class and tell the other students what they did with polyethylene in their work. I learned more useful information from the students than from the professor. It was his style of involving the students that made the course interesting and valuable. We learned from the expertise of the professor as well as from student peers. I heartily recommend this teaching technique.*]

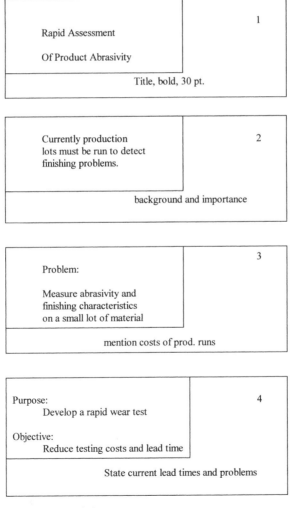

Fig. 14.4(a) Use of 3 by 5 cards to plan visual aids

Teaching should be fun. Young minds are usually more open when the atmosphere is friendly and comfortable. Talk with the students before and after class. Get to know them. Ask the class at the start if anyone has any good news to share. Interject personal experiences that may relate to the subject at hand and ask the students for comments on laws and theorems that you discuss. Never belittle a question or questioner.

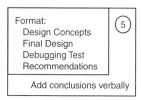

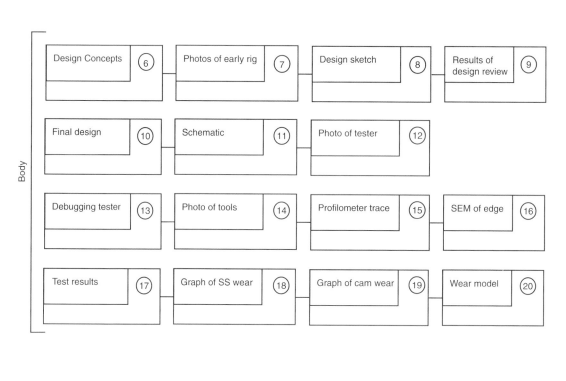

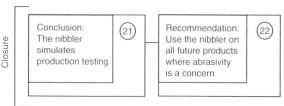

Fig. 14.4(b) Visual aids for the body of the talk

Our final suggestion on effective teaching is to test frequently and give students many opportunities to make up for a poor test grade. If grading is based on ten quizzes, a midterm, and a final, students have more opportunity for success. Give students ample opportunity to do well. Try to be fair and impartial in grading and let students know how they are doing throughout the course.

If you are only teaching a one or two session class, some of these testing techniques may not apply. The other suggestions still apply, especially the one about being prepared. Know your subject and present it in a logical, clear way.

Entertaining. The preparation of an entertaining talk is outside the scope of this text, like teaching is. However, here are some simple suggestions:

- Do not tell jokes unless you are good at it.
- Do not make sexual or off-color remarks.
- Do not say anything nasty or libelous.

Very few people are effective joke tellers; be sure that you are one before using this presentation technique. Never use profanity or off-color remarks to get a laugh in front of a group. Do not denigrate individuals. It is acceptable to highlight a person's idiosyncrasies, but do it in a light-hearted manner. Self-deprecation is the safest way to get a laugh when you are making an entertaining professional presentation.

Our final advice for preparing an entertaining talk for a peer's retirement or similar affair is to be prepared. Plan what you will say and when. Rehearse it. Make what you say appear spontaneous, but have it scripted and planned. When you make your presentation, throw away your script. Have fun yourself; go through your plan as if you are saying whatever comes into your head. Be loose and make clean, humorous remarks that will not offend anyone.

14.3 Visual Aids

Options. The purpose of visual aids is to get the attention and increase the information retention of your audience. There are studies that show that people retain about 50 or 60% of what they see, but only 10% of what they hear. [*Is this why politicians never use visual aids? Maybe they do not want the voters to remember their preelection promises.*]

Presentations are given to persuade, update, address a problem, instruct, or entertain. Technical presentations are usually given for the first four reasons. In all of these instances, get the attention of the audience and have them remember your message. Using visual aids can help you attain these objectives in conveying a message.

Which visual aids should you use? The choices are outlined in Fig. 14.5. Each has advantages and disadvantages. Most people have access to chalkboards, flip charts, transparencies, or slides. Chalkboard talks fit schools. They do not work well in industry because you are wasting people's time while you write on the board. In 2000, some universities have progressed to a method of studio teaching, where each student has a computer and the instructor writes on an electronic screen. The writing is displayed on large monitors that replace chalkboards. The computers at each student's station are used for subject demonstrations, virtual labs, and homework.

Flip charts are unwieldy to prepare ahead and carry to a presentation. Slides are ideal for a formal talk at a large gathering, but slides may be more difficult to prepare than transparencies. Word slides can be made from transparencies placed over a colored sheet of paper and photographed with a copy camera. Graphic arts vendors have computer systems for generating slides with elegant colors and fonts, but they tend to be expensive.

There is, of course, the option of generating slide shows with various types of computer programs. With some overhead systems you can display the visual aids directly from computer files on the company network Intranet (or from the laptop that you may use as a notebook when traveling). This is not always available, and so transparencies should always be produced from computer-generated visual aids. Most conference rooms in industry and schools contain an overhead projector, and transparencies with a very professional look (including photos) can be easily made with computer software. Color transparencies can also be produced from computer-generated visuals, if you have a color printer (and color copier, too, if you cannot print transparencies from a computer printer).

Visual aids presented directly from a computer can be impressive, but there are some considerations to be addressed if you are contemplating this. The biggest consideration at a conference is the time required to set up the necessary equipment. Most users have a laptop computer loaded with 10 or 20 "slides." They toggle the slides from the laptop, but the laptop must

Type of visual aid	Equipment required	Suitable for
Chalk talk	Chalk board	Teaching, team meetings
Flip charts	Easel and charts	Team meetings, problem solving sessions
Transparencies	Overhead projector, screen	Any type of gathering
Slides	Projector, screen	Any type of gathering
LCD/computer	Overhead projector, screen LCD display PC or laptop	Any type of gathering
Video	Video projector of VCR screen of monitor	Any type of gathering
Demonstration	Props	Small gatherings

Fig. 14.5 Spectrum of visual aids

be connected to power and then to the video projector. Many conferences are held in hotels. Video projectors are rented from audio/visual companies, and special arrangements must be made to ensure that all equipment is in place. Because a computer slide show takes setup time, it is usually necessary to move your talk to a time after a break so that the break time can be used for setup. If all works well, these visual aids can be spectacular. You can fade in and out; you can show a short video. You can have music. However, when the computer acts up, you may have no visual aids.

[*At a recent conference that I attended, the session chairman was setting up to use a PC for his talk. There was not a good place to set his laptop, so he ended up balancing it on the overhead projector. It fell off and stopped working. In a dither, the author took off for the conference office to seek repair help. He returned two and a half hours later. He spent all that time on the phone with the laptop manufacturer. They were guiding him through circuit board tests over the phone. He got it working by switching chips and gave his talk after most of the audience left.*]

Needless to say, there are risks with the use of video projector/laptop visual aids. Very few work as planned because stuff happens. This risk must be considered when contemplating the use of this system. It is always prudent to produce transparencies as back up, even if the computer presentation is being done with a proven internal system within an organization.

Videos and movies are effective visual aids if you have the necessary time and equipment to make them. Demonstrations are wonderful if you have the time to do them, but they usually can only be done for a small group, unless you speak from a stage and have big props. In summary, the choice of a visual aid medium is up to you. Every visual aid in Fig. 14.5 has been and can be effective. They all work if used properly. Pick the medium you are the most comfortable with and obtain the necessary equipment to do it right.

Word Visuals. As just mentioned, word visual aids like title, purpose, objective, format, and so forth can be made with any word-processing program. Select a font size that will fill the width of the visual aid with five or six words and only use five or six lines on each visual. Usually, a font size of 24 to 30 points is adequate. Violating this rule defeats the purpose of visual aids. You want people to see the visual.

RULE

Never use a sheet of normal 12 point text as a visual aid.

Small fonts cannot be seen by many. Bold fonts are preferred, but do not use non-standard fonts; they are distracting and hard to read. Do not write text in colors. Use black or a single dark color. Remember that 15% of the world is colorblind. They may not distinguish a green line from a red line. Do not use giant fonts. They are just as distracting as small fonts. Again, planning visuals on index cards is a great way to design them.

Caution is also recommended when designing computer generated visuals. Many presentation programs allow a wide array of visual features and cute options. It is easy to have a gray cobweb or quilt pattern behind your words. They may look nice on your computer monitor, but when projected, the background may obliterate words or be a distraction. Some computer artists also like to insert borders and logos, which are often unnecessary and may just be another distraction. If you must use these programs, try to find the "plain and simple" command in the software menu.

RULE

Visual aids—keep them simple.

Graphs. The preparation of graphs is discussed in Chapter 8, "Using Illustrations," and the same approach applies when making graphs for presentations. The following list summarizes some guidelines for graphs in reports as well as presentations:

- No more than two variables
- Black on light color background slides lights up the room and speaker.
- Use bold fonts on all numbers and words.
- Project in anticipated size room to see if font is large enough.
- Clearly label axes.
- Only use tick marks—not full lines—for graduations.
- Use rational scales on graph axes.
- When comparing a number of graphs of the same properties (different materials), use the same graph scales.
- Select graph scales such that the data occupies most of the graph area.
- Do not use colors that cannot be seen by colorblind people.
- Use error bars to show statistics on data points.
- Make graphs simple.
- Use detailed captions.
- Avoid acronyms.

Do not use three-dimensional graphs; just use graphs with an x-axis and a y-axis. Do not use color unless colorblind people approve it. The most common error made for visual aids is to prepare graphs on a spreadsheet with default line widths and typing. These are fine for reports, but they are not suitable for visual aids. They usually cannot be seen past the speaker. They need bold type and a font size larger than 12 points. Make the lettering as large as possible. Of course, also observe the guidelines in "Using Illustrations," Chapter 8, on labeling axes, spacing tick marks, and showing error bars. Always put a caption above or below your graphs, and do not use acronyms. Descriptive legends and captions are better without acronyms, especially for visual aids. They act as a "crib sheet" and allow the audience an opportunity to read what the slide is about as well as listen to you tell them. This helps retention by audience participants.

Some people prefer to use tables in place of graphs. It is much better to avoid using tables—period. If you must use them, make the numbers and letters bold and at least 14 points in size. This limits what you put on a table. [*I use tables as visual aids for things like listing chemical compositions of metals and the like. There are really no graph alternatives for this type of data.*] When there is no suitable alternative for tabular data, just make sure type is larger than normal.

RULE

Use fourteen point, boldface type minimum on tables used in visual aids.

Photos. It is often desirable (and sometimes even expected) to present some photographs during a talk in some professions. Photographs can add interest and attract attention. Photos are often the very best visual aid, and their use is recommended where appropriate. For example, always try to use photos when describing machinery or difficult-to-describe situations or structures. Photos are still worth a thousand words.

Color is acceptable and desirable for photos, which can be converted to overheads on a color copier. This produces better quality than a black and white copier with photo mode. Scanners give overheads a quality similar to the color copier. Digital cameras are useful for photos of equipment and large specimens. Digital photos can be printed directly on transparency media with acceptable quality.

Slides are the best type of photo for projection to groups. Nothing can compete with a good color slide as a visual aid. The highest success can be attained with slides made without a flash on the camera. Flashes usually yield a bright spot that obliterates the subject. Try to use high-speed film and flood lamps for uniform subject lighting. Slides should be slightly overexposed rather than underexposed. They project better. Of course, digital photos or scanned silver halide photos can be projected, too.

Videos. A video camera can be used to make a video tape for projection to a group. The risks of using this system are similar to using a laptop computer for a talk. You must arrange for a meeting room with special equipment, that is, a video cassette recorder and a large monitor on a high stand. Sometimes, this is not a problem; other times it is.

Another consideration in using videos is the amateur quality of many videos recorded with handheld cameras. If you are making a video of a test by yourself using a handheld camera, you may end up with a "jiggle-ridden" video that does not impress your audience. [*Recently, a peer made a video of the operation of a new precision measuring device. He held the camera himself and talked while he was operating the machine. That video is now the team joke. It was hilarious, but he intended it to serve as an end of project report to management.*] A failed visual aid can have a negative effect on the purpose of a presentation.

Videos combined with a computer display program can be very effective. You do not need the video cassette recorder (VCR). The video can be made on a personal computer and projected on a video projector. In engineering, it is becoming popular to show motion of devices as videos that are part of a presentation. Again, the equipment risks are present.

In summary, videos can be of great value in an electronic report or presentation. However, do not make the video methodology the objective of the study. Also, if you are going to use a video in a presentation, make it to professional quality. A poorly executed video works like a poorly executed joke in a presentation. It can have such a negative effect that you would have been better off not using any visual aids.

RULE

If you use videos, make them with professional quality.

14.4 Presentation

Making an oral presentation is often intimidating to people, but oral presentations only intimidate the first few times. They quickly lose their power to incite fear and nervousness. The ability to speak in front of an audience is acquired with practice. Always be prepared, and nervousness will dissolve after your opening sentence. Know what you are going to say; have effective visual aids. Look the audience in the eye and present your message with confidence.

The nervous mannerisms to avoid include:

- Lack of eye contact
- Talking to your visual aids
- Blocking the projected image or screen
- Shuffling file cards or papers
- Reading a text
- Lack of enthusiasm for your subject
- Mumbling
- Nervous pauses (uhs)
- Machine gun delivery
- Lethargic delivery
- A death grip on the podium
- Standing motionless
- Looking at only one person in the audience
- Nervous tics
- Shaking when you use a laser pointer
- "One-second" residence time of visual aids on the screen
- Monotone delivery
- Too brief
- Too long—talking over the allotted time

The last admonition is probably the difficult one to avoid. People who have shed all nervousness in making oral presentations can develop a related malady—the inability to end a talk. The cure for talking too long is to decide exactly what you want to say ahead of time; time your presentation and do not deviate in delivery. It is usually unintentional adlibs that cause overruns in a presentation. Keep to your plan.

RULE

Keep oral presentations within allotted time. Talk concise.

Nervous mannerisms are probably the second most common bad habit, following the problem of speaking too long. Another very common practice is to partially cover an overhead during a talk. This is totally unnecessary, and you will appear as if you are talking down to the audience.

RULE

Never partially cover an overhead. It insults the audience's intelligence.

All of the bad habits in speaking are annoying to the audience, and they can be eliminated easily if you know that you are doing these things. Often, you do not. Ask an associate to check your talk to see if you have any of these undesirable mannerisms when speaking.

Many other undesirable things can also happen during a presentation. You can minimize them by doing some checks before the presentation:

- Learn how to control room lighting.
- Check the seating and make sure people can see your visuals.
- Check that equipment works.
- Check for pointers.
- Check the microphones.

Equipment failures are always possible. Try to have a contingency plan. [*One time I was the first speaker at a large meeting. There were about 250 people in the room. My slides were in the projector, and I checked that it worked and was in focus. I brought up my title slide, and the smoke showed on the screen. The projector was overheating. The projectionist from the audio-visual company shut down the projector and ran out of the room to get another. I stood at the podium for what seemed to be an eternity while the audience talked amongst themselves. After about 10 minutes a new projector arrived. It did not work. Without recourse, I gave my talk with no visual aids. I described my elegant graphs and drew them in the air with my finger. The audience was very compassionate. Ever since that event, when I use slides, I bring my own projector (in working order). Another preemptive action that I take on presentations to out-of-town groups is to carry two sets of visual aids. One is in my luggage and another on my person. I had a briefcase stolen from my hotel room at a conference. I had already given my talk, but I got the message that talks are not safe locked in a hotel room.]*

Summary

Formal reports are often orally presented to the sponsors. The people who paid for the work may want a face-to-face presentation of results. An oral presentation allows intercourse not possible with just a written report. Tailor your presentation to the audience; use appropriate visual aids. Talk in a concise way without nervous mannerisms and those pausing "uhs;" show confidence and enthusiasm in your delivery. The ability to make an effective oral presentation is a must for a successful technical career.

Some parting suggestions on oral presentations follow:

- Never apologize at the start of a talk for your visual aids, poor English, or lack of preparation. Just give what you have with confidence.
- Make a conscious effort to analyze your audience.
- Make no jokes.
- Your talk should contain the same sections as your written report.
- Make the purpose and objective of your presentation evident to the audience.
- Always have a proper closure (conclusions, recommendations).
- Make visual aids "seeable." Do not even think of using normal size (12 pt) type in a visual aid.
- Only 5 or 6 words on each line of a visual aid
- Only 5 or 6 lines on each visual aid
- Any type of visual aid can be effective if well executed. Select the type that you are most comfortable with.
- Eliminate nervous mannerisms with observation and practice.
- Never read a talk.
- When speaking in your native language in a country that speaks another language, speak slowly and distinctly, not louder.
- Use good visual aids, body language, and eye contact to convey your message to the audience.
- Use presentation evaluation sheets distributed to the audience to hone your presentation skills.

Important Terms

- Video
- Nervous tic
- Joke
- Transparency
- Movie
- Visual aids
- File cards
- Chalk talk

- Audience
- Monotone
- Overhead projector
- Delivery
- Eye contact
- Slide
- Jargon
- Nervousness
- LCD display
- Media
- Flip chart
- Demonstration
- Attention
- Uhs
- Site review
- Verbal abstract

For Practice

1. Compare the advantages and disadvantages of slides and transparencies for presenting the results of a big project to your company management.
2. Describe how you would prepare for a ten minute project update presentation to your coworkers.
3. Describe how you would prepare a 30 minute presentation to management on a major project.
4. Describe how you would prepare to present a technical paper at an international conference.
5. Describe how you would prepare a talk for presentation using an overhead projector.
6. Cite 10 things that could go wrong when you give a talk to a local technical organization. How would you deal with them?
7. Outline a presentation to your teammates on your progress on the laboratory safety audit.
8. Cite six presentation flaws and how to correct them.
9. Describe how you would prepare word slides and graphs for a 30 minute talk to company management.
10. Describe how a presentation is a verbal abstract.
11. You are giving a talk on a department improvement project (whatever you want to improve). Prepare the introduction visual aids.
12. Cite six displays of nervousness in giving a talk. How would you correct them?

To Dig Deeper

- C.T. Brusaw, G.J. Alred, and W.E. Oliw, *Handbook of Technical Writing,* 4th ed., St. Martins Press, 1993
- S.E. Lucas, *The Art of Public Speaking,* 5th ed., McGraw-Hill, 1995
- N.A. Pickett and A.A. Larter, *Technical English,* Longman, New York, 1996
- R.L. Sullivan and J.L. Wircinski, *Technical Presentation Workbook,* American Society of Mechanical Engineers, 1996

CHAPTER **15**

Getting It Done

CHAPTER GOALS

1. *Understand how to focus on a writing task*
2. *Know how to deal with writing impediments*
3. *Know the importance of measuring your writing effectiveness*

THIS BOOK is mostly dedicated to the details of technical communication: how to develop writing strategy and a writing style, how to make illustrations, and how to write reports and other documents that pertain to technical fields. This final chapter is about doing the actual writing. How do you as a technical person find time, energy, motivation and resolve to sit down and produce a technical document? How do you deal with all of the work and nonwork factors that complete with writing time? It does take time; it does take focus. Writing requires concentration. How do you clear your plate and make it happen?

Hopefully, by this point, you are convinced that writing technical documents and making oral presentations are important parts of a technical career and that you need to do these things. The overall objective of this book is to make you well equipped to meet the communication needs of your organization, and the topic of this Chapter is writing discipline—the ability to produce timely technical documents that meet the needs of you and your organization.

The format or arrangement of this Chapter is to discuss the factors that [*from my experience*] are impediments to writing. There are many excuses for not writing, and this Chapter describes them and suggests ways to overcome them. Specific factors that influence the discipline required in writing are motivation, making time to write, timeliness, dealing with interruptions, dealing with data, dealing with writing skills, dealing with collaborators, and dealing with computer substitutes for writing. A discussion on monitoring your communication effectiveness and maintaining your communications skills concludes the Chapter.

15.1 Impediments to Writing

Writing is not easy, and many obstacles may make the writing process more difficult. Some of the skills needed for writing are:

- Making the time to write
- Dealing with interruptions
- Resisting the urge for additional study
- Dealing with weak writing skills
- Dealing with writing mechanics
- Dealing with computers
- Setting deadlines

Making the Time to Write

With downsizing, everybody in industry is required to do more. Some people say they do not have time to write technical documents. So how do you get extra time for technical writing? The short answer is that you make time. Schedule a portion of any project for writing. Most engineers and scientists work from project to project. You make a proposal for a machine design, solve a repetitive part failure, develop software—whatever. Most technical supervisors require submission of a plan for the project as part of the approval process. One way to get time to write a report is to schedule the time.

Designate time in the project plan for writing a report. Figure 15.1 shows an example of budgeting time for report writing. [*My personal method is like that, but more generous on the report writing phase. We are only*

Research Plan Task	Completion Time*
1. Develop elastohydrodynamic model	3.5 months
2. Incorporate microtopography of both surfaces in the model	5
3. Perform laboratory tests on bearing designs arrived at from modeling	5
4. Compare test results with results predicted by the model	6
5. Prepare final report	7

*months after start of project

Fig. 15.1 Project plan designating time for report writing

funded for one fiscal year on a project. The fiscal year coincides with a January to December year. My project plans always show all test work completed by the end of August. This leaves four months for data analysis, report writing, proposal writing, symposia, and vacation. I plan on the fourth quarter of the year to wrap projects up. The technicians who perform tests for me are aware of my modus operandi, and they also plan on completing all testing by the end of the third quarter.]

Another hint on making time for writing is to be organized and meticulous in gathering data throughout a project. You should have a data collection system that you are comfortable with as a tool for report writing. It does no good to schedule a month of report writing in November if you cannot find the test results from February. Engineering and science projects take significant amounts of calendar time. Develop a system that ensures that essential data are not lost. [*Cubby holes, a project journal, a computer database, and infallible technicians work for me.*] Find a system that works for you.

Dealing with Interruptions

Many time-management seminars are periodically offered in most major cities. They present some useful suggestions on how to deal with interruptions that interfere with report writing and other work. For example, the office space for technical staff is often some sort of cubicle. If you sit such that you can make eye contact with passersby, there is a strong likelihood that they will talk with you and create an interruption. The solution proposed in some timesaving seminars is to rearrange your cubicle so that you never make eye contact with aisle traffic.

Having a cubicle with a guest chair is an additional problem. This chair creates an open invitation for passersby to stop and chat, again creating interruptions. The time-management people suggest storing heavy objects like a broken printer on the chair. Keeping a briefcase on a chair works quite nicely, too. Only remove it if you really want to talk to the interrupting person.

The time management people also have suggestions on how to limit the extent of interruptions. If a visitor is overstaying his/her welcome, make a remark like, "Thanks for sharing this with me; I've got to make a call now." If they still do not get the hint, stand up. Stay standing until they leave. Or just walk out to get a supply or look in your mailbox.

Telephone calls are probably by far the most destructive interruptions. Develop a way to limit their invasion into your workspace. [*I have my own system for dealing with them. Firstly, I hate phones, I never answer the phone at home, and I probably would not have a phone if I lived alone.*] A phone by definition is an interruption. Somebody is intruding unannounced into your intellectual space. Certainly all people need timely communication with others, especially clients. This is why e-mail was invented. It does not intrude (unless you want it to). It is timely (if you answer messages at midday and at day's end). E-mail has made the telephone totally unnecessary in many businesses. One system for preventing phone interruptions is presented for your consideration in the following list:

- Never wear a pager unless answering the phone is your primary job.
- Never answer your office phone when you have a visitor.
- Have a 10 second maximum message on your answering machine.
- Take and return phone messages no more than twice a day at designated times; state this on your answering machine message.
- Never talk with unsolicited sales callers. Say, "No thank you," and hang up.
- When you are away from your desk and hear your phone ring, never dash to take the call. The machine will do it.
- Say, "I have to go to a meeting now" to phone callers who want to chat.
- Never call anybody unless it is about business. Chatting at home with out-of-town relatives and friends is excepted.
- Never answer the phone when you are on private time (outside working hours).
- Do not even think of buying a cellular phone.

This phone management system may not be for all technical people, but it has the approval of Phoners Anonymous (PA). It prevents phone addiction, and it helps cut down interruptions that cut into your contemplative and report writing time.

Writing requires quiet time to focus on your report message. All technical persons need to develop systems to evade interruptions so they can focus on review of data and packaging them in a coherent message. Find a quiet place for writing. Empty conference rooms work wonders, as do home offices. Conference rooms usually have schedules, and it is almost always possible to find one that is open for a one or two hour period. Once you reserve it, pack up your data and spend time focusing on summarizing data, distilling it into conclusions, and making recommendations in a report. Needless to say, working at home requires supervisor approval. [*I have never been denied it.*] It is also possible to do "phone work" at home. Most answering systems can be accessed from your home office, so you can work at home and still have access to the communications at work.

Each person's work and home situation is different, so it is difficult to present guidelines that fit these many situations. The overall idea, however, is to establish a quiet-time system that provides uninterrupted time to add intellectual value to work and put it in a report form that can be shared with others.

Resist the Urge for Additional Study

The urge for more study is a widely practiced excuse for not writing a report. There are always results if you work on a project. They might not be the expected or desired results, but there will be some results that should be shared with others. Status reports should be issued on projects that need additional studies. In fact, yearlong studies (or longer ones) need written quarterly reports. If you carry a significant number of long-term projects, a quarterly report on each could consist of a bullet statement followed by a paragraph describing the current project situation:

Status Report Example

3rd Quarter 1998
Development of New Slitter Knife Materials

- The cemented carbide experiment in the first quarter was unsuccessful. Four knives with a brazed carbide surfacing were fabricated in the knife room, but radial cracking of the surfacing precluded continuation of this venue.
- A quotation has been received for a cemented carbide insert reaction bonded to a stainless steel substrate. Four insert knives will be tested by the end of October. In addition, carbide lower knife prototypes will be coupled with upper knives made from a new proprietary knife material. It is anticipated that all of these knife systems will be tested on the laboratory slitter by year end.

Film Scratching Test

Five different films were scratched with four different laboratory test rigs and it was determined that the ball-on-plane test produced repeatable results that correlate with field assessment of scratching susceptibility (Figure 1). The other tests will not be used in future studies.

The elements of a project status report are:

- Title
- Description of work completed
- Problems
- Comparison to plan
- Adjustments to plan

The above examples did not show any significant problems. If problems need to be addressed or if project plans have changed, these types of issues need to surface in status reports.

The beauty of status reports is that they are more informal than many of the other reports described. They are like an abstract with data (where additional supporting data may be attached in the form of illustrations or visual aids. These informal status reports can serve as the basis of a formal report on the entire project.

In summary, if it is felt that more work must be done on a project before a report on results can be issued, address this in a status report. This negates the "needs additional work" argument for not writing a report. As an aside, bosses seem to love these kinds of reports. No doubt their brevity makes them easy to read and digest. In addition, they feel more assured because they know the status of the projects in the department. Make status reports a part of your report writing repertoire. You will not be sorry.

Dealing with Weak Writing Skills

Many people in technical jobs are unsure of their writing skills, and this is a frequent reason why many technical people never write reports. They feel that they are not good at writing. Their spelling or grammar is not good; they are not proficient at formulating a logical report or constructing good sentences. In general, they are intimidated by the task.

The fear of writing is almost a catch-22 situation. People are not comfortable writing reports because they do not write reports and develop writing skills. Like any skill, writing skills are developed and maintained by practice. For example, poor spelling is a common problem that makes people reluctant to write. There has always been a stigma attached to poor spelling. Spell checkers on computers have made it easier to prevent misspelling, but the legitimate fear of a misspelling still requires careful attention and review. Always read a document over on the screen and fix these kinds of word mistakes. Then print the document and read it again. Invariably, you will find other spelling, grammar, or punctuation errors.

RULE

Print out a hard copy and proofread it before electronic distribution.

Maybe some people can make a document perfect without seeing it in hard copy, but many lesser mortals need a hard copy to find all of the mistakes. This proofreading procedure, in addition to spell checking, will prevent poor spelling as a writing impediment. We know that spell checkers are limited. For example, the machine would not complain if I typed "no" for "know" in the preceding sentence.

Lack of typing skills may be another excuse for not writing reports. Computers have forced technical people to personally type their own reports, spreadsheets, and so forth. Department secretaries no longer do this job. What should you do if you are a poor typist? The official answer is to take a typing class and learn to type. [*I did, but I did not. That is, I took a typing class, but I learned that I have a physical/mental disability that precludes my becoming a satisfactory typist. I did not learn very well. I handwrite and outsource the typing to people who are good at it. I pay for this from personal funds if it is personal work, and I use project funds for work-related reports.*] You should budget for typing in project plans and outsource it if you are a poor typist. It certainly is cheaper for your organization to pay $10 an hour for 10 typed pages from a professional typist than to have you spend an hour typing two pages at a corporate labor rate of $50 an hour. Never let poor typing skills keep you from writing technical documents. There is always a way around this problem.

Dealing with Writing Mechanics

Report Structure. Any language has rules on how words and phrases should be arranged in sentences. Word use and the structure of sentences are part of report structure, but many technical people have trouble putting their message in a logical report structure. The report "recipe" presented in just about every Chapter of this book is good way to make writing easier with a consistent report structure:

- Introduction
- Body
- Conclusions
- Recommendations

This simple structure works for all types of technical reports. The only real work is deciding what goes into these report sections.

Grammar. Grammar includes all aspects of writing: word choice, forms and structures of words, arrangement of words into phrases and sentences, and the rules for a particular language. Grammar rules are different for every language. What is grammatically correct in the English language may be totally wrong in French, for example.

Poor grammar can be an impediment to active report writing. This book is not going to show readers how to make their grammar correct. This is (was) the responsibility of primary education. Some grammar issues are addressed in preceding Chapters, and additional guidelines on avoiding common grammar mistakes are given in Appendix 6. Figure 15.2 illustrates what they used to teach in elementary schools about parts of a sentence 50 years ago. They probably have computer programs that teach this now, and most word processing programs come with a built in Notre Dame nun who underlines grammar errors in green as you make them. Voila! There is no need to be apprehensive about writing because you forgot some of the basic rules of grammar. In addition, some organizations still have professional editors that review papers and similar technical documents that will be in the public domain for any kind of error in grammar, spelling, or punctuation, as well as company proprietary information.

In summary, grammar questions should not pose an impediment to active technical writing. Reasonably correct grammar is a must, but it does not need to be perfect. The guidelines in Appendix 6 can help to make your grammar reasonable.

Punctuation. Punctuation marks are standardized symbols to separate sentence parts and to end sentences. They are intended to make the sentence (message) easier to understand.

Without Commas

The samples were abrasive blasted with silicon dioxide a known carcinogen in a sealed booth.

With Commas

The samples were abrasive-blasted with silicon dioxide, a known carcinogen, in a sealed booth.

There are only eleven of these little things, and they certainly should not intimidate potential report writers. The guidelines presented in Appendix 6 hopefully will answer most questions that may cause apprehension and

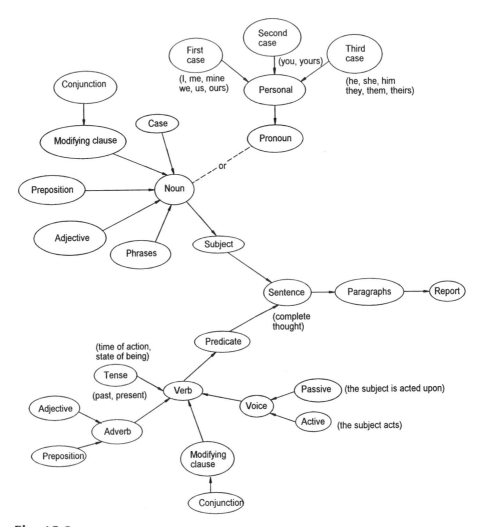

Fig. 15.2 Parts of a sentence

discouragement in writing. The following is a basic summary of punctuation:

- Periods signal the end of a sentence (complete thought).
- Commas signal changes in thought and keep words and clauses separated.
- Semicolons join closely related sentences.
- Hyphens join closely associated words and broken words.
- Dashes are used to signal a sudden change in thought. (They are different from hyphens; they are longer. Use two hyphens with no space between them.)
- Question marks end a question.
- Parentheses signal minor digressions.
- Exclamation points signal surprise and do not belong in technical writing.
- Apostrophes show possession.
- Quotation marks separate nonstandard words/thoughts and the work of others.
- Colons are used to start letter salutations and to introduce significant lists (often a complete sentence).

Punctuation can be part of your writing style, but it is probably not in your best career interest to use wrong punctuation. Do not let the methodology of punctuation get in the way of a message. Keep in mind the preceding definitions and use the punctuation references elsewhere in this text for troublesome punctuation situations. Strive for a reasonably correct presentation that does not detract from your message. The message is what is important. Errors in writing mechanics can always be explained as part of your style. You will never get arrested for a comma omission, but you could lose a job for not documenting your research.

Dealing with Computers

In this new millennium, computers are almost always used in some way for writing technical documents. You will probably do all your writing on a computer; your illustrations may be generated on a computer. Computers may control your test rigs and collect experimental data.

How will and should computers affect technical writing? The simple answer is that they should make writing easier. Computers have made corrections and rewrites less painful. Spelling and grammar checks have made dealing with language more palatable. Standardized document formats have simplified dealing with report mechanics. However, as with all good things, there are some potential negatives. The biggest threat posed by computers to technical writing is the use of transient computer output to replace formal technical documents (complete with experimental details, test data, conclusions, and references—an information package suitable for archiving).

Too often, technical people try to use electronic transfer of experimental data to replace reports. Teams can have "team suites" on a server someplace in the organization. Process verification or control information similarly is stored on a server that can be accessed by many individuals. Most will not allow modification of the data by browsers, but workers are sometimes encouraged to store raw data in these computer team suites. Do not allow these data storage tools to serve as writing substitutes.

Proper communication of technical information requires interpretation as well as the details of what was done. Figure 15.3 is an example of data sent to a network file for study by team members. It appears to be the output of some statistical software package but is devoid of captioning and identification fields. Graph axes are unlabeled or labeled in acronyms, data cannot be rearranged, unwanted data clutters fields and in general, a nonuser of the software is ignorant of the usefulness or interpretation of the data. A technical report is needed to tell others what happened in the experiment. A data dump like the one in Fig. 15.3 is not a useful substitute to a report, even if it is shared on a computer server.

Response: 4SIGMA - VERTICAL STEADINESS

Summary of Fit

RSquare	0.255427
RSquare Adj	0.089966
Root Mean Square Error	1.814453
Mean of Response	12.16696
Observations (or Sum Wgts)	23

Effect Test (Type III)

Source	Nparm	DF	Sum of Squares	F Ratio	Prob>F
Std DELTA_A	1	1	2.692918	0.8180	0.3777
Std A	1	1	1.922607	0.5840	0.4547
PRODUCT	2	2	19.717798	2.9946	0.0754

Sequential (Type 1) Tests

Source	Nparm	DF	Seq SS	F Ratio	Prob>F
Std DELTA_A	1	1	0.561348	0.1705	0.6845
Std A	1	1	0.050212	0.0153	0.9031
PRODUCT	2	2	19.717798	2.9946	0.0754

Parameter Estimates

Term	Estimate	Std Error	t Ratio	Prob>|t|	VIF
Intercept	3287.2671	4286.845	0.77	0.4531	0
DELTA_A	6023.3451	6659.964	0.90	0.3777	1.4306758
A	-2378.991	3113.101	-0.76	0.4547	2.0814843
PRODUCT[2383-2388]	-0.044936	0.664509	-0.07	0.9468	1.8427624
PRODUCT[2386-2388]	1.6983234	0.845326	2.01	0.0598	1.7363901

Fig. 15.3 Example of experimental data sent to a team suite

There are computer software packages intended for every imaginable purpose, and the output should be designed with the ultimate users in mind. A better example of computer output from a statistical process control (SPC) program is shown in Fig. 15.4. Axes are defined, and the graphical presentation aids interpretation for those familiar with SPC methods. Templates can also prompt a writer to use a standard format for a technical paper; there are templates and software for every type of document. For those in research, there is an electronic laboratory notebook software to serve as a repository of experimental data, test procedures, and development. It may take the place of the laboratory notebook, that was advocated in Chapter 3, "Performing Technical Studies."

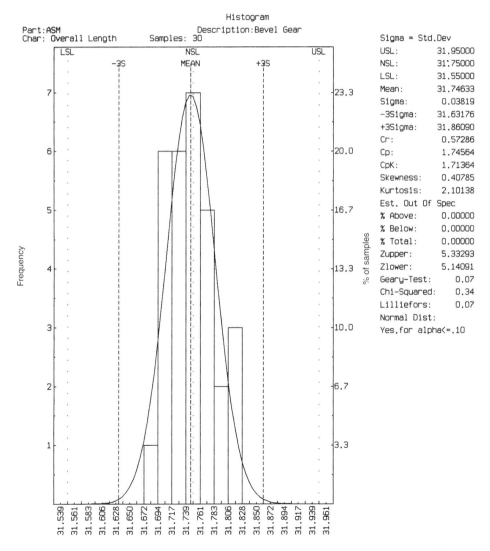

Fig. 15.4 Example of statistical data with better captioning and description

How do proficient technical writers deal with the various writing aids available with computer software? Computers should be used to help you, but hard-copy reports are still used for two main reasons: (1) computer software systems are transient and (2) long, involved, complicated documents are difficult to read on a computer monitor. Reason 1 is based upon many years of computer and software flux. Most companies update software and hardware on a periodic basis, and chances are likely that you will lose valuable work. [*I had expert systems that cost in excess of $10,000 on my computer that were discarded by the company computer police with the hardware. My $1,000 graphing program had a similar fate. At home, I keep my 1982 computer in the attic so that I can retrieve documents that I stored in the system in the 1980s. I suspect that I will die before I go through the hassle of reconfiguring this now-obsolete system. Our company library now requires electronic and hard copy for archived documents. I suspect that the hard copies will be the only form retrievable ten years hence.*] Hard copy is a very reliable method for long-term storage and access.

Readability of computer monitors is another reason for continued use of hard copy. Most computer monitors cannot accommodate the equivalent of a typical typed page. A typical formal report may take 20 screens. Most document readers do not have the patience and eye power to read 20 screens. Monitors have lower resolution than a typed document. The image is not completely stable, and the glare and limited viewing position make reading on the computer difficult. Attention span is greatly reduced. Reader friendly computer documents should not exceed two or three screens. This precludes this medium for formal reports and papers that require significant verbiage and data.

Hard copy is still the preferred medium for significant documents. Newspapers, books and magazines have been available on the computer since the early 1990s, but the hard-copy versions of these modes of communication are still preferred by a vast margin. Hard copy is just easier on the reader. This of course may change in the future, as more and more professionals conduct their activities with electronic devices. [*They never carry a pad of paper, only their pager, cell phone, and laptop*].

Setting Deadlines

As noted in previous Chapters, technical writing needs to be timely. Proposals need to be written during budgeting season. Investigation reports need to be written as soon as the work is done. The report is the vehicle for presenting the investigation results. The same situation exists for formal reports on research projects and most other forms of technical writing.

The way to achieve timeliness and to help succeed in writing technical documents in general is to write to a deadline. If you are a manager trying to get a report from an employee, assign a deadline. If you determine your own working priorities, set a deadline for yourself. Make it a specific date and track progress toward the goal. Humans are imparted with an innate need for goals; people always seem to do better if they have goals. They focus their energies, and this helps them achieve goals. A deadline is a fantastic help in any task. In fact, if projects do not have deadlines, they often never reach completion. There is something in human essence that needs and wants goals.

So, do not fight your essence; set a deadline for writing all technical documents and meet it. Book publishers always give authors a deadline for every step in the writing process. Similarly, technical journals have very rigid deadlines for abstract submission and final manuscript submission. Most professional writers work to deadlines. [*I am writing this book to a deadline that I set with the publisher seven months ago. I will force myself to meet this deadline. I want to meet this deadline to meet my promise to the publisher, but, more importantly, I want to finish and take a vacation. I will reward myself for meeting this goal. Maybe this is why human nature likes deadlines. Meeting a deadline usually has rewards associated with it.*] There is a definite psychological reason why deadlines work.

In summary, set a deadline for each technical document. Make it timely and realistic; above all, meet it. When you do, reward yourself; give yourself a dip in the hot-tub, a long shower, a martini—whatever you find pleasurable. A deadline is a very definite aid in getting the job done.

RULE

Set a deadline for a document; meet it and reward yourself.

15.2 Maintaining Writing Skills

One of the main reasons many engineers and technical people are poor report writers is that they do not write enough to develop and maintain writing skills. Some engineers use e-mail notes as a substitute for a written documentation. E-mail notes can serve as interim reports or for answering technical problems. They are acceptable for informal meeting minutes and similar transient communications but not for investigations or projects. A technical document almost always needs illustrations and data tabulation, even photos. If an e-mail message is detailed enough to include supporting data, you may as well make it a formal or informal report that can be archived. The format and content should always conform to the guidelines presented in this book.

Technical writing must be practiced, regardless of the way a document is prepared. Like any acquired skill, continual practice is needed for good writing. How do you practice technical writing? The easiest way is to write for every project—make it a habit to conclude all projects and assignments with a written report. If typical projects take six months or a year, write interim reports. Essentially, write reports or other technical documents daily and weekly. The longer the interval without writing, the greater the likelihood of losing writing skills.

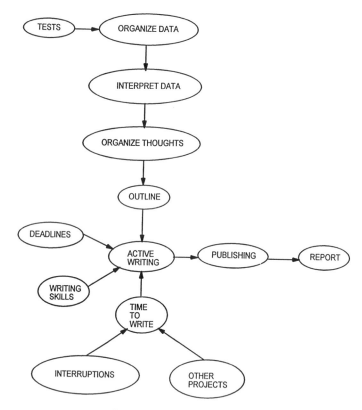

Fig. 15.5 Report writing steps

Meeting minutes provide another opportunity to maintain writing skills while also making meetings more efficient. Circulate the meeting minutes to attendees. Make them complete, and try to include the basic elements of all technical reports as described in Chapters 7 to 11. Whatever your situation may be, opportunities for regular report writing probably exist. Search out the opportunities and establish a writing program that will keep you proficient. Establish a goal of at least one report per week. Remembering the steps and guidelines for report writing (Fig. 15.5) will help you build your career.

15.3 Measuring Report Results

When you write a report for an archival journal, there is usually a review of your work by three or four peers. These reviews can be a measure of how your writing style matches what others believe to be suitable for technical writing. If reviews suggest major revisions, this is an indication that something is amiss in your style. If your reports come back from peer review with just minor changes, this indicates that you are doing things right.

Most archival journals give authors 50 reprints of their paper free of charge. The number of requests for these reprints is a measure of the popularity of the topic addressed. [*I have several file drawers full of reprints. I never gave away 50 reprints on any paper; however, I found reprints very valuable in consulting. Somebody may ask a question that was covered in a published paper. I just give them a copy of the reprint.*] Thus, requests for paper reprints can serve as a measure of the popularity of a technical document.

Published books are the easiest technical documents to get feedback on. Book royalties depend on book sales (usually 10 to 15%). The larger the royalty check, the larger the readership. A successful technical reference, one not suitable for teaching, may be considered successful by some publishers if 1000 copies were sold. [*This is the reason why highly technical books on narrow subjects must be a labor of love. Royalty earnings are such that the authors may be getting 10 cents an hour for time researching and preparing a book. This is my fourth book, and so far, each has taken five years to research and write.*]

A simple way to gage your writing proficiency is to compare yourself to department peers. Most organizations keep a central list of reports issued within the organization or department. Periodically review the list of document titles and see how you compare with others in writing volume. Do you write less or more than others? It is also helpful to get copies of reports written by peers and read them for report quality. Is their style better or worse than yours? You may find that you are an expert writer compared to your peers (or not). If you uncover something missing in your work, take action and improve it. If you find your peers using nonapproved writing formats or illogical formats, also take action. Suggest standard formats [*like the one proposed in this book.*]

Evaluation forms attached to your document can be helpful in getting feedback on the effectiveness of your technical writing. This type of form is often used on newsletters to keep address lists up to date. It may contain some questions like:

- Did you find this information useful?
- Was the information timely?
- Is the report format acceptable?
- Is the report length right?
- What is your overall rating of this document?

This kind of feedback works best if it is done throughout an organization. If you post technical documents on your department website, the website hits could serve as a measure of the effectiveness, or at least the popularity, of your topic.

Our final suggestion on gaging writing is to ask one's boss what he or she thinks about your technical writing. Ask if your technical documents conform to department format. Determine if you are writing enough or too much. Do your documents fulfill their intended purpose? Does the boss feel that your documents are well written? Are there any areas where you could improve? Your supervisor determines your pay and promotions and your writing can be a factor in a performance evaluation. Make sure that it meets his or her expectations.

Summary

This Chapter gives some suggestions on how to deal with distractions and work situations that get in the way of technical writing. The impediments to proper documentation of work are many. The distractions are countless, and sometimes the rewards are fleeting. However, documentation of technical work is essential. As a technical person, your product may be only words on a page or computer screen, but to your employer, they are intellectual property. They are why you earn a salary. They are the result of your schooling, the result of experimentation. These words and figures may be used to make profits, to create jobs, to solve health problems, to build cities. Writing is important.

This text started as class notes for an informal writing course for newly hired technical people and transfers to a central engineering department for a large manufacturing concern. Many of the new people either were not familiar with the technical writing procedures of the organization or were not active writers for a variety of reasons. The course essentially presented the division writing expectations along with suggestions on how to make writing tasks easy and still meet company requirements.

In the ten years or so that this course was offered, there has been a steady stream of technical people who completed the course and moved on to a career with emphasis on technical writing. Some of the reasons why this text has been effective for users are:

• They know how to write various types of technical documents.
• They know how to organize, present, and interpret their data.
• They strive for concision—all documents are as short as completeness will allow.
• They have accepted that technical writing is a part of a technical profession.

The final recommendation is to write on a regular basis. Establish the attitude that the job is not done until the documentation and communication to others is complete. Write immediately after completing the work; use the suggested planning and outline process, format, style, and tone, and have your work reviewed by at least one trusted editor. Make the document as correct as possible, and follow up to make sure that your writing intentions were satisfied. Good technical writing will make your job and career easier and less stressful. A good report/document puts closure on a project or task. Then you can go on to your next challenge.

Good luck in your career and don't forget the simplified approach to technical writing.

Important Terms

- Time management
- Additional study
- Project plan
- Writing skills
- Cubicle
- Spell check
- Data management
- Typing skills

- Record keeping
- Grammar
- Interruptions
- Punctuation
- Phone management
- Skill maintenance
- Timeliness
- Report metrics

For Practice

1. Write a project plan that includes budgeting time for reports.
2. State five ways to prevent unwanted interruptions at work or school.
3. Describe how you would organize and record test data on a one-year long project to develop a new product, an automatic napkin dispenser, for restaurants.
4. Describe why report timeliness is important.
5. Write a status report on this course to your spouse, friend, roommate, and so forth.
6. State three ways to overcome poor writing skills.
7. Describe how you would check grammar and punctuation in a formal report.
8. Describe how you would determine if your reports were effective.
9. State five long-term aids to effective report writing.
10. Describe how you intend to implement the learnings of this book.

To Dig Deeper

- A. Eisenberg, *Writing Well for the Technical Professions,* Harper and Row Publishers, 1989
- S.E. Pauley, *Technical Report Writing Today,* Houghton Mifflin Co., Boston, 1979
- W.S. Pfeiffer, *Technical Writing, A Practical Approach,* Macmillan Publishing Co., 1991
- P.P. Sageev, *Helping Researchers Write—So Managers Can Understand,* 2nd ed., Battelle Press, 1994
- M.E. Skillen and R.M. Gay, *The Chicago Manual of Style,* University of Chicago Press

Example of a Laboratory Test Report

MATERIALS ENGINEERING LABORATORY REPORT

REPORT NO. 99073

TO:	P. Buch, Polymer Processing Division, 2/35/KP
TITLE:	Evaluation of Tenter rail materials
AUTHOR:	K. G. Budinski, KP Materials Engineering Laboratory
CONTRIBUTORS:	M. Kohler, C. Mroczek

Problem:

The Polymer Processing Division requested that Materials Engineering Laboratory to evaluate the friction and wear characteristics of a new phenolic material as a candidate for improved wear life on tenter rails in polyester sheet manufacturing machines. Hardened chain links (60 HRC) slide against these rails at high sliding speeds (1 to 3 m/s) and at elevated temperatures (up to 260 C). The current rail material is an aramid-reinforced phenolic that occasionally fails producing wear debris that may contaminate the product. The purpose of this study is to determine if a new grade of carbon-fiber reinforced phenolic can provide improved serviceability over the present rail material. The objective is to obtain a rail material that does not fail in service.

Investigation:

Two samples (2 cm x 6 cm x 100 cm) were submitted for wear testing; one was identified as sample 1 (current B317 phenolic), the other as sample 2 (new rail material). The test plan was to perform a block-on-ring wear test on each material and assess the wear on both members. The blocks would be made from the test plastics and the rings would be hardened steel. The tester is shown schematically in Fig. A1.1. The test procedure used was ASTM G77 with the following testing parameters:

normal force – 44.48N
sliding distance – 10,000 m
sliding velocity – 1 m/s
plastic temperature – 120 °C
ring roughness – 0.1 μm

Three blocks were made from each plastic; matching 4620 rings were fabricated and hardened to 60 HRC. The blocks were machined so that an as-molded surface was the test surface.

Friction was continuously recorded throughout the test. Evaluation of wear behavior included comparison of running friction, volume loss from the test blocks and volume loss from the test rings. Three replicates were tested from each material. Profilometry was used to measure wear scar volumes.

Results:

The wear test results which are presented in Fig. A1.2 show that there is not a statistically significant difference in the wear behavior of the two materials. The average wear of the new rail material was slightly lower. Neither material produced measurable damage to the hardened steel counterface. The friction test results (Fig. A1.3) show similar results. There is not a significant difference throughout the test.

Conclusions:

The new rail material candidate did not have significant improved friction or wear characteristics over the current rail material.

Recommendations:

If the new rail material is comparable in cost to the present rail material, we recommend production trial of a length on 307 machines where most failures occurred. The slight improvement detected in our laboratory tests may become more significant under active service conditions.

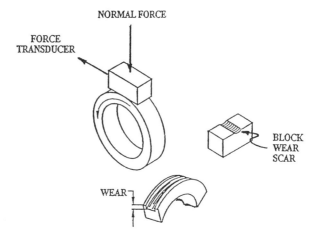

Fig. A1.1 Schematic of block-on-ring wear test

Rail Identification

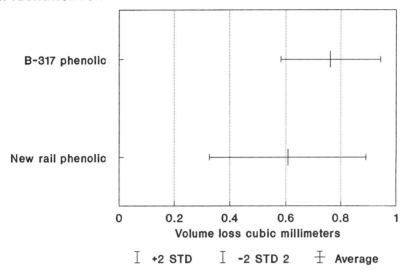

Fig. A1.2 Average volume loss for B-317 phenolic and new phenolic. Error bars are for ±2 standard deviations (STD).

Rail Identification

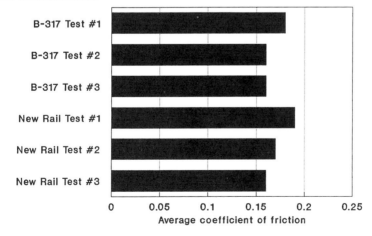

Fig. A1.3 Coefficient of friction with steel for the current rail phenolic (B-317) and a new phenolic. Test was performed on a LFW-1 block-on-ring tester with a 4620 steel ring with the following conditions: normal force of 44.48 N, ring rotational speed of 1 m/s, 10,000 meters of sliding.

Example of a Formal Report

BUD LABS – DESIGN ENGINEERING DIVISION

Subject:	Carbide Flex Seals for Gel Reactors
Report No.	227A42
Date:	7/20/98
Review Date:	7/20/98
Author(s):	K.G. Budinski
Contributors:	R. Swartz, F. Grant

Abstract:

Seal leakage problems in the support manufacturing facility prompted a study in the Materials Engineering Laboratory to determine if there are materials for face seals that will last for two years compared to the present life of about four months.

Laboratory tests were conducted to screen candidate materials for improved wear life. Cemented carbide self-mated appeared to have the best performance in the wear tests. Three prototype seals were manufactured with a solid carbide stationary member and a composite carbide/stainless steel for the rotating member. It was determined that the braze-clad carbide surface was not successful because of cracks.

A project extension is requested to investigate alternate ways of bonding carbides on the rotating flexure.

Approved By: _____ Date: _____

Edited By: _____ Date: _____

Introduction

Gel reactors in the Manufacturing Support facility had twelve shutdowns in 1997 because the face seals at the agitator feed-through (see top of Fig. A2.1) leaked. Each shutdown costs approximately $20,000 in lost production and $1800 in seal replacement costs. Nine of the twelve 1997 failures were caused by wear of the seal faces. Another cause of leakage was the sticking of the spring. The spring load is used to accommodate the waviness and wobble that exist in the system due to machinery tolerances and tolerance buildup.

Manufacturing Support requested the Materials Engineering Laboratory to investigate material improvements to increase the wear life of seals to at least two years. A department engineer then proposed a way to eliminate the spring (Fig. A2.1 bottom). The seals have the same general shape as the existing seal, but the design uses flexible web design [1] to provide the spring load on the seal. This system was put into production to solve leakage problems resulting from spring sticking and breakage, but the hardened contacting surfaces did not provide the desired service life of 500 hours. They wore significantly and leakage started at about 200 hours.

It is the purpose of this project to screen candidate seal couples and select a couple that will improve service life. The objective of this work is to eliminate the costs of frequent seal failures. This report summarizes the work. Laboratory tests are described, followed by an investigation of improved manufacturing techniques, discussion of results, and recommendations.

Laboratory Tests

Examination of worn seals shows that leakage occurs when the male member of the seal face wears a significant groove into the stationary member. The rotating member does not wear exactly the same as the stationary member and when the wear groove gets to a depth of about 0.5 mm leakage starts. It was decided to screen various candidate materials with a laboratory wear test and then field test the winning candidates in the pilot plant.

There are a variety of standard tests used to investigate metal to metal wear characteristics [2-5]. The crossed-cylinder test of ASTM G 83 was selected, because it produces significant wear in hard materials in a relatively short test time. The test method places a rotating cylindrical pin at 90° against a stationary pin (Fig. A2.2). A "divot" or worn area is removed from the stationary pin, and a groove is formed on the rotating pin. The wear volume on both members is measured from mass changes during the test. The test parameters were:

- speed = 0.22/sec,
- normal force = 200 N,
- test duration/sliding distance = 2h - 20,000m.

The test couples included hardened steel, ceramics, and cemented carbides. Three replicate tests were conducted on each couple and the average wear volumes were used in comparison graphics.

Results

The wear test results (Fig. A2.3) indicate that a number of candidate couples had lower system wear than the existing couple, self-mated type 440C stainless steel. The couples with less system wear were:

> Self-mated C2 cemented carbide
>
> 440C stainless vs. C2 cemented carbide
>
> D2 tool steel vs. C2 cemented carbide
>
> M2 tool steel vs. C2 cemented carbide
>
> C2 cemented carbide vs. A 11 tool steel
>
> M4 tool steel vs. C2 cemented carbide

Any of these couples should provide a wear improvement over the present system. The lowest wear rate in the screening test occurred with type C2 cemented carbide self-mated.

Implementation

It was decided to implement the wear results by building three prototype seals out of C2 cemented carbide. The complication in doing this is keeping the flexure design on the rotating member of the seal. Cemented carbide is about three times as stiff as the

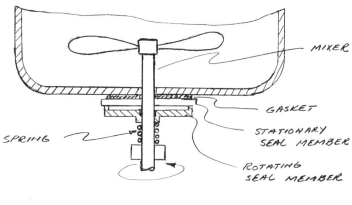

EXISTING DESIGN

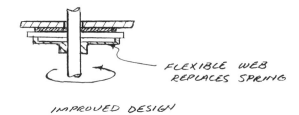

IMPROVED DESIGN

Fig. A2.1 Proposed design improvement for face seal

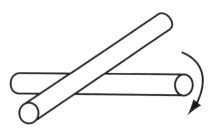

Fig. A2.2 Cross-cylinder wear test

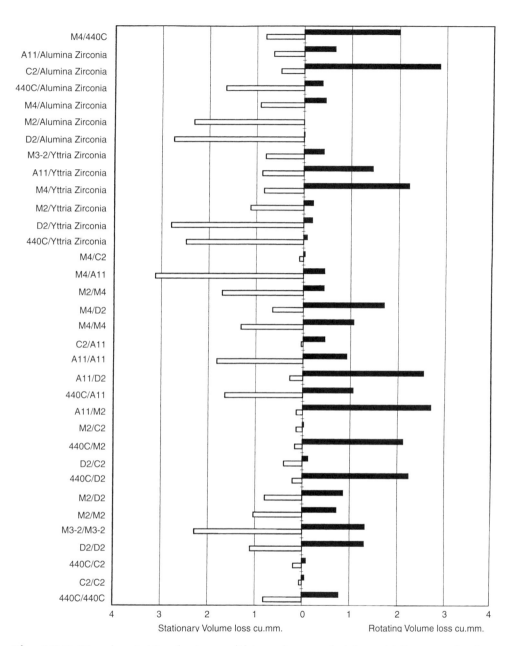

Fig. A2.3 Wear characteristics of various candidate couples; wear of rotating and stationary members in crossed-cylinder tests

stainless steel that is currently used in the flexible member of the seal. In the current design, the flexible part of the seal is only one millimeter thick. To get the same flexing action from cemented carbide, the member would be only 0.3 mm thick. According to several carbide suppliers, it would not be possible to fabricate the lower seal with a flexure this thin. Cemented carbide is too brittle. It would crack.

An alternative solution would be to clad the seal face on the rotating member with carbide. There are a variety of ways to accomplish this, but we opted to clad the rubbing surface with cemented carbide braze cloth. This technique has been used by others [5, 6] to clad rubbing surfaces and it appeared that it would work in this application.

Three prototype seals were fabricated with solid cemented carbide for the stationary member; the flex member of the seal was made from type 316 stainless steel with a one-half millimeter-thick cladding of carbide braze cloth applied to the rubbing surfaces of the seal. A cross-section of the clad seal surface is shown in Fig. A2.4. The edge and surface appearance is shown in Fig. A2.5. There were cracks and severe edge roughnesses that were uncovered by finish grinding. Both would make the seal prone to leaking.

Discussion

The braze-clad cemented carbide was selected over thermal spray processes for this application since previous investigations [7, 8] have shown problems exist in using these porous coatings for seal surfaces. The braze-clad carbide is not porous. We did not anticipate cracking in the clad layer. These results suggest that if we implement a carbide-to-carbide couple on the gel reactor seals, we will have to invent a solution. A new method needs to be developed for affixing cemented carbide to the rotating flexure.

Conclusions

1. A seal made from cemented carbide self-mated should provide a significant improvement in life over the current couple of 440C stainless steel self-mated.
2. Application of carbide to the flexure member of the seal by braze cloth failed due to cracking of the carbide overlay.

Recommendations

The wear tests conducted in this investigation suggested that there are six candidate mating couples that will provide improved seal life. It is recommended that this project

Fig. A2.4 Surface condition of seal

Fig. A2.5 Cross section of clad seal face

be funded for another four months at a cost of $40,000 to develop an alternate way of bonding carbide to the seal flexure member and to perform pilot plant test another candidate couple (M4 versus cemented carbide).

References

1. R. Swartz, "Use of flexible seals to reduce leakage," Technical Report Number A 2798483, Rochester, NY: Bud Labs, 1994.

2. ASTM Standard G 99, <u>Test Method for Wear Testing with a Pin-on-desk Apparatus,</u> West Conshohocken PA: ASTM, 1995

3. ASTM Standard G 83, <u>Test Method for Wear Testing with a Crossed-cylinder Apparatus,</u> West Conshohocken PA: ASTM 1998.

4. ASTM Standard G 132, <u>Test Method for Pin Abrasion) Testing,</u> West Conshohocken, PA: ASTM, 1995.

5. P. Orsat, "Use of Braze Cloth for Hardfacing Impellers," Journal of Fluid Mechanics, Vol. 27, No 6, Sept. 1995, pp 203-10.

6. H. Magee, "Cemented Carbide Surfacing," Surface Engineering, Vol. 2, No 4, Oct. 1997, pp 201-203.

7. H. Moore, "Plasma Arc Coating Rotary Seals," Machine Design, Vol. 1, No 13, August, 1995, pp 103-7.

8. D. Duncan, "Surface Engineering Processes," Technology, Vol. 3, No. 8, pp. 192-202

Example of an Informal Report

MATERIALS ENGINEERING LABORATORY

REPORT NO. 20178

To: R. Aponte, Web Support Department, 3/12/BP Date: 7/24/98

Subject: Reactor Seal Failures

From: K. G. Budinski, Materials Engineering Laboratory 2/10/BP

Contributors: M. Kohter, C. Moczek

Problem

The Manufacturing Support department is experiencing premature agitator seal failures of 12 DTA reactors. Seals are lasting only about 6 months resulting in an annual maintenance cost in excess of $250,000. The Materials Engineering Laboratory was requested by Manufacturing Support to investigate new seal materials that would produce a seal life of 2 years as opposed to the average 6 month life that presently exists.

It is the purpose of this report to:

- present laboratory test data on screening candidate materials for the seal couples
- give the results of prototype tests on one of the new couples using self-mated cemented carbide face seal

The objective of this report is to obtain additional funding to continue tests.

Investigation

Procedure. Metal-to-metal screening tests were conducted on carbide seal couples using the crossed-cylinder wear test per ASTM G-83 (speed 0.22m sec, normal force 200 N, total sliding distance 20 km). The crossed-cylinder test was selected because it produces significant wear in hard materials in a relatively short test time. The test method places a rotating cylindrical pin at 90° against a stationary pin (Fig. A3.1). A "divot" or worn area is removed from the stationary pin, and a groove is formed on the rotating pin. The wear volume on both members is measured from mass changes during the test. The wear volume on both members is measured from mass changes during the test.

The test couples included hardened steel, ceramics, and cemented carbides. Three replicate tests were conducted on each couple and the average wear volumes were used in comparison graphics. The test parameters were:

- speed = 0.22/sec,
- normal force = 200 N,
- test duration/sliding distance = 2h - 20,000m.

Test Results. The wear test results (Fig. A3.2) indicated that a number of candidate couples had lower system wear than the existing couple, self-mated type 440C stainless steel. The lowest wear rate in the screening test occurred with self-mated, Type-C2 cemented-carbide couples.

Discussion. Although self-mated cemented-carbide couples had the lowest wear rate in the screening tests, these prototype seals were made using a brazed carbide cladding (2 mm thick) on the flexible member of the seal. It was determined that this cladding process was not suitable for obtaining a carbide surface on the flexible seal member. It developed radial cracks and chipping in the finishing operation.

Summary

This project has identified several seal couples that show promise of significant improvement in service life over the present couple of self-mated Type-440C stainless steel. Brazing of the carbide material was not suitable for the flexible backing, but we have a standby plan to develop a second carbide prototype based on deposition methods. This approach has a good likelihood of success, and additional funding of $40,000 is requested for a pilot-plant test of another material candidate and prototype. An elapsed time of four months is anticipated to complete all work up and including installation of one seal in a production reactor.

Attch: Fig. A3.1 and A3.2

cc: A. Black

    D. Chattergee

    R. Roads

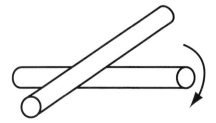

Fig. A3.1 Cross-cylinder wear test

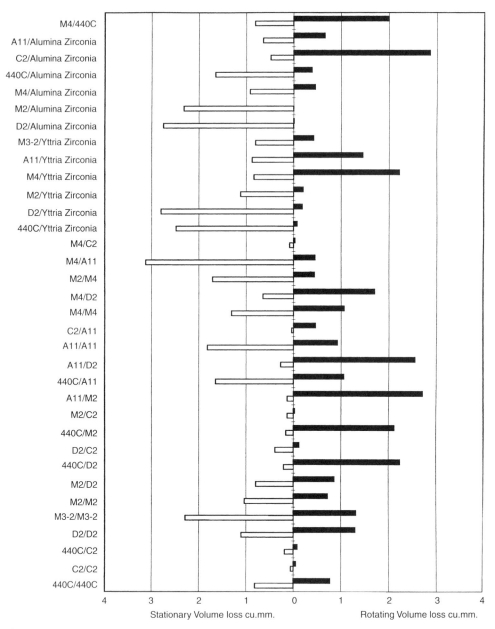

Fig. A3.2 Wear characteristics of various candidate couples; wear of rotating and stationary members in crossed-cylinder tests

Example of Visual Aids
for a Verbal Presentation

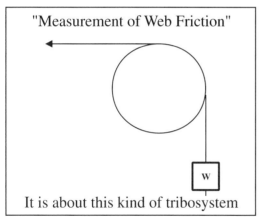

"Measurement of Web Friction"

It is about this kind of tribosystem

Overhead 1

Purpose of Study:
Simulate web transport with friction test

Objectives:
- **Improve control of conveyance processes**
- **Eliminate conveyance problems**

Overhead 2

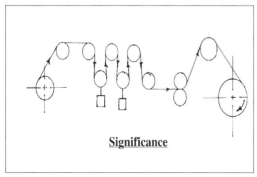

Significance

Overhead 3

Presentation Format:

1. **Fundamentals of capstan friction**

2. **Development of a standard apparatus**

3. **Effect of testing variables**

Overhead 4

Basic Elements of Capstan Tester

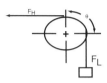

$$\frac{F_H}{F_L} = e^{\mu\theta}$$

F_H = Higher tension
F_L = Lower tension
 μ = Coefficient of friction
 θ = Wrap angle

Overhead 5

Test Set Up

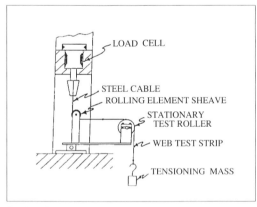

LOAD CELL
STEEL CABLE
ROLLING ELEMENT SHEAVE
STATIONARY TEST ROLLER
WEB TEST STRIP
TENSIONING MASS

Overhead 6

Conditions of capstan friction test

· Test at room humidity
· Normal force, 1 lbf
· Web width of 1 in.
· Velocity, 6in./min.

Overhead 7

Effect of temperature

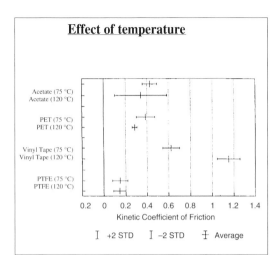

Overhead 8

Test Variability:
• Coefficient of variation
• Within-lab = 6.5%
• Between-lab = 9.4%

Process Capability:
• Day-to-day = 0.07
• Single day 0.04

Overhead 9

Conclusions:

• The capstan test is a reasonable simulator of a web conveyance tribosystem.

• Its use is advocated for applications with a reasonable match with this tribosystem

Overhead 10

Some Meanings of Questionable Phrases

THE FOLLOWING table was adapted from *A Glossary for Research Reports* by C.D. Grahan, Jr., which appeared in *Metals Progress,* Volume 71, Issue 5, May 1957 and is reprinted with permission:

Phrase	Meaning
"It has long been known…"	I have not bothered to look up the original reference.
"Of great theoretical and practical importance…"	Interesting to me
"While it has not been possible to provide answers to these questions…"	My experiment did not work out, but I wanted to publish it anyway.
"Extremely high purity"	Composition unknown except for exaggerated claims of the supplier
"Three of the samples were chosen for detailed study."	The results on the others did not make sense and were ignored.
"Accidentally strained during mounting"	Dropped on the floor
"Handled with extreme care during experiments…"	Not dropped on the floor
"A fiducial reference mark on the specimen"	A scratch
"Although some details have been lost in reproduction, it is clear from the original micrograph that…"	It is impossible to tell from the original micrograph.
"Typical results are shown."	Best results are shown.
"Agreement with the predicted curve is:	
excellent…"	Fair
good…"	Poor
satisfactory…"	Doubtful
fair…"	Imaginary
"Correct within an order of magnitude"	Wrong
"It is believed that…"	I think
"It is generally believed that…"	A couple of other guys think so too.
"It might be argued that…"	I have such a good answer for this objection that I shall now raise it.
"It is clear that much additional work will be required for a complete understanding of…"	I did not understand it.
"Thanks to Joe Glotz for assistance with the experiment and to John Doe for valuable discussions."	Glotz did the work, and Doe explained it to me.

Grammar and Punctuation

GRAMMAR refers to the rules of language. It includes the forms of words, word usage, and the arrangement of words in sentences and sentences into paragraphs. In many cases, word usage and rules are from earlier languages. Some English words are the same in French (resume and repertoire); some have Germanic origins (stein). Others are Italian (pizza and spaghetti), and many have Latin origins. The word belligerent comes from the Latin "bellum gero," which means to wage war. Sometime long ago in antiquity, rules were established on how to put words in sentences and that words have different forms depending on usage. Latin has forms of pronouns and rules on placement of modifying words. It has genders for words and singular and plural forms of words. This is the basis for many rules of grammar.

Throughout this book, it has been assumed that you, the reader, are aware of the basic rules of English grammar. Nonetheless, this appendix reviews some basic rules to help you achieve good grammar in report writing. Topics include:

- Parts of speech
- Parts of a sentence
- Constructing reasonable sentences
- Constructing reasonable paragraphs

A6.1 Parts of Speech

The English language contains eight types of words or parts of speech.

- Nouns—places, persons, and things that are subjects in sentences
- Pronouns—words that take the place of nouns
- Verbs—action words
- Adjectives—words that modify nouns
- Adverbs—words that modify verbs
- Conjunctions—words that join sentences or phrases
- Prepositions—words that connect nouns or pronouns
- Interjections—exclamation words like "help," "wow," "stop," and so forth

Every sentence must contain a subject (noun) and predicate (verb). Pronouns and verbs also have several forms that should not be mixed within a sentence or a series of sentences.

Pronouns are defined as first-, second-, and third-person pronouns with several forms of usage as the subject, predicate, or a modifier (adjective) within a sentence. The basic singular and plural forms of pronouns are as follows:

Person	Subjective	Objective	Possessive
Singular Forms			
First	I	Me	My, mine
Second	You	Your	Your, yours
Third	He, she, it	Him, her, it	His, hers, its
Plural forms			
First	We	Us	Our, ours
Second	You	You	Your, yours
Third	They	Them	Their, theirs

Make sure that you use the same person throughout a sentence and a related sequence of sentences.

Mixed Person

I have been conducting these tests for many years, and *we* feel that the drum test is most appropriate.

Person Agreement

I have been conducting these tests for many years, and *I* feel that the drum test is most appropriate.

Verbs have different tenses. Tense pertains to when the action happened. Select a tense for action words and be consistent in the use of tense in a section.

Tense	Example
Present	Complete
Past	Completed
Past perfect	Had completed
Present perfect	Has completed
Future	Will complete
Future perfect	Will have completed

Like pronouns, the tenses of verbs should not be mixed in a sentence (or even in a series of sentences, if possible):

Mixed Tense

We *completed* the project, and we *will dispose* of the test samples.

Tense Agreement

We *completed* the project and *disposed* of the test samples.

Adjectives and adverbs are modifiers. When multiple adjectives and/or adverbs are used as a modifier, the words need to be connected together by a hyphen. These multiple-word modifiers are known as compound adjectives:

Examples of Compound Adjectives

The high-alloy structural steel was used for the high-strength container.
The strain-rate sensitivity can be evaluated from the shock-wave test results.

Adverbs used in combination with adjectives generally are not hyphenated:

Hyphenation not Required

The overly eager engineer was still not familiar with the complex design procedures.

Adverbs often act as a modifier of an adjective, and so hyphenation is unnecessary. This is the case in the above example. Also note that the words "complex" and "design" are not hyphenated either. The word "complex" describes the procedure, not the design. Be very careful when you hyphenate. Excessive use of hyphens can be a distraction and even confusing. Just remember that the hyphenated words must act together as a distinct adjective.

Conjunctions (and, but, or, nor, yet, and so forth) can be used to connect separate sentences or independent clauses. The word "because" may also act like a conjunction, although it implies a closer connection between two thoughts. For example, consider the following two sentences:

> Sentence 1
>
> The work on the slitter was delayed, and the support staff turned their attention to the cinching process.
>
> Sentence 2
>
> The work on the slitter was delayed, because the support staff turned their attention to the cinching process.

The sentence has the same basic structure, but the conjunction defines the form of the connection between two thoughts.

Prepositions (about, above, after, among, at, before, by, for, into, in, like, on, since, of, because of, until, and so forth) should be used to connect a subject to a phrase. For example, consider the following examples of the same basic thought to the rest of the sentence.

> Version 1
>
> This project was closed because of a low ranking after review.
>
> Version 2
>
> This project was closed due to a low ranking after review.
>
> Version 3
>
> This project was closed after review.

Each version is adequate and correct. However, the following version is not good:

> Improper use of the Word "Since"
>
> This project was closed, since the review process gave it a low ranking.
>
> Preferred
>
> This project was closed, because the review process gave it a low ranking.

The preferred use of the word "since" as a preposition should convey the passage of time like some other prepositions such as "after" or "until." In the preceding example, the real meaning is to convey a cause-and-effect relationship between two events. With statements about cause and effect, use the word "because." Since should be used as a preposition to convey passage of time:

> Preferred use of the Word "Since"
>
> Since the last team meeting, several organizational changes have delayed progress.

A6.2 Sentence Construction

The building block of any document is the sentence. It is a complete thought, and it requires a subject (what the thought is about) and a predicate (what action occurred or will occur). The subject can be a noun, pronoun, or a complex group of words that define a person, place, event, or thing:

Simple Noun as the Subject

The beaker was agitated.

A Complex as the Subject

Titration by acidification and neutralization is the normal way to test for glucose.

The subject (and predicate) can have modifiers, but they remain as the basic elements of a sentence. Sometimes, in technical writing, you can get so engrossed in the technical terms that a subject or predicate is omitted. The result is an incomplete sentence in need of correction:

Incomplete Sentence

With a chemical composition of 0.1%C, 4.0%Mn, 2.7%Si, 0.8%P and vacuum ladle degassing.

Modifying adjectives, adverbs, and clauses should be placed as close as possible to the nouns and verbs that they modify. One of the most common errors in sentence structure is to start sentences with modifying clauses. These clauses may not technically violate rules of grammar and they may modify a noun or verb. They usually, however, add unnecessary words and detract from the clarity of the sentence.

Awkward Sentence

As part of a study into the effect of paint slivers, we conducted tee-bend paint adhesion tests.

Preferred—Direct Statement

We conducted two tee-bend paint adhesion tests in our study of paint slivers.

Five words were saved in the preferred example, and the sentence has exactly the same meaning. It is almost always possible to move the clause into the sentence with fewer words. Make this a priority in writing.

RULE

Do not start sentences with clauses (at least try not to).

Also try to make your sentences proactive. Negative sentences may not be grammatically incorrect, but they usually add words, and a negative tone is always undesirable.

Negative Sentence

The use of low pH solutions turned out to be a wrong direction in our study.

Preferred—Positive Statement

Our study indicated not to use low pH solutions.

Active and Passive Voice. Sentences can be written in the active or passive voice. You must make the choice. It is almost always preferred to write in the active voice (when the subject does the action) rather than in the passive voice (when the subject is acted on). The active voice makes statements sound more authoritative, and, usually, writing in the active voice aids concision.

Passive Voice

Statistics were applied to the first group of experiments.

Preferred—Active Voice

We applied statistics to the first group of experiments.

In the preceding example, using the active voice leads to the use of a personal pronoun. As mentioned earlier, some journals prohibit their use. It may also be undesirable to use personal pronouns in formal reports. However, you can still write in an active voice without the use of personal pronouns. For example, the same example could be written as:

Active voice without personal pronoun

Procedure Description

Statistical analysis checked for any correlation of variables within the first group of experiments.

Procedure Results

Statistical analysis of the first group of experiments verified the predicted correlation.

Procedures are intended to do things and produce results. Therefore, try to say what procedures and processes do; this can give your document an active voice without resorting to personal pronouns.

However, the descriptive nature of most technical documentation is often easier to express in the passive voice. The passive voice means that you, as the writer, are making observations about what was done and what happened. This passive voice is sometimes unavoidable and indispensable in describing the methods and results of a technical study. Observation is an underlying foundation of the scientific method. If it is better to write in the passive voice, then so be it. This is why a passive voice is prevalent in many formal reports and technical papers.

Personal pronouns and an active voice are more common for informal reports. Saying "I did this" or "we did that" is a conversational tone that can be acceptable for informal reports to teammates, colleagues, or immediate superiors. Such a tone is easier to read and can still be objective without seeming too aloof. The recommendation is to use the voice that feels most appropriate for the writing situation at hand. However, try not to mix voices, tenses, or moods in a sentence or a section of a report.

Mixed Voice

We completed the hardness study, and the results were reviewed.

Preferred—Same Voice

We completed the hardness study and reviewed the results.

Active Voice without Personal Pronouns

Review of the hardness test results completed the evaluation and determined specification conformance.

Mixed Tense

We will buy a new microscope, and we bought a polarizing stage as an accessory.

Preferred—Same Tense

We will buy a new microscope and an accessory polarizing stage.

Mixed Mood

I will take action on the budget overrun, but it would be well to consider review of the entire budget process.

Preferred—Consistent Mood

I will take action on the budget overrun and review the entire budget process.

Parallelism in a sentence means that words and sentences should have similar forms if they are dealing with the same subject.

Lack of Parallelism within a Sentence

The stylus profilometer was evaluated for suitability, and we purchased one for use on paper.

Preferred—Parallel Structure

The stylus profilometer was evaluated for suitability and purchased for use on paper.

Lack of Parallelism between Sentences

Tensile tests on the problem metal indicated a lower than normal yield strength. Because yield strength is a critical parameter, we rejected the lot.

Preferred—Parallel Sentences

Tensile tests on the problem metal indicated a lower than normal yield strength (a critical parameter). The lot was rejected.

If you are writing a particular way, do not change it within a sentence or between sentences. Use the same voices and tenses.

A6.3 Paragraphs

Paragraphs are groups of sentences that express a thought. When you change thoughts or subjects, start a new paragraph. There are, however, guidelines on the makeup of paragraphs. The first guideline is to not make the paragraph too short. There is a rule in the English language to designate a new paragraph by indenting the first word of the new paragraph, or in block format, by skipping a line between blocks of writing. This text uses the former, but both techniques are acceptable.

Paragraphs that are too short, like one sentence, do not look or read well. It suggests that your thought process is vacillating. You can only hold a particular thought for one sentence. Some sentences are paragraphs in themselves; this is common in legal writing. A long sentence is a complex sentence. Complex sentences are hard to read and prone to misinterpretation. Keep sentence length reasonable. There is no set definition of reasonable length, but if you have one go on for more than two typed lines, you are taxing the limit. Likewise, long paragraphs appear foreboding to the reader. It suggests rambling. The maximum length for a paragraph is arbitrary, but a length from ¼ to ½ typed page is a reasonable guide.

Parallelism is also a concern in paragraphs. It is not preferred practice to write a few sentences in the active voice and then switch to the passive voice. The same applies to tense, mode, and style. As with sentences, try not to change the way you are writing in the middle of a paragraph. It is annoying when people do this in conversation. A person will be talking about a baseball game and drift into a discussion of his or her lawnmower. All have encountered this. This style of writing can evoke less than a favorable reaction from readers. Thought drifting is especially undesirable in technical writing, because technical writing is usually asking for action on the part of the reader. You may not get what you are seeking.

The last issue is composing transitions in paragraphs. Stand-up comedians and public speakers work at mastering transitions from one subject to another. This is part of effective speaking. This applies to technical writing as well. When you move from one paragraph to another, try to use transitional words. If you are discussing a particular chemical reaction, let the reader know if the following paragraph is still concerning this reaction or you are moving on to another subject.

Use of a Transition Sentence

The corrosion of cemented carbide in electrical discharge machining dielectric was assessed by 30 day immersion tests in agitated dielectric. Four replicate polished samples with dimensions of $10 \times 30 \times 40$ mm were immersed in 250 ml of dielectric at room temperature. The weight of each sample was measured and recorded prior to immersion. The mass change in testing was determined by weight change measurements and inductively completed plasma atomic mass spectroscopy for tungsten and cobalt in the corrodent.

Additional corrosion information was obtained by optical microscopy of adjacent surfaces for pitting and other forms of surface damage. Pitting was quantified by atomic force microscopy of pit depth in ten, 20-micrometer square fields on each sample. Average pit depths were compared to show corrosion differences in the three candidate cemented carbides in testing.

The first sentence in the second paragraph lets the reader know that this paragraph is a continuation of the corrosion test procedure. This is the transition statement.

A6.4 Punctuation

Every language has punctuation symbols, which are the standardized marks in writing to separate or connect letters, words, or sentences. The most common punctuation marks in English are:

Symbol	Name	Purpose
.	Period	To end sentences that make a statement, in numbers
?	Question mark	To end a sentence that asks a question
;	Semicolon	To end a sentence that is closely related to the subsequent sentence
:	Colon	To introduce a list or formal statement
,	Comma	To separate sentence parts, to clarify
" "	Quotation marks	To denote a quotation
!	Exclamation point	To end a bold sentence or statement
—	Dash	To separate parts of a sentence (made from two hyphens)
-	Hyphen	To join related modifying words
( )	Parentheses	To add a statement, reference, or clarifying term to a sentence, for mathematics
[ ]	Brackets	To denote reference numbers
'	Apostrophe	To denote possession

Other languages have their own standardized marks. Although anyone writing a technical document should be familiar with the use of punctuation, this section reviews the basic uses of punctuation.

Use of Periods

To denote the end of a sentence that makes a statement

The project was completed on time and within budget.

To end an indirect question

We were asked if we could work overtime on this project.

To end a polite command

Please exit to the left.

To end abbreviations that could be read as a word

Fig. = Figure

Decimal point in numbers

35.007 inches

Use of Question Mark

To end a sentence that asks a question

Have you completed your weekly report?

Exclamation Point

To end a sentence or word that expresses emotion

Eureka! I found it.

Factorials in statistics

$5! = 5 \times 4 \times 3 \times 2 \times 1$

Use of Semicolon

To separate closely related sentences

Quench hardened steels have microstructures composed of martensite and retained austenite; the amount of each depends on the steel and quench rate.

The samples were placed in the ultrasonic cleaner; acetone was used as the cleaning solvent.

To separate listings with lots of commas

Alloy A: 0.2-C, 4.0-Cr, 7-Ni; Alloy B: 0.1-C, 5.0-Cr, 6-Ni

Use of Colons

Introduction of a list within a sentence or following a sentence

Three chemicals are candidate solvents for the coating: acetone, methylene chloride, and toluene.

Introduction of a formal statement

Summary:

The experiment did not produce the expected result of time dependence. However, it did illustrate that the rate of dissolution was significantly reduced by the presence of bromine ions. This knowledge will be applicable to the phase II work involving…

A letter salutation

Dear Mr. Jordan:

To introduce a quotation

The project manager said: "The project must be completed by November 30 or we will lose the Borden contract."

Use of Commas

To set apart an introductory statement from the main message of a sentence

During the corrosion test, the monitoring device inadvertently shut down.

Before a conjunction (and, but, however, etc.) linking complete sentences

This project produced significant value for the company, and it served as a model for project execution.

The test failed, because the monitoring device shut off.

For large numbers

We had over 66,000 attendees during the exhibit.

To separate items in a list within a sentence

This project will investigate rate dependence, temperature dependence, the role of additions, and optimum concentrations.

For clarification

Thirty, 12-person subgroups were formed.

To separate parenthetical expressions or clauses

The jig grinding machine, a new K4000 computerized model, was moved from the laboratory to the prototype center.

To separate transition words (however, consequently, for example, in fact, etc.) within a sentence

The project results were valuable; however, the test procedure did not address all the objectives.

To separate multiple modifying words

Rapid, cryogenic freezing of the specimens produced quick, precise, and accurate microtome results.

To separate dates, addresses, and titles

January 4, 1987

Rochester, New York, USA

H. Shapiro, UP

R. Schwartz, Director of Research

Use of Quotation Marks

To separate a quotation within a sentence

George Bernard Shaw once stated: "The only way to keep from being miserable is to never have enough free time to wonder if you are happy."

To denote a word, or phrase that requires special attention from its normal usage in a sentence

The "scum" on the bottom of the vessel was supposed to be a white, fluffy precipitate.

To refer to a title in a sentence

Roger Noor's book, "The Analasticity of Viscoelastic Solids," is a significant work on the properties of rubbers.

To denote the use of nonstandard work/phrase

The workers on the shop floor termed management's raise schedule "un grande mess."

Use of Dashes

To add a change in thought in a sentence

The vessel emptied in record time—without air pressure.

To shift tone in a sentence

We tried to resolve our disagreement amicably—but I never forgot it.

To make a separation more distinct than a comma

Fundamental—yet misunderstood—concepts are the reason why…

Use of Hyphens

To connect words as a compound adjective

The eight-foot ladder was used to paint the ceilings.

To separate words in a compound noun

Twenty-nine is a number, and it should be spelled out when it is the first word in a sentence.

To separate repeated letters in compound words

We had to re-examine our results.

To break up words (at syllables) that do not fit in a line

The length of the coating chamber was extended to the anticorrelation department.

Use of Parentheses

To separate a clarifying word or statement in a sentence that is not part of the central theme of the sentence

The supervisory ranks (director and up) were evaluated for effectiveness in the third and fourth quarter.

To direct the reader to an illustration

The corrosion rate increased monotonically for the duration of the test (Figure 3).

To enclose numbers in lists

Three test samples were submitted: (i) as received, (ii) surface hardened, (iii) hardened , and (iv) oxide treated.

Use of Brackets

To enclose reference attribution numbers

Charmers [3] performed this experiment with recycled aluminum scrap.

To separate items already in parentheses

The situation in the shop division (welding, sheet metal [light], and piping) needs to be resolved.

Use of Apostrophes

To denote possession

Mark's hammer was found on the truck fender.

To make contractions

I didn't do it.

Summary

This Appendix briefly skims some basics on English grammar, which is more fully covered in entire books on the subject. For purposes of engineers writing reports, writing an accurate and useful report on a job is more important than having it grammatically perfect. All people with English as their main language have been subjected to the rules. In most cases, they have been drilled into us in grammar school. The few grammar items discussed in this Appendix reflect common grammar flaws that have been observed in technical writing.

This Appendix concludes with some recommendations of additional books on writing. Some general suggestions are:

- Refrain from use of slang expressions and figures of speech.
- Keep one tense and voice in sentences and paragraphs.
- Use verbs that convey action or results by the subject in a sentence (active voice).
- Use the passive voice in formal reports when you, the writer, want to describe your own work.
- Use passive voice in summaries and abstracts.
- Refrain from use of personal pronouns in formal reports and published papers.
- Delete unnecessary words; strive for a concise document.
- Try to write positive sentences.
- Try to start paragraphs with a word or sentence that reflects the topic of the paragraph.
- A plural verb must be used for a plural subject.
- Avoid run-on sentences and sentences with no separation (semicolon or conjunction) between complete thoughts.
- Be consistent in style, tone, voice, tense, and so forth in a document.
- Use the past tense for work done in the past.
- Try not to start sentences with a modifying clause.

Important Terms

- Subject—what a sentence is about
- Predicate—the verb and its modifiers in a sentence
- Noun—a person, place, thing
- Verb—a word denoting action
- Clause—a group of related words with a subject and predicate
- Sentence—a complete thought
- Adjective—a word that modifies a noun
- Adverb—modifies a verb
- Singular—referring to only one
- Plural—referring to more than one
- Passive voice—the subject is acted upon (The boy was passed.)
- Active voice—the subject does the action (The boy ran fast.)
- Tense—indicated present, past, or future time
- Imperative—words giving direction
- Antecedent—precedes
- Gender—male, female, neuter
- Personal—type of pronoun (pertaining to people)
- Infinitive—clause or verb starting with "to"
- Gerund—the "ing" form of a verb
- Possessive—belonging to
- Pronoun—a word used in place of a noun (He walks to work.)
- Personal pronouns
  —singular first person: I, mine, me, my
  —second person: you, your
  —third person: he/she, him/her, his/hers
  —plural: we, us, ours, yours, they, them, their

To Dig Deeper

- C.T. Brusew, C. J. Alred, and W.E. Alin, *Handbook of Technical Writing,* St. Martin's Press, New York, 1993
- J. Feierman, *Action Grammar,* Fireside Publications, New York, 1995
- L. Rozakis, *The Random House Guide to Grammar, Usage, and Punctuation,* Random House Inc., New York, 1991
- W. Strunk, Jr. and E.B. White, *The Elements of Style,* Allyn and Bacon, Needham Heights, MA, 1979

Additional
Report Mechanics

GOOD PREPARATION of a document ensures reports with proper appearance and preferred presentation techniques. Additional guidelines discussed in this appendix are:

- Page layout
- How to handle abbreviations
- Capitalization
- Contractions
- Indentation
- Italics
- Numbers
- Page numbers

A7.1 Page Layout

An important part of technical writing is to have the product look professional and, more importantly, be readable. Appendixes 2 and 3 are examples of formal and informal reports, respectively. If you are writing a paper for a journal, the format must follow author instructions for that journal.

If you are writing a formal report that will reside in your organization, the cover sheet should contain pertinent filing information: title, date, report number, author, contributors, key words, and revision date (if put in a library). Do not address the report to an individual. Use a distribution list on a detachable cover sheet. (See Fig. 9.5 in Chapter 9, "Formal Report: The Outline and Introduction.") Do not number the title page of a formal report. Start on page two and call it "two." This is accepted practice. You must decide on the font, type size, and placement of hierarchy of headings in the body of your report. Are you going to need three or four types of headings? How will you make heading levels distinguishable?

Informal reports should look like the ones in Appendix 1 or 3. Most word processing software systems allow generation of a template that can be consistent throughout an organization. This produces consistency in appearance and forces correct layout. Informal reports are usually only one page long, but they can be longer. Short or long, they must contain sections on introduction/problem description, investigation, results, summary (or conclusions and recommendations, if you have some).

A7.2 Abbreviations/Acronyms

Avoid using acronyms in a title. Acronyms should be defined at first use, but it is awkward to define a term in a title. [*I am categorically against acronyms. Nothing confuses readers more than an unfamiliar acronym. There is plenty of time in life to state complete titles or names of organizations. This is my opinion.*]

If abbreviations appear to be needed in a document, use them sparingly. The only exceptions are the many abbreviations used for physical units, such as the units specified by the Le Système International d'Unités (SI). Some common SI abbreviations are:

Measurement	SI Units	Non-SI
Distance	m, meters	ft, foot and in, inch
Time	s, second	h, hour and min, minute
Force	N, Newton	lb, pound [*slug is correct*]
Pressure	PA, Pascal	psi, pounds per square inch
Area	m^2, meter	in.^{2}, square inch
Mass	g, gram	lb, pound

The complete list of SI units and abbreviations is available in most libraries in publications such as the American National Standards Institute (ANSI) ZZ10.1-1973 or ASTM E 380 Metric Practice Guide, ASTM, West Conshohocken, PA, 1972.

A7.3 Capitalization

Capital or upper case letters were invented to draw attention to a word. The convention in the English language is to capitalize the first word in every sentence and names of people and places.

Words to Capitalize

Proper nouns

James Day, General Electric

The first word after a colon, but only if a complete sentence follows

The lab will be cleaned in a particular sequence: We will clean the corrosion lab, then the wear lab, then the metallographic area.

Institutions and Organizations (department, associations)

National Science Foundation, Shops Division

Specific places

Syracuse, New York, Materials Engineering Laboratory

Specific periods in time (but not seasons: winter, spring, summer, and fall)

October, The Industrial Revolution, The Great Depression, The Reformation, and Lent

Recognized ethnic groups, not economic or social groups (factory workers, economists, etc.)

American, Native Americans, Italians, Poles

The first word of the salutation and close of a letter

Dear Mr. Johnson:

Regards,

Sincerely yours,

Trade names

The bearing surfaces were separated by a one-millimeter thick sheet of a fluoro-carbon (Teflon®).

RULE

Never capitalize all letters in a document or computer message. All caps are difficult to read. Use upper and lower case.

A7.4 Contractions

Contractions are abbreviated words with an apostrophe replacing the deletion of several letters:

- Won't = Would not
- Can't = Cannot

Contractions should never be used in technical writing [*my opinion*]. They are essentially a form of slang, and slang expressions are often not appropriate in technical writing. Some technical writing teachers disagree with this. They believe that writing without contractions is stiff. Use the appropriate practice in your organization. Use contractions sparingly, if at all.

A7.5 Indentation

Indentation of five typed spaces can be used to denote a new paragraph. Some books are written that way. [*This is what I was taught 50 years ago in grammar school.*] Another way of separating paragraphs is to skip a line to start a new paragraph. This is a block page layout. This is often a more effective visual break than indenting, and it requires fewer keystrokes. Either way, you must create a visual break for the reader.

Indentations are appropriate for listed items:

- Lists of numbers
- Conclusions
- Recommendations
- Special instructions

A7.6 Numbers

There are other conventions that apply to the use of numbers in technical documents. In the United States, large numbers are denoted with a comma, for example, 3000, 10000, and 200000 should be written as 3,000, 10,000, and 200,000. Most people prefer commas in numbers over 1,000. In some countries, a comma is used in place of a decimal point. Keep this in mind in international documents. Other guidelines include:

- Never start a sentence with a number; spell it out.
- Write out numbers less than ten.
- If you use a variety of numbers in a sentence, be consistent in how you write them (even if they are lower than 10): "We have 460 parts in stock and only 4 are on backorder."
- Put a zero before numbers less than one.
- Do not add insignificant digits to numbers.
- Spell out numbers that are prone to misinterpretation: We ordered four 2×4 boards.
- Use numbers for decimals and fractions.

- Use decimals for complex fractions.
- Place a hyphen between the number and a word, when it is part of a compound adjective: "The 156-page report was poorly written and inconsistent."

A7.7 Fonts

Computer word processing programs have scores of fonts, but most are unnecessary or difficult to read for technical documents. Never italicize an entire document or use strange fonts. Acceptable fonts are those used in textbooks, newspapers, and magazines. Never use bold on large sections of text. Only use bold text for headings and similar text that needs to be set apart.

Type size also varies. Some word processing software packages comes with a 10-point type size as the default. This is too small for normal eyesight. Similarly, type size larger than 12-point type is too large for an entire document. Set the default to the font and type size of preference. An 11-point or 12-point type size is generally adequate. Type size depends on the font, but for the common fonts used in technical documents (like Times New Roman), 11- or 12-point type is easy to read.

A7.8 Appendixes

Appendixes serve a valuable function in technical reports. They can be used for supporting data that would clutter a report. This makes review of supporting data optional. For example, if you have six spreadsheets of data on a manufacturing process that serve as the basis of a report, append them to the report. However, do not even put them in the appendix if it is likely that one of the report readers will use the spreadsheets. Sometimes, customers of service organizations prefer to do their own statistics on test data, or they republish results in their reports. They need raw data for this. Append it.

Other Types
of Documents

THIS APPENDIX describes some writing hints on other types of documents besides reports. Examples include:

- Executive summary
- Meeting minutes
- Newsletters
- Resumes
- Patent
- Biographical sketch

A8.1 Executive Summary

An executive summary can be mandatory on big projects in some organizations. It is a synopsis of a formal report containing more detail than an abstract and including sections corresponding to the sections in the formal report. Essentially, each section is abstracted. The purpose of an executive summary is to digest your information for management. The philosophy behind this type of report is that managers are too busy to read a thirty-page report, but they need the information for decisions. The executive summary should be no longer than 10 percent of the formal report, and it does not contain graphs and illustrations unless one or two are absolutely essential to understanding the message in the report.

[I do not use executive summaries because my employer does not use them. However, my former employer used the following format for formal reports:

- *Title*
- *Introduction*
- *Conclusions*
- *Recommendations*
- *Body*

The introduction had to be very thorough. The combination of Introduction, Conclusions, and Recommendations constituted an executive summary. Busy executives read only the first three sections. An abstract was not necessary because the introduction contained most of the information in an abstract.]

Writing executive summaries is not endorsed here, but if you work for an organization that requires them, by all means use them. The following example can be used as a guide; essentially, you abstract each section and delete figures. The executive summary shown below condenses a 27-page report into 3 pages. Note that there are subheadings and closure matter (conclusions and recommendations). It is obvious that this type of document can be read more quickly than the full report. In some organizations, the executive summary follows the title page; others put it between the abstract and the report introduction.

Example of an Executive Summary

Tool Wear Crisis

October 5, 1998

Author:

K.G. Budinski, Materials Engineering Laboratory

Problem:

In the second quarter of 1998, a significant reduction in perforator tool life with Brand A film was compared with that of Brand B film. The manifestation of excess punch and die wear was excessive film dirt in the form of skiving or slivers that could lead to customer complaints. Finishing assembled a team to address this issue. It is the purpose of this summary to present the status of work conducted in the Materials Engineering Laboratory (MEL) aimed at improving tool materials to solve this problem. The objective of this document is to report where we stand as of this date in using improved tool materials to solve this problem. Work performed in five areas is discussed: 1. fundamental studies, 2. improved cemented carbide tools, 3. cermet tools, 4. tool coatings, 5. tool surface treatments.

Fundamental Studies:

In the first quarter of 1998, the Materials Engineering Lab initiated a project to simulate the type of tool wear that has been observed in production perforating punches and dies. A crater forms on the die top and on the end of punches. It starts about 40 µm away from the cutting edge and grows in depth and width until it consumes and deteriorates the cutting edges. MEL developed a ball-impact test that simulates the fluid-like behavior of estar support under the compressive stress as encountered in perforating. It was learned that cratering around cutting edges is produced by erosion resulting from radial movement of film on the tool surfaces. Ball-impact tests on various tool materials suggested that the observed erosion included a material removal component from abrasion, adhesive wear, and chemical action from coatings on the films.

Improved Carbide:

All of the current perforator tools are made from cemented carbides which are tungsten carbide particles held together with a cobalt binder (from 6 to 15 percent). Because the ball-impact test suggested that there was a component of corrosion in the mechanism of material removal on perforator dies, it was decided to investigate replacing cobalt-binder carbides (C101) with a grade with a more corrosion-resistant nickel/chromium binder (C805). Corrosion tests were conducted on both grades by immersion in backside chemical and emulsions and by dry film contact tests.

These tests indicated that the C805 grade had significantly better resistance to these chemicals than the current cobalt binder carbide, C101. We also conducted abrasion tests to compare the abrasion resistance of C805 to C101. C805 demonstrated 20% lower abrasion resistance than C101, but this was not deemed to be significant since we have successfully used high cobalt carbides in production. These materials had abrasion resistance almost three times lower than that of C805.

The remaining concerns regarding C805 as a replacement for C101 are edge chipping and corrosion of polished die tops in the electrical discharge machining (EDM) cutting of die holes. Tests performed on the MEL nibbler indicate microscopic edge chips on a tool tested for 10^6 nibbles (emulsion up). The die top corrosion does not occur with the old tooling. We have not identified the reason for this corrosion or the solution, but it may be solvable with EDM machine controls (antielectrolysis).

Cermet Tooling:

Concurrent with our efforts to make tools from a corrosion-resistant carbide is work aimed at making tools from a steel-bonded carbide or cermet. There is currently only one cermet under test, S-40, which is composed of 45 percent by volume titanium carbide in a type 440C stainless steel matrix. Test sample material is on hand and corrosion and wear test samples are in various stages of machining or polishing. It is too early to comment on how this material will compare with C101 and 805. Corrosion and nibbler tests will be done by year-end.

Coatings:

A possible tool wear solution is to apply a thin, hard coating to the existing tools. A robust coating will add wear resistance and possibly corrosion resistance. Current candidate coatings include titanium nitride, titanium carbonitride, and tungsten carbide/carbon. The potential problems with the use of coatings are: 1. We cannot use thick coatings (greater than 1 μm) because of possible interference with punch-to-die clearances. 2. Many PVD coatings form nodules in deposition that could scratch film. 3. Coatings must be tested for adhesion. All of these problems are solvable, but it may take until the end of the year to select a suitable coating and supplier. This effort is proceeding.

Surface Treatments:

There are some surface treatments that may improve erosion resistance of tool surfaces without adding dimension, for example, ion implantation, diffusion treatments, and cryotreatment. We are considering two processes for this tooling problem: 1. Use of as-EDM surfaces, 2. The Toyo diffusion process (TD). Nibbler tests have been performed on the former with good results and a set of C-perforator tools has been TD treated and sent to B58 for production tests. This process diffuses vanadium into the carbide to produce a vanadium carbide skin about 10 μm deep that is very abrasion resistant.

Conclusions:

1. C101 produced more erosion than C805 in MEL ball-impact tests with Supra film.
2. C905 has better resistance than C101 backside chemical and emulsions.
3. C805 may have problems with die top corrosion in EDM and cutting edge chipping that need to be solved.
4. We do not have the data at present to decide on the feasibility of cermets, coatings, and surface treatment solutions.

Recommendations:

1. Continue fabrication of production tooling from C805 with emphasis on getting test punches and dies running as soon as possible.
2. Survey the wear of the C805 tools by removing a punch and die for MEL testing after reasonable production increments.
3. Start fabrication of cermet tools for production testing.
4. Coat a quantity of 20 punches and dies with 1 μm titanium nitride and test in production if coated surfaces are nodule-free.
5. Fabricate 20 punches and dies with fine finish EDM surfaces (<5 microinch roughness) and test in production with periodic wear surveys like #2.
6. Continue the search for other alternate carbide grades with MEL nibbler and abrasion tests.

A8.2 Meeting Agendas

A meeting agenda, by definition, is a list of topics to be discussed. A meeting without an agenda is usually an unproductive meeting. Like any task, you need to focus, work at it, and complete it. An agenda should serve as a plan, and it should contain the following elements:

- Subject
- Time
- Location
- Invitees
- Name of person who called meeting
- Expectation
- Specific items to be discussed (a list with times)
- Closure (conclusions, actions, recommendations, etc.)

Most published agenda contain elements 1 to 5 and 7, but expectations and closure are often ignored. If you state what you hope to achieve in the meeting before discussions start, the chances of staying on track are much better. If you are calling a meeting to select a motor/gearbox for a new machine under design, state this. Introductory remarks should also make it clear that your expectation is to leave the meeting with a size, type, and supplier for the drive system. The meeting closure is the documentation on what will be purchased.

Figure A8.1 shows an agenda with all the suggested elements. Creating an agenda takes very little writing and very little work, but the payback can be significant. A published agenda distributed to all invitees well ahead of the meeting (a week is reasonable) is recommended. If you want people to come prepared on certain subjects, state this. Sometimes, for continuing team meetings, you may not be aware of what various team members have to discuss at the meeting. In this instance, send out an e-mail asking for agenda items from team members. As the meeting leader, it is your

MEETING AGENDA

Date:	March 27, 2000	**Time:**	9:00 A.M. to 10:00 A.M.

Subject: New backside antistats

Location: Room 42, Building 43, Main Office

Invited: R. Brad, P. Know, R. Tyse, J. Tres, M. Monsel

Called by: Ralph Swartz, PCDM

Expectations:

1. Decide on test plan for co-extruded backing.

2. Decide on disposition of Chem 180.

Agenda Items:

1. D. Spring report on laser oblation of C coating (10 min.)

2. Support issues – Frank Klee (5 min.)

3. Coating of matte particles – Anne Worst (5 min.)

4. Plan for pilot tests – All (10 min.)

5. Review co-extrusion test data – All (5 min.)

6. Chem 180 – Bill Fink (10 min.)

7. Action items – All (5 min.)

8. Summary – Ralph Swartz

Fig. A8.1 Example of meeting agenda

responsibility to put these items into a coherent meeting with expectations and tangible results.

E-mail has made it extremely easy to publish agenda item requests and completed agenda. Use of this communications tool makes meetings more productive and usually less boring if you enforce topic time limits. This simple tool could make a significant improvement in your work life.

A8.3 Meeting Minutes

Figure A8.2 presents an example of minutes from a meeting. If the organization votes on issues like consensus standards, there are rules that may apply to recording minutes. For example, *Robert's Rules of Order* requires:

- Group, date, time and location
- Attendees, type of meeting
- Minutes of previous meetings read and approved or revised as approved
- Issues addressed
- Votes taken and vote count
- Description of actions
- Time and date of actions
- Secretary signature

Meeting minutes for a team meeting in your organization need not be as formal. However, if minutes are to add value to an effort, they need to document completed tasks and assigned tasks. They should serve as the record of the group of what transpired and as the to do list for members. Essentially, minutes should summarize what transpired in various areas. Make each subheading coincide with the topic discussed and summarize what transpired in each area. If you are assigned to writing meeting minutes, use these guidelines and make statements that pinpoint action items. Let minutes serve as a to do list as well as a historical record. Concision should apply to minutes. Try to summarize statements. Most team members do not appreciate minutes that take up six or seven computer screens.

A8.4 Newsletters

Formal meeting minutes can be long (and boring) to read and to listen to. One remedy is to write a newsletter that replaces the minutes. [*I have been writing the newsletter in Appendix 9 for more than 15 years.*] The format of the sample newsletter in Appendix 9 is more than just an abstract of what happened in the committee or task group meeting. The newsletter also summarizes the technical programs of the committee, future meetings, and miscellaneous news. Newsletters can be mailed to members of record in hard copy; it is also common to post them on web sites.

Minutes of the Meeting of Main Committee G02
Held on 7, May, 1998 at Atlanta Hilton, Atlanta, GA

The following members and visitors were in attendance: Robert Allen, Sharon Black, Ray Mayer, Peter Blain, Tim Summer, Ken Budinski, David Chambers, Frank Hirsch, Brian Camp, Maggie Johns (Chair), Floyd Woodson, Visitor: Lawrence Moore

I.   Call to Order
       The meeting was called to order by the chairman at 5:10 P.M.

II.   Circulation of Attendance List
       The chairman welcomed members and visitors to the meeting. The quorum requirements were checked and satisfied.

III.   Minutes of last meeting, held at ASTM Headquarters were approved.

IV.   Staff Manager's Report
       The Staff report was given in the Executive Meeting.

V.   Membership Secretary's Report
       The Membership Secretary's report was given in the Executive Meeting.

VI.   Vice Chairman Report
       The Vice-Chairman's report was given in the Executive Meeting.

VII.   Future Meeting Sites
       The future meeting schedule was presented in the Executive Meeting.

VIII. Report on Liaison Activities
       Liaison Activities were given in the Executive Meeting.

IX.   Subcommittee Reports
       A.   G02.10 Erosion
            No report was given
       B.   G02.20 Computerization
            G117 was approved. The new "Standard for Digital Data Acquisition in Wear and Friction Measurements" will be submitted for ballot.

Signed: _____

Fig. A8.2 Meeting minutes

Newsletters have whatever format their editors want. There are no rules in writing them other than the basic rules on style, tone, punctuation, and grammar. [*I have an aversion to clip art in newsletters, because it makes the document whimsical. If this is the intent, then it may be acceptable.*] As the author, decide on style and what the newsletter contains. Above all, make it useful. Make sure that it contains information that helps your readers in some manner.

A8.5 Resumes

Resume writing is, of course, a significant part of the employment scene. Effective resumes are described in a multitude of books, pamphlets, and career newsletters, and a systematic treatment is beyond the scope of this book. For detailed guidance, refer to resume writing books.

Figure A8.3 is a sample resume that reflects the popular style for a young person entering the workforce in the year 2000. Resumes are only one or two pages long, and they are customized for each employer. They show why the employer needs your particular skills. Most resume writing books suggest that a resume contain the following elements:

- Position sought/career objective
- Education
- Skills
- Work experience (last three employers with the most recent one first)

The purpose of a resume is to get you an interview so you can sell yourself in person. Too much or too little information on a resume may not get you an interview. Write everything that you feel is important. If it is four pages long, abstract and consolidate to two. You can say less about work experience further in the past. [*Yes, delete your four years as a newspaper delivery person in high school if you had three jobs since high school.*] However, be careful not to leave any gaps in the chronology of your career. Gaps always raise the question. What happened between 1992 and 1995?

Probably the most common problems with resumes are spelling, grammar, or punctuation mistakes. Needless to say, these errors shed a bad light on job prospects.

RULE

Always have your resume proofread by others.

Some additional hints from experts are:

- Do not use "cute" special paper—just quality bond.
- Do not use acronyms.
- Do not try humor.
- Use bullets where appropriate.
- Be concise.
- Use concrete examples.
- Try to demonstrate a fit to the advertised job.
- Use active, strong words.
- Be honest.
- Be consistent in style.

After you have written, rewritten, and had your resume proofread at least twice, then write a cover letter. This letter introduces your resume and demonstrates your writing skills. The letter should be brief but contain several paragraphs. It should contain these elements:

Stacey M
street address
city, state Zip

phone number

Objective

To obtain employment as a co-op (cooperative education) under the major of chemical engineering, preferably to work winter and summer quarters.

Qualifications

- **Leadership:** Active member of high school student government; cheerleading coach; management skillls in current place of employment.
- **Career Related Skills:** Great personal skills, hard working, self motivated, and quick learner.

Work History

Shift Supervisor-1995–present
Bruster
Opened and closed store, banking, scheduling, customer/store relations, and supervision of crew

Sales Representatitve
Upton's Department Store
Opened and closed store, experience on register, and floor sales.

Cashier, Sales Representative
Agway
Cashier, floor sales, arrangement of floor merchandise

Education

- **1997 East Cowet High School:** Advanced and AP Courses. GPA: 3.75
- **Present Georgia Institute of Technology:** expected graduation 2001

Volunteer Experience

Rotary Interact, Habitat for Humanity, Peachtree City Packers Cheerleading Coach

---

References

Ken Budinski, affiliation, address, phone

Fig. A8.3 Example of resume

- Address the letter to a specific person (usually in the personnel department).
- The first paragraph should state the job you are interested in and how you learned of the opening.
- The second paragraph should briefly highlight your qualifications that match their job description. Say something about the company, or in some way indicate that you know what they do. Why should they hire you?
- Refer to your enclosed resume and state how you wish to be contacted.

The following is an example:

Example Cover Letter for Job Inquiry

100 Fox Road
South Fork, Montana 12143

October 26, 1998

Ms. Joan Smith
Human Resources
Alfred Manufacturing
Alfred, Montana 12753

Dear Ms. Smith:

I am submitting my resume for your consideration for the metallurgist position that Alfred Manufacturing advertised in the June 1998 issue of Metal Progress.

I am currently a senior in Metallurgical Engineering at Montana School of Mines, and I believe that my interests and background match your job description. I am proficient in finite element analyses, C-language, and CAD as stated in your ad. In addition, I have always been a fan of Alfred tractors. I was raised on a farm where we have two of them. I loved pulling a 4-bottom plow with my dad's Alfred 610.

In summary, I appreciate your consideration for your metallurgist opening. I am sure that I can offer Alfred Manufacturing skills that will contribute to maintaining Alfred as the premier tractor manufacturer in the U.S. Thank you for your consideration.

Very truly yours,

(Signature)
(Typed Name)

Home Phone: 212-407-2314
School Phone: 212-406-4318
Fax: 272-401-3218
e-mail: rdf@cobalt.org

A8.6 Biographical Sketch

If you make a presentation to a technical society or other organization outside of your place of employment, you may be asked for a brief biographical sketch. The following is an example of a system consisting of three elements: present position, education, and expertise. There are no universally accepted standards for a biographical sketch, but in situations where your talk will be advertised in a bulletin or the like, you probably should keep the length to less than 200 words. State what you think the audience should know about you to make your message believable.

Example of a Biographical Sketch

Kenneth G. Budinski

Current Position

Technical Associate, Materials Engineering Laboratory, Kodak Park Division Eastman Kodak Company (1964 – present)

Education

- BS Mechanical Engineering, 1961 General Motors Institute, Flint, MI
- MS, Metallurgical Engineering, 1963 Michigan Technological University, Houghton, MI

Expertise

Materials selection for machine and product design; friction of solid/solid tribology systems, wear of tool materials, tribology testing, failure analysis

A8.7 Patents

A patent is a very special type of technical document. Patents may be the best archived and the most structured and defined type of reports. Patents can be good for seventeen years; they are available to all. They are kept available for searches indefinitely. They are the ultimate example of archiving and are permanently on file in the country in which the patent was issued. Patents are given a number that is unique; if you want to protect an invention worldwide, you need to get patents in other countries. There are laws surrounding the procedures for filings, and the rules for writing these documents are very detailed.

Patents disclose the details of an invention to the public. The granting of a patent allows the inventor exclusive rights to the invention for a period of years (17 years for a utility patent). The format and language of the document is specified by the issuing agency such as the United States Patent and Trademark Office (www.uspto.gov). Patents are very rigid in style and form to ensure readability by others. All patents look alike. They may look foreboding, but they are not difficult to read because the section headings are clearly marked. They contain a title section, references, an abstract, claims, and, in most cases, drawings. The basic contents of a patent include:

- A heading with filing statistics
- Prior art-related patents
- Assigned to (inventors/owner)
- Abstract
- Body containing field, background, summary of the invention, embodiment (what is the advantage), examples of the invention
- What is claimed (legal part drafted by an attorney)

Writing a patent is usually best left to a patent attorney, but anybody can write one. There are significant fees for the various steps in the process. These fees are forfeited if the document is rejected because it has not met filing or writing requirements. There are many references to guide you through the process of writing a patent if you are up to the challenge.

Every organization should have a strategy on when to apply for a patent. [*Our company, Bud Labs, has adopted the strategy to only patent inventions that are in some way associated with the corporate mission statement, which is to be the worldwide leader in the development and manufacture of tribology equipment.*] A general rule-of-thumb regarding the patentability of a process or device is that the invention must be new, useful, and not obvious (i.e., not shown to be true by common experience):

- A patent is shown to be new by searching issued patents. This step is required when you submit a patent request, and patent searches can be done electronically (e.g., see the web site for the United States Patent and Trademark Office at www.uspto.gov).
- You must demonstrate that the invention is useful. It must work and produce some desirable effect.
- A patentable invention must be nonobvious. This means that you cannot predict the outcome from common experience. For example, you design a light bulb that automatically dims and brightens from the ambient light. It is not obvious that a light bulb can do this.

In summary, writing a patent or a body of work is an option to consider, but make sure that it is in keeping with company strategy and it meets patent criteria. A patent lawyer should be contacted if a patent appears feasible and in the best interest of your organization.

To Dig Deeper

- R.C. Levy, *Inventing and Patenting Sourcebook,* 2nd ed., Gale Research Inc., Detroit, 1992

Example of a Technical Newsletter

WEAR NEWS

Vol. 2, No. 10

Meeting of ASTM G2 Committee, ASTM Headquarters, Nov. 5-7, 1997

Conference Reports

The Fall meeting of the G2 Wear and Erosion Committee was held at ASTM's new headquarters in West Conshohocken, PA. This newsletter usually begins with a report on our most recent technical program. There was no symposium or workshop in conjunction with the committee meetings, so we will begin with reports on some fall tribology conferences.

First World Congress in Tribology - K. Budinski

This may have been the biggest tribology meeting ever held. There were 410 formal papers and 480 poster papers. Attendees represented 51 countries and about 1000 people pre-registered. The conference spanned four and one half days with a fancy evening banquet in the middle. The conference was held at Westminister Central Hall which is directly across the street from Westminster Abby. Some sessions were held at the headquarters of the Institution of Mechanical Engineers, (I Mech E) one block from Central Hall and abutting the park that is part of the Buckingham Palace complex. Quite an interesting meeting location.

The Congress was sponsored by many organizations, but the organization that ran the event was I Mech E. The Westminster Central Hall location was chosen because 60 years ago, it was the site of a similar "significant" conference on wear. One person who attended the 1937 conference also attended this one.

There were five concurrent sessions each day so it was not possible to get the flavor of each session. The following list of sessions (and the number of papers in each), hopefully, will present a picture of what areas of research the organizing committee feels are important enough to warrant a session.

Contact Mechanics (20)

Wear and Friction Testing (36)

Lubricant Chemistry and Rheology (36)

Surface Engineering (36)

Solid Lubrication (20)

Thick film Bearings and EHL (36)

Wear in Metallic Systems (20)

Wear in Ceramic Systems (18)

Bio-tribology (20)

Rolling Bearings (18)

Wear in Polymer Systems (18)

Manufacturing and Maintenance (20)

Environmental Issues (8)

Tribology in Extreme Environments (8)

The conference seemed to be heavily attended by academicians. A review of the attendance list presented a surprise in that more than half of the U.S.A. attendees were from industry. The following is a list of the top ten countries in attendance and the percent age of attendees that listed a corporation as their affiliation.

Country	Number of Attendees	Percent from Industry
UK	256	33
US	180	57
Japan	134	18
France	50	8
Sweden	47	29
Germany	36	25
Russia	28	7
Poland	24	4
Yugoslavia	20	30
Switzerland	18	11

The Congress was opened by an admonition from Professor Peter Evans (the author of the 1960 report that lead to the word "tribology") to revise the current trend in tribology research towards contact mechanics and modeling practiced by many academicians. "Tribology must regain user participation."

Another "cosmic truth" in Professor Evans' remarks was that tribologists need to quantify the savings produced by their work and make them known to those who control purse strings. Politicians understand cost savings, not asperity contact models. Professor Evans' "scolding" was followed by one of Professor Duncan White's always-interesting reviews of tribology happenings. He cited examples of tribology advances in the last few decades and he claimed that they were comparable to the industrial revolution. Automobile engines that run a hundred thousand miles without an overhaul were non-existent in 1950. People in the U.S.A. now buy used cars with that many miles. We have done well according to Professor White.

In summary, the technical sessions at the First World Tribology Congress suggest that the focus of universities is in developing models and first principles that explain wear and friction observations. Industrial people are close-lipped about their tribology advances. There appears to be little progress in relating material properties to wear behavior. We cannot calculate wear life from measurable material properties. The calculations for fluid film thickness, however, are quite well-established for many tribosystems. Research in lubricated systems has moved on to the atomic and molecular scale; they are probing how lubricants adhere and how they separate surfaces.

The materials people seem to be looking towards ceramics, composites, and coatings for tribosystem improvements. There were no talks about steels, bronzes, cast irons and babbitts that are the workhorse of tribomaterials. Everybody is using expensive devices (AFM, SFM, SIMS, XPS, etc.) that poke at the outer 100 nanometers or so of surfaces.

These studies do not appear to be aimed at tangible near-term benefits to industry. Biotribology is a long overdue activity. Participation and funding is increasing and these efforts may lead to prosthetic joints that can last for a lifetime. Interest in molecular dynamics seemed to be declining compared to its introduction in the early 1990's.

Overall ownership of tribology appears to be in the hands of mechanical engineers. They are addressing wear and friction from the physics/mathematics approach. Material engineers and chemists are doing less research in tribology. This may be the reason why the UK is only realizing a fraction (1/3) of the 1.5 billion Pounds annual savings that they (Inst. of Mech Engr.) claimed that tribology research would yield when they founded the field in the mid 1960's. (The last sentence is a biased editorial comment from a material's engineer; Mea copa.)

US - Eastern Europe Conference on Nanotribology and New Tribological Materials - K. Budinski

This was a most unusual conference since papers were presented in several languages in three locations in two countries. The conference started on September 1 in Minsk Belarus; then moved by bus to Grodno Belarus about 100 km of Minsk, then on to Warsaw. About 50 participants went on the bus and additional participants showed up at the various institutes that were visited. Most of the 48 or so papers that were presented to the traveling group were given at the Belarus Academy of Sciences in Minsk. The touring participants also visited laboratories at a research institute in Minsk, the Polytechnic Institute in Byalstok Poland and at the Warsaw Polytechnic University.

The U.S.A. delegation (13 strong) was sponsored by the National Science Foundation (NSF) and included participants from government, industry, and academia. Each US delegate presented a paper. The papers from the entire conference will be published in the Journal of Friction and Wear (a C.I.S. publication). The Conference topics included surface engineering, tribochemistry, testing, monitoring, modeling, plastics, and biotribology. The research interests of the former USSR countries (C.I.S.) appeared to be aligned with those of the U.S.A. and U.K. However, the effort has been greatly reduced, since the breakup of the USSR. We were told that in 1990, there were over 2000 tribologists in the USSR. There are now between 600 and 1000 in about 150 tribology centers in Russia, Belarus, and the Ukraine. Belarus still has about 25 centers.

The stated purpose of the conference was information sharing between U.S. and the former USSR Countries in Tribology. The object was to improve joint research efforts and establish new ones. Traveling and taking meals with tribologists from seven former USSR countries certainly allowed us to become familiar with their work and vice versa. My overall impression of the tribology situation in the former USSR is that they have many competent researchers and reasonably well-equipped laboratories, but they are working with government funding that has been substantially reduced. Factories making military hardware no longer need tribology help. Research requests from the private sector have not reached expectations. It appeared that they were soliciting research funding from the U.S. One institute in Belarus that we visited offered to work on any materials/tribology or manufacturing problems. Labor costs are lower in Belarus and it may be possible to get a lot of value for research dollars. However, their border bureaucracy appeared (to me) to significantly reduce the potential for U.S. - Belarus alliances. A border crossing from Belarus to Poland may take 30 hours. It took us 3 hours even though our hosts had a letter from the agency chief asking them to give us VIP treatment. The governments of Eastern European countries need to learn how to be visitor friendly. I suspect that only then, will their research institutes become viable contenders for outside research dollars.

Leeds-Lyon Symposium, London - Peter Blacke

This Leeds-Lyon Symposium, held September 4-6 1997, was the 24th in a series of late summer meetings which normally alternate between Leeds University in England and two universities in Lyon, France. In Lyon, the meeting has been held under sponsorship of Ecole Centrale de Lyon and INSA (National Institute for Science and Applications). Due to the World Tribology Congress (WTC), the venue was moved from Lyon, whose turn it was this year, to London. In London, agreement was made with Professor Brian Smithe to host it at Imperial College the week before the WTC. Each conference in the L-L series carries a theme related to tribology. This one was "Tribology for Energy Conservation." There were 139 registered attendees of which 10 were from the USA.

Throughout the conference, there were seven invited lecturers and 12 parallel sessions of contributed papers. The invited papers were as follows:

"The Role of Tribology in Life Cycle Design," Prof. R. Doe, University of Surrey, UK

"Energy losses by Friction and Wear /Improved Fuel Economy," Prof. W. J. Browne, Technical Academy of Essengler, FRG

"Energy Conservation Through Friction Reduction," Dr. S. Smithe, Ford Motor Company, USA

"Challenges in Tribology Posed by Energy Efficient Technology," Dr. H. Whitee, Imperial College, UK

*Energy Conservation Through Extended Component Life," Dr. M. H. Joones, University College of Swansea, UK

"The Impact of Tribological Issues on Energy Conservation in Metal Forming Operations," Dr. J. H. Smithe, University of Sheffiel, UK

The technical sessions comprised fifty-four contributed presentations. The sessions were: Friction Reduction (2 sessions), Lubricants (2 sessions), Wear, Hydrodynamics, Elastohydrodynamics, Surface Roughness, Manufacturing, Component Life, Condition Monitoring, and Automotive.

The short (15 minutes) time allotments for each of the contributed papers presented problems for both native English speakers and non-native English speakers. You could sense the stress on some of the speakers who were struggling to finish their talks on time. As a result, some of the presentations seemed disjointed and rushed. A set of abstracts for all of the talks was handed out to the registrants. A symposium proceedings, containing all papers and titled the same as the symposium, will be published in hard cover by Elsevier Publishing Company. Publication typically takes a year, so look for the proceedings in September or October of 1998.

On Friday afternoon, there was an optional, student-led tour of the tribology laboratories at Imperial College. Much of the work there was on elastohydrodynamics lubrication and rolling contact (at least two of the faculty specialize in this area). Funding was from oil companies like Exxon, Vavoline, Shell, and Mobil. One test method has been developed for the evaluation of diesel fuel lubricity additives. Apparatus were constructed to study continuously variable transmissions and constant velocity joints. A small, bench-top fretting rig was available for use on coatings. There was also research being done on a rather large simulator to investigate the wear of splines in helicopter gearboxes.

The theme of the conference, to be held 8-11 September 1998, will be "Lubrication at the Frontier (thin film and boundary lubrication)."

Fall G2 Committee Meeting - K. Budinski

It has been many years (probably more than 8) since the G2 Committee met at ASTM headquarters. This meeting provided an opportunity for attendees to see the new headquarters building and meet some of the staff that we deal with over the phone, but seldom meet. In lieu of a symposium or workshop, we toured the building, had a tutorial on standards writing, and on Wednesday afternoon we toured LNP Corporation's plastic research lab which was nearby.

The ASTM building is impressive. It is on the banks of the Schuylkill river in a wooded valley that was formerly a sleepy rural community. It is not so sleepy now. The headquarters building is situated at the confluence of at least four major commuter routes, and it is a real challenge to cross one of these streets on foot. The building has four floors, and it sits on a ramp garage. Visitors who come by car can park under the building, and they can fend with the traffic better than visitors on foot. There are 150 or so staff at headquarters, and it appears to be a very pleasant place to work.

Our meetings started on Thursday morning. We visited LNP in the afternoon and held some meetings in their conference room. The remaining meetings were held Friday morning at headquarters. The following are summaries of what transpired at the various subcommittee meetings:

Terminology Activities - Peter Blacke substituted for Chairman, Frank Doe. Negatives on the recent ballot on the definition of wear coefficient and friction were discussed. The negative on wear coefficient was ruled persuasive, and the item must be reballoted. An editorial change was recommend for the friction term. Frank Doe submitted a classified index of wear terms for consideration for inclusion in the G 40 terminology standard. Some attendees suggested that this document be reformatted as a guide on the use of tribological terms. Others wanted more terms defined. The disposition of the index was deferred to Frank.

Peter Blacke proposed new definitions for inclusion in the G 40: two-body and three-body abrasive wear. Peter's strawman definitions were submitted to Frank Doe for ballot consideration.

Abrasive Wear Activities - Jim Evans (White Rock Engineering) agreed to assume the chairmanship of this subcommittee. One negative was received on a recent ballot on the G 105 standard. It was revised to add key words. The discussion at the meeting resulted in the withdrawal of the negative (the owner was present).

Grindability test - Peter Blacke discussed the proposed pin-on-belt test that was developed for ceramics. There are two versions of the test but basically, it involves abrading the end of a ceramic pin with a 220 grit diamond belt (10m/s, 10N force). A discussion on the need for making this test a standard resulted in the consensus that it should go forward. Peter will continue to champion this test.

Ball Cratering test - This activity is centered in the UK, with task group chairman, Mark White. Nobody was present at the meeting to report on the status of standardization, but John Evans agreed to contact Mark and try to keep the test progressing. It is widely used to measure the thickness of this PVD coating and many labs have the equipment.

Miller test - Task Group chairman, Jim Evans, reported that he would like to update the slurry abrasion test to include the use of a gold rider for testing slurries of low abrasivity (Miller Number less than 10).

Non-Abrasive Wear activities - The subcommittee meeting was chaired by chairman, Brian Dow (Falex).

Block-on-ring test - Task group chairman, Ray Brown (Consultant), reported results on recent inter-laboratory tests on three different plastics. The data acquired to date suggest that the test time needs to be increased from the present 2 hours to 20 hours. Ray proposed taking the plastic method out of the G77 standard and making a new standard for plastic testing. Ray will proceed with this course of action with a goal of January for balloting of the plastic-free G 77 test.

Reciprocating test - Task Group chairman, Peter Blacke, reported that a research report has been submitted to ASTM on this test. It contains all of the details of the test development and inter-laboratory tests.

Paint test – Dr. Roll from Ford proposed standardization of a paint durability test at our last meeting. To date, there has been no word of activity to this end. Peter Blacke will contact him to get the issue resolved.

New Business - Bill Rand (NIST) suggested an activity aimed at development of a guide on coating wear tests. Bill will advertise on our Web page for collaborators.

Erosion Activities – Chairman, Howard Larm, has retired from the University of Kansas and from the erosion subcommittee. He is currently working at the National Research Lab in Vancouver, but will be moving out of the country mid 1998 for retirement in Italy or Hungary. Bill Rand agreed to serve as interim chairman. A symposium on high temperature wear is still being planned for 1999. We thank Howard for his contributions to G 2. We will miss him, and we wish him a great retirement.

Cavitation - It was reported that there are five European scientists interested in participating in inter-laboratory tests on the stationary specimen test. We need samples for testing (ceramic and nickel aluminides). If any one can help with a donation of test coupons, contact John Doe for details. The test coupons are needed for test on the stationary sample test. The G 32 vibrating specimen test is also up for its five-year re-approval.

Friction Activities - Subcommittee chairman, Ken Budinski (Kodak), presented a draft of a standard on friction testing web material for the presence or absence of lubricants. This is an ISO inclined plane test that has been used for over 20 years to determine if wax is present on photographic films. Inter-laboratory tests will be conducted with Peter Blacke and Ray Brown as participants. In our standards writing tutorial on Wednesday, we learned about ASTM's new software that will allow comment on a ballot on the Internet. This will be tried in the development of this standard. If any G2 members wish to participate in the document writing and balloting on the Internet, contact Ken at 716-477 2027.

Computerization Activities – Chairman, Bill Rand, reported ballot results on a Guide for Digital Acquisition of Wear and Friction Measurements. There were no negatives and the ballot will be moved to Main ballot. The G117 standard on calculating and reporting measures and precision of inter-laboratory test data is up for re-approval. Bill will revise the standard to include ways of dealing with outlying data and then reballot.

Web site - Bill offered to develop a new home page for the G2 committee. He will include new activities and who to contact for various existing activities.

Long Range Planning – Chairman, Peter Blacke, discussed ways that G2 may increase membership and participation:

1. Present papers to other groups on standards development.
2. Promote G2 to co-workers and suppliers.
3. Notify local technical societies of symposia in their localities.

Other suggestions included wear tutorials with symposia, web site advertising of new activities, and expansion of our activities into newer areas of tribology such as nanotribology.

Chairman's Message - John Evans

It was my pleasure, on behalf of the G 2 Committee to present Peter Blacke the Frank Brautigam Award in recognition for his outstanding leadership as the task group chairman in the development of ASTM G 133- Linearly Reciprocating Ball-on-flat Sliding Wear. This method can evaluate the wear behavior of ceramics, metals and other materials of interest. Round robin testing evaluated the wear resistance of silicon nitride.

A symposium on Wear Processes in Manufacturing has been finalized for 1998. It will be held on May 6 in Atlanta in conjunction with the Spring G 2 meeting. Fourteen papers covering a variety of wear issues in manufacturing will be presented. Please plan to attend. In the fall, our meeting site has been changed to Oak Ridge National Labs. This should be an interesting meeting site. You will be able to visit an excellent research facility and spend a few fall days in the Smokey Mountains of Tennessee.

Final Notes

Future meetings:
*   May 6,7 & 8, 1998, Atlanta GA, with a Symposium on "Wear Processes in Manufacturing"
*   Fall, 1998, Oak Ridge National Laboratory, Oak Ridge, TN
*   April, 1999, Atlanta, GA, with the Wear of Material Conference

Peter Black gets Bronson Award -   A highlight of the meeting at ASTM headquarters was granting of the Bronson Award to Peter Black for his outstanding work at the task group level in developing the reciprocating wear test. Headquarters has a large congregation area with the walls lined with the permanent awards given by various committees. The Bronson Award is G2's only permanent award, and for once, the recipients had an opportunity to see their names on the plaque that resides at headquarters.

Example of a Journal Article*

* Reprinted from *Wear*, Volume 203–204, Kenneth Budinski, Resistance to Particle Abrasion of Selected Plastics, 1997, p 302–309, with permission from Elsevier Science

Reprinted from

WEAR

Wear 203–204 (1997) 302–309

Resistance to particle abrasion of selected plastics

Kenneth G. Budinski

Eastman Kodak Company, 3177 Latta Road, Suite 146, Rochester, NY 14612-3092, USA

WEAR
An International Journal on the Science and Technology of Friction, Lubrication and Wear

Scope
This international journal is dedicated to the rapid publication of high quality papers on the important subjects of wear and friction together with papers on related aspects of lubrication. Thorough refereeing of all papers, in accordance with procedures for international journals in science and engineering, ensures that Wear is an international forum for multidisciplinary communications on topics such as: the fundamental knowledge of wear phenomena as applied to engineering and science, including nanometre and atomic scale aspects of tribology, design and materials; wear of natural biological and implanted materials; contact phenomena; control and prevention of wear processes; new materials and their characteristics in terms of surface structure, wear, corrosion, oxidation resistance, friction and high temperature strength; new engineering concepts and innovation in materials (plastics, composites, ceramics); controlled wear by abrasion; wear of cutting and forming tools; lubricants and mechanisms of lubrication controlling wear and friction; surface phenomena, their physics, chemistry and practical consequences. Papers concerned with these fields may be submitted. Facilities also exist for the prompt publication of short Research Reports and Letters subject to editorial approval.

Publication Information
Wear (ISSN 0043-1648). For 1997 volumes 201–212 are scheduled for publication. Subscription prices are available upon request from the Publishers. Subscriptions are accepted on a prepaid basis only and are entered on a calendar year basis. Issues are sent by surface mail except to the following countries where Air delivery via SAL mail is ensured: Argentina, Australia, Brazil, Canada, Hong Kong, India, Israel, Japan, Malaysia, Mexico, New Zealand, Pakistan, PR China, Singapore, South Africa, South Korea, Taiwan, Thailand, USA. For all other countries airmail rates are available upon request. Claims for missing issues should be made within six months of our publication (mailing) date.

Orders, Claims, and Product Enquiries
Please contact the Customer Support Department at the Regional Sales Office nearest you:

New York
Elsevier Science
P.O. Box 945
New York, NY 10159-0945, USA
Tel.: (+1) 212-633-3730
[Toll free number for North American customers: 1-888-4ES-INFO (437-4636)]
Fax: (+1) 212-633-3680
E-mail: usinfo-f@elsevier.com

Tokyo
Elsevier Science
9-15 Higashi-Azabu 1-chome
Minato-ku, Tokyo 106
Japan
Tel.: (+81) 3-5561-5033
Fax: (+81) 3-5561-5047
E-mail: kyf04035@niftyserve.or.jp

Amsterdam
Elsevier Science
P.O. Box 211
1000 AE Amsterdam
The Netherlands
Tel.: (+31) 20-4853757
Fax: (+31) 20-4853432
E-mail: nlinfo-f@elsevier.nl

Singapore
Elsevier Science
No. 1 Temasek Avenue
#17-01 Millenia Tower
Singapore 039192
Tel.: (+65) 434-3727
Fax: (+65) 337-2230
E-mail: asiainfo@elsevier.com.sg

USA mailing info: *Wear* (ISSN 0043-1648) is published monthly by Elsevier Science S.A. (P.O. Box 564, 1001 Lausanne). Annual subscription price in the USA is US$ 4753.00 (valid in North, Central and South America), including air speed delivery. Second class postage rate is paid at Jamaica, NY 11431.
USA POSTMASTER: Send address changes to *Wear* Publications Expediting Inc., 200 Meacham Avenue, Elmont, NY 11003. **AIR-FREIGHT AND MAILING** in the USA by Publications Expediting Inc., 200 Meacham Avenue, Elmont, NY 11003.

Advertising Information
Advertising orders and enquiries may be sent to: **International**: Elsevier Science, Advertising Department, The Boulevard, Langford Lane, Kidlington, Oxford OX5 1GB, UK. Tel.: +44 (1865) 843 565. Fax: +44 (1865) 843 952. **USA and Canada**: Weston Media Associates, Dan Lipner, P.O. Box 1110, Greens Farms, CT 06436-1110, USA. Tel.: +1 (203) 261 2500. Fax: +1 (203) 261 0101. **Japan**: Elsevier Science Japan, Advertising Department and **Customer Support Department**, 1-9-15 Higashi Azabu, Minato-ku, Tokyo 106, Japan; Tel: +81-3-5561-5033; Fax: +81-3-5561-5047; E-mail: KYF04035@niftyserve.or.jp

ELSEVIER

Wear 203–204 (1997) 302–309

WEAR

Resistance to particle abrasion of selected plastics

Kenneth G. Budinski

Eastman Kodak Company, 3177 Latta Road, Suite 146, Rochester, NY 14612-3092, USA

Abstract

Accelerated abrasive wear of plastic parts in a piece of production machinery prompted a laboratory study to find a material with better abrasion resistance. The abrasion occurred in a machine that compacted 'sand-like' particles of an inorganic compound. The abrasion resistance of a wide variety of plastics and different durometer polyurethanes (21 materials) was tested with a modification of the ASTM dry-sand rubber wheel three-body abrasion test. Only one material, a polyurethane, had better abrasion resistance than the material that was currently in use. Hardness, friction and scratch tests were conducted on the test materials to try to understand the role of material properties in this type of abrasion. None of these correlated with the wear data. Previous investigators of plastic abrasion related abrasion resistance to the fracture energy and friction. The wear data developed in this study did not correlate with the specific model proposed by Ratner. However, it was possible to obtain a reasonable correlation with a deformation factor that included the friction of the abrasive on the plastic and a term that related to the energy required to deform the material plastically. A test similar to a Brinell hardness test was used to arrive at the deformation energy of the 21 test materials. The more easily the material deforms in contact with a particular abrasive, the better the abrasion resistance. © 1997 Elsevier Science S.A. All rights reserved.

Keywords: Abrasion resistance; Plastics

1. Introduction

This study was prompted by an equipment problem that was occurring in the production of crystallites of an inorganic compound used in the preparation of photographic emulsions. The material was similar to ordinary table salt in size, appearance and compressive strength. The problem to be addressed by this study was the accelerated wear of plastic guides that directed the powder into a briqueting press. The powder was directed into compacting rollers with two parallel plastic plates of ultrahigh molecular weight polyethylene (UHMWPE) that were about 6 mm thick. The plates were only lasting about two weeks before they had to be replaced because of excessive abrasive wear from the powder rubbing on the plate as it entered a roller nip. The plates were not particularly expensive, but replacing these plates was a very expensive operation. They were 'buried' well into the machine and their replacement required the loss of up to two days of production. The plates were made from plastics because metals may introduce contamination (wear debris) and because ceramics were too brittle to withstand the flexing that occurs in the plates. Our assignment was to conduct laboratory tests to determine whether another material would provide improved service life over the UHMWPE.

Many studies have been conducted on the wear of plastics mated to metals, but the studies on the abrasion resistance of plastics are fewer and less conclusive. Most reports seem to recommend additional studies [1–10]. There are some standard tests for the abrasion resistance of plastics; one of the most widely used tests is the Taber Abraser test, ASTM D 4060 [11]. This test involves abrasion of a flat sample of plastic with rotating rubber/abrasive wheels or with sandpaper wheels. Another test, ASTM D 1242 Procedure A [12], uses loose abrasive distributed on a rotating platen. The loose abrasive is pressed into the rotating sample. ASTM D 1242 Procedure B [13] uses what is essentially a belt sander to abrade specimens on a conveyor that cycles in and out of contact with the sandpaper. ASTM G 56 [14] has been used to measure the abrasion resistance of plastic to paper. A ball rider is rubbed on a large paper-covered drum. ASTM G 132 [15] is an abrasion test in which the ends of vertical pins rub on a large sandpaper-covered drum. Although this test was developed for metals, the concept has been used by others to test plastics [16]. All of these tests were considered as candidates for a laboratory test to screen materials to address the above production problem.

The test selected to rank materials was yet another ASTM test, ASTM G 65, the dry-sand rubber wheel abrasion test [17]. This test, which is illustrated in Fig. 1, was developed

K.G. Budinski / Wear 203–204 (1997) 302–309

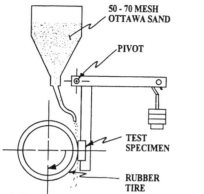

Fig. 1. Schematic diagram of the ASTM G 65 dry-sand rubber wheel abrasion test.

Table 1
Plastics, elastomers and metals selected for testing

Base polymer	Filler (wt%)/reinforcement
Polyphenylene sulfide (PPS)	40% carbon fiber (CF)
Polystyrene (PS)	None
Epoxy (EP)	40% woven glass
Polyphenylene sulfide (PPS)	PTFE/chopped glass (CG)
Phenolic (PF)	Woven aramid fiber (NOM)
Polytetrafluoroethylene (PTFE)	Chopped glass (CG)
Polyoxymethylene (POM)	PTFE
Acrylonitrile/butadiene/styrene (ABS)	None
Epoxy (EP)	Woven cloth (Cot)
Phenolic (PF)	Woven cloth (Cot)
Polyetheretherketone (PEEK)	None
Polytetrafluoroethylene (PTFE)	None
Polyimide (PI)	None
Polyethylene (HDPE)	None
Polyamide (PA)	MoS_2
Polyurethane (PUR) 55A	None
Polyurethane (PUR) 90A	None
Polyurethane (PUR) 75D	None
Polyurethane (PUR) 85D	None
Polyethylene (UHMWPE)	Oil
Polyethylene (UHMWPE) Control	None

Reference metals: AISI type 316 stainless steel (hardness 92 HRB), Stellite 6B (hardness 43 HRC), type A2 tool steel (hardness 60 HRC).

to rank the abrasion resistance of both hard and soft metals. The material to be tested is line-contact loaded against a rubber wheel and silica sand is metered into the nip. Wear is assessed by measuring the volume of material (by mass change) removed from the specimen in a fixed period of rubbing. This particular test is one of the most used abrasion tests and the ASTM committee responsible for this test is continually working on improvements and new test procedures to increase the applicability of the test. An experimental test procedure developed by the subcommittee for ranking the abrasion resistance of coatings was selected as the most appropriate test. The wearing counterface is vertical with parallel abrasive flow (three-body abrasion), like the application. It was felt that this test produced the closest simulation of the real tribosystem.

It is the purpose of this paper to describe the abrasion resistance ranking of 17 plastics and 4 elastomers for application as wear plates in a particle briqueting machine. The objective of this study was to identify a material that would provide longer service life than the UHMWPE that was currently in use. The format of this paper is to describe the laboratory testing, to present the test results, and then in the discussion to describe the tests that were conducted to find plastic properties that correlated with the test results.

2. Procedure

The dry-sand rubber wheel tester uses a 228 mm diameter chlorobutyl rubber wheel (60Shore A) as an abrader. The wheel is 12.7 mm wide and runs at a single speed, 20.9 rad s^{-1}. The test samples are from 4 to 12.7 mm in thickness, 25 mm wide, and 76 mm long. The wear surfaces are the 25×76 mm^2 faces. The loading force of the specimen against the wheel can be up to 140 N. The abrasive is 215 to 300 μm silica. All test users purchase test sand from the same source. The sand flow is in the range 300–400 g min^{-1}. There are three procedures in the ASTM test standard that employ dif-

ferent normal forces and test durations. The test procedure used in this study employed a 45 N force and a test duration of only 200 wheel revolutions (60 s). This procedure was used successfully by the ASTM G02.3 Abrasive Wear Subcommittee in 1990 to conduct interlaboratory tests on polymeric coatings. The test materials included the plastics and elastomers listed in Table 1. These materials were selected because of their successful performance in other plant operations.

All test samples were made from bulk materials. The samples with woven reinforcements were tested on their flat face. Where possible, the abrasion tests were conducted on as-molded surfaces. Wear volumes were calculated from mass changes during the test.

3. Test results

The average volume losses are compared in Fig. 2. Typical wear scars on the plastic specimens are shown in Fig. 3. Only one material, a 90Shore A durometer polyurethane, had better wear resistance than the UHMWPE plastic currently in use. An oil-lubricated UHMWPE was comparable in wear resistance to the control material as were several other hardnesses of polyurethane elastomer.

The glass and carbon fiber-reinforced plastics had the worst abrasion resistance of the reinforced materials. Cotton and linen-reinforced materials had better abrasion resistance than the glass-reinforced materials. The worst wear on a non-reinforced plastic was on polystyrene. There appeared to be a preferred hardness for polyurethane—the top of the Shore

K.G. Budinski / Wear 203–204 (1997) 302–309

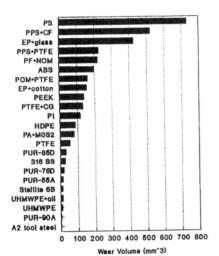

Fig. 2. Wear volumes for plastic candidates in a modified ASTM G 65 abrasion test.

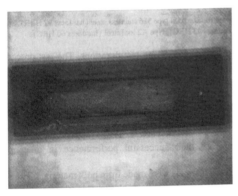

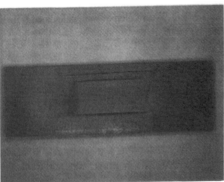

Fig. 3. Typical appearance of wear scars on plastic specimens.

A scale. Harder (Shore D) and softer materials did not wear as well. All of the plastics tested, except UHMWPE and polyurethane, had lower abrasion resistance than the soft stainless steel reference material. A2 tool steel at 60 HRC

was more abrasion resistant than all of the test materials, but it was not significantly more abrasion resistant than a 90 Shore A polyurethane. These tests suggest that under certain conditions, some plastics and elastomers can have particle abrasion resistance comparable with that of hard steels.

4. Discussion

The objective of this study was barely met. We sought to identify materials that were significantly better than the UHMWPE in abrasion resistance. Only one material, a polyurethane with a Shore A durometer of 90, had better abrasion resistance than UHMWPE, and it was only about two times better; the desired improvement was ten times the current service life. The plastics included in this study represented some of the 'premier' plastics from the standpoint of plastic-to-metal wear resistance. Carbon fiber-reinforced polyphenylene sulfide (PPS) is a very high-performance plastic bushing material. Cotton-reinforced phenolic has been used for difficult wear problems for decades. It is used in automobile gears and heavy duty rolling mill bearings. Epoxy/glass composites have extremely high compressive strength. One would think that out of 21 high-performance plastics/elastomers several would be significantly more abrasion resistant than UHMWPE, which is an ordinary plastic that has been used for abrasion applications for many years. This was not the case. The remainder of this discussion is directed towards understanding why UHMWPE and polyurethane (PUR) have better abrasion resistance than engineering plastics that are much stronger and stiffer (and more expensive).

4.1. Role of hardness

What controls the abrasion resistance of plastics? The classic relationship for abrasive wear of metals is that wear is inversely proportional to the hardness of the metal [18]. The harder the metal, the lower the abrasion rate. Unfortunately there is no single plastic hardness scale that is suitable for the wide range of plastics/elastomers included in this study. Nonetheless, Shore D and recoil hardness tests were conducted on the test plastics to explore the plausibility of a hardness–abrasion relationship. As shown in Fig. 4, the hardest plastic, glass-reinforced epoxy, was harder than UHMWPE by a factor of about 1.4, but their abrasion rates varied by a factor of about 60.

It was thought that the resilience of the plastics may play some role in resisting indentation and scratching by hard particles. There is a commercially available hardness tester that uses the rebound velocity of a spherical-ended sabot to measure the hardness of metals. The harder the metal, the greater the rebound velocity. This device was used on the test materials. As shown in Fig. 5, these rebound hardnesses did not show an apparent correlation with the abrasion volume losses.

K.G. Budinski / Wear 203–204 (1997) 302–309   305

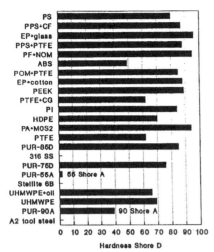

Fig. 4. Shore hardness of test materials.

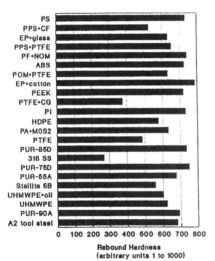

Fig. 5. Rebound hardness of test materials (based on the recoil velocity of a spherical ended sabot).

4.2. The role of scratch resistance

For many years the Taber Abraser has been used to rank the abrasion resistance of plastics [1,6,16,19]. As mentioned previously, this device can be used for two-body or three-body abrasion. This type of test is not unlike a scratch test where the abrader is an abrasive-filled rubber or a sandpaper-covered wheel. Fixed sharp particles are imposed on the test surface. In an attempt to simulate this type of material removal, scratch tests were conducted on the test plastics using a 60° included angle diamond cone stylus with a tip radius of 200 μm. Yamaguchi [16] and Briscoe et al. [20] suggested that scratch hardness is a factor in abrasive wear

of plastics. The scratch hardness is measured by the width of the furrow produced by the scratching stylus:

$$H = P/b^2$$

where H is the scratch hardness, P is the stylus load, and b is the furrow width.

The test plastics were scratched for a distance of about 50 mm at a speed of 10 mm s^{-1}, with a mass of 1 kg on the stylus. The force required to produce the scratch was recorded continuously and this force was converted into static and kinetic friction coefficients by dividing by the normal load. Typical scratches are shown in Fig. 6.

As shown in Fig. 7, the materials with the highest scratch hardness, such as the epoxies and phenolics, had the poorest abrasion resistance. Two of the elastomers, PUR-55 A and PUR-90 A, did not scratch at all. The various plastics appeared to scratch by different mechanisms (Fig. 8). The reinforced plastics displayed ragged edges on the scratch furrow. The unfilled polymers tended to produce smooth scratch furrows; some produced furrows that varied in width suggesting what appeared to be a stick–slip type motion during surface deformation. Four metals and a cemented carbide were scratched with the plastics/elastomers to see whether they responded 'properly' to hardness differences. The

Fig. 6. Scratches in test specimens after scratching with a 60° diamond at 100 magnification: (a) cotton-reinforced phenolic, (b) polyamide + molybdenum disulfide.

K.G. Budinski / Wear 203–204 (1997) 302–309

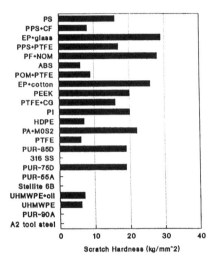

Fig. 7. Scratch hardness of test materials.

scratch hardnesses of all of the plastics were lower than the hardnesses of the metals (Table 2).

The metals displayed scratch hardnesses that generally correlated with their hardness, but the relationship was not suc-

cinct. These results suggest a friction effect. Stellite 6B did not scratch even though it is much softer than the tool steel and cemented carbide. Overall, the scratch tests did not lead to a relationship that clearly related the scratch hardness to the observed volume losses in the abrasion tests. A second and third series of scratch tests were conducted with carbide styli with larger radii (2 and 10 mm diameter balls). Only a few of the test materials scratched with the 2 mm stylus and none scratched with the 10 mm stylus. The scratch tests were concluded and the scratch force data were evaluated for possible correlation with abrasion rates.

4.3. Friction considerations

The force measurements obtained in the scratch tests suggest that when a scratch stylus produces plastic deformation/fracture of the surface, the force is probably a reflection of the energy required to produce deformation and removal of material. The friction 'coefficients' μ measured in scratch tests on the test plastics are given in Table 3.

These results suggest that the friction coefficient depends to a significant degree on the degree of plastic deformation produced in the scratching operation. The wear test results did not correlate with any of these scratch test results. In the

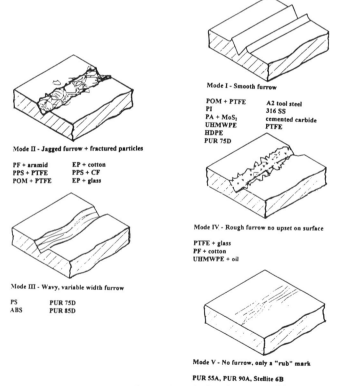

Fig. 8. Appearance of 60° diamond scratches in various test materials.

K.G. Budinski / Wear 203–204 (1997) 302–309   307

Table 2
Scratch hardness of four metals and cemented carbide

Material	Scratch hardness ($kg\ mm^{-2}$)	Vickers hardness
6061-T6 aluminum	25	47
316 stainless steel	100	143
Stellite 6B	No scratch	435
A2 tool steel	400	625
Cemented carbide (C2)	10 000	1854

Table 3
Coefficient of friction μ measured in scratch tests on plastics with various scratch styli

Stylus/loading mass	Average μ	σ [a]	r range
60° diamond/1000 g	0.58	0.14	0.2–0.9
2 mm diameter carbide/1000 g	0.21	0.2	0.2–1
10 mm diameter carbide/1000 g	0.14	0.13	0.05–0.7

[a] Standard deviation.

abrasion test, however, the scratching material is 50–70 mesh silica sand. In an effort to determine whether scratching with silica makes a difference in scratch results, a scratch stylus with silica grains on the rubbing surface was fabricated. Friction coefficients were calculated from the force measurements made with a silica sand stylus. The steel stylus was 6 mm in diameter and a monolayer of sand was adhered to the hemispherical end of the stylus with a cellulosic lacquer. As shown in Fig. 9, the friction coefficient of the plastic/sand couple did not correlate with the wear test results. The average friction coefficient for the sand/plastic couples was 0.32 ($\sigma = 0.22$, $r = 0.1$ to 1.1). Most test samples were permanently scratched by the sand stylus with a normal force produced by a 500 g mass on the stylus. The scratches and the friction coefficients were smaller than for the diamond stylus. These data suggest that the friction coefficient of this tribosystem includes a measure of surface deformation as proposed many years ago by Bikerman [21].

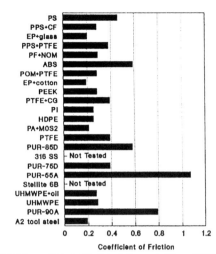

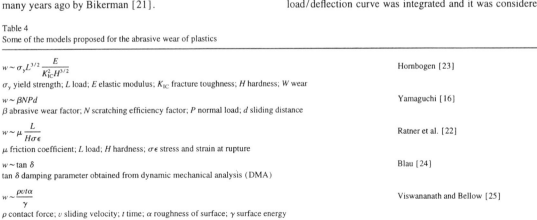

Fig. 9. Friction coefficients (kinetic) of silica sand sliding on test materials.

4.4. Abrasion models

Up to this point, the traditional tribological properties of friction and hardness have failed to correlate with the abrasion results. Some additional plastic abrasion models from the literature were reviewed (Table 4) for direction. One model that seemed to be quite reasonable was that proposed by Ratner et al. [22], where the rate of material removal was said to be inversely proportional to the product of 'stress and strain at rupture'. It was not clear how stress and strain at rupture could be obtained from the G 65 abrasion blocks that were available as test materials. It was decided that the load/deflection curve for plastic deformation of the surface by a hemispherical indenter may be a predictor of at least the ability of a material to deform plastically as in scratching. With this reasoning, the abrasion test coupons were indented to a fixed depth of 1.25 mm with a 6 mm diameter indenter in a universal tension/compression tester. The area under the load/deflection curve was integrated and it was considered

Table 4
Some of the models proposed for the abrasive wear of plastics

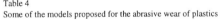

$w \sim \sigma_y L^{3/2} \dfrac{E}{K_{IC}^2 H^{3/2}}$	Hornbogen [23]
σ_y yield strength; L load; E elastic modulus; K_{IC} fracture toughness; H hardness; W wear	
$w \sim \beta NPd$	Yamaguchi [16]
β abrasive wear factor; N scratching efficiency factor; P normal load; d sliding distance	
$w \sim \mu \dfrac{L}{H\sigma\epsilon}$	Ratner et al. [22]
μ friction coefficient; L load; H hardness; $\sigma\epsilon$ stress and strain at rupture	
$w \sim \tan\delta$	Blau [24]
$\tan\delta$ damping parameter obtained from dynamic mechanical analysis (DMA)	
$w \sim \dfrac{\rho v t \alpha}{\gamma}$	Viswananath and Bellow [25]
ρ contact force; v sliding velocity; t time; α roughness of surface; γ surface energy	

K.G. Budinski / Wear 203–204 (1997) 302–309

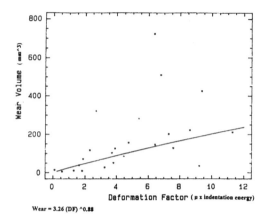

Wear = 3.26 (DF) ^0.88

Correlation Coefficient = 0.7273

Fig. 10. Effect of ease of surface deformation (deformation factor) on the volume of abrasive wear.

as a measure of the energy required to deform plastically the test material.

Wear data were plotted vs. the reciprocal of this energy term as proposed by Ratner. There was poor correlation. After manipulating the energy data in a variety of ways it was determined that the best correlation existed with a deformation factor that included the friction coefficient of the sand on the test material:

$$w \sim \mu(Se)$$

where w is the abrasion rate, μ is the friction coefficient of sand on the plastic surface, and Se is the area under the load/deflection curve from the ball indent test.

The lower the product of friction and deformation energy, the lower the abrasion. As shown in Fig. 10, the correlation is much less than perfect (correlation of 0.73), but this correlation is certainly better than the correlation obtained with the hardness and scratch parameters. This relationship also seems reasonable. The deformation factor for elastomers is low and they have good abrasion resistance. The same can be said about the UHMWPEs. The contribution of friction coefficient is thought to be that low friction can reduce wear because the abrasive (in three-body abrasion) is less likely to dig in and form a plow mark or scratch. However, a high-friction material such as PUR has good abrasion resistance because the abrasive grains tend to roll through the wear interface rather than become fixed on one member and plow a furrow. The correlation seemed plausible and further work with the other models was decided against.

5. Conclusions

1. Polyurethane with a durometer of 90Shore A has more abrasion resistance to AFS 50–70 silica in a three-body abrasion test than UHMWPE.

2. The hard, reinforced and filled engineering plastics had relatively poor abrasion resistance to silica sand in the three-body test used in this study.
3. The plastics that deform easily when acted on by loose abrasive particles (e.g. silica) are less likely to produce material removal by scratching/fracture.

6. Summary

Twenty-one plastics/elastomers were subjected to a three-body abrasion test to find an improved material to solve a production problem. The tests did not identify a material with the desired 10-fold increase in abrasion life. The improvement was only of the order of two times, much less than anticipated. However, this study reconfirmed that UHMWPE and high Shore A polyurethanes have better abrasion resistance than most other plastics and elastomers.

The explanation for their excellent abrasion resistance appears to be their ability to deform easily and their favorable friction characteristics against most other materials. The model suggested by this study needs more development, but it is felt that it may have correlated better if this study did not include such diverse plastic/elastomer systems (glass-reinforced composites, injection moldable commodity plastics, carbon fiber-reinforced engineering plastics, and elastomers). They were not from similar groups or families. Future studies need a less diverse group of test materials. Also, it appears that the abrasion model should include a fracture toughness term since the more brittle materials lost material by brittle fracture in the scratch tests. The elastomers may need a term relating to their resilience (branching) and the neat materials may need a more well defined friction term. In 1981, Bartenev and Lavrentev [19] concluded the chapter on abrasive wear in their book on the friction and wear of polymers with the statement: "To the present there is no theory of abrasive wear of polymers." This situation appears to prevail, and it may not be possible to improve on the UHMWPEs and polyurethanes for abrasion resistance until a better model is deduced.

References

[1] J.M. Thorp, Abrasive wear of some commercial polymers, *Tribol. Int.,* *15* (1982) 59–68.
[2] J. Larsen-Basse and T. Ahmad, Slurry abrasion of polymers under simulated submarine conditions, *Wear, 122* (2) (1988) 135–149.
[3] D.H. Buckley and R.L. Johnson, Friction wear and decomposition mechanisms for various polymer composites in vacuum to 10^{-9} mm Hg, *NASA Tech. Note D-2073*, December 1963 (NASA, Washington, DC).
[4] G.F. Cole and R. Traviesco, Wear by paper on nylon matrix composites. In L.H. Lee (ed.), *Advances in Polymer Friction and Wear*, Vol. 5B, Plenum, New York, 1974, pp. 689–702.
[5] I.M. Hutchings, *Tribology*, CRC Press, London, 1992, pp. 156–162.
[6] G.E. Power and J.H. Dumbleton, An improved method for determining the wear of polymeric coatings, *Wear, 25* (1973) 373–380.

[7] B.J. Briscoe and P.D. Evans, The influence of asperity deformation conditions on the abrasive wear of irradiated PTFE. In K.C. Ludema (ed.), *Wear of Materials 1989*, Vol. 2, American Society of Mechanical Engineers, New York, 1989, pp. 449–457.

[8] R. Kaneko and E. Hamada, Microwear processes of polymer surfaces, *Wear, 162–164* (1993) 370–377.

[9] D.G. Bellow and N.S. Viswanath, An analysis of the wear of polymers, *Wear, 162–164* (1993) 1048–1053.

[10] S. Bahadur and D. Gong, Formulation of the model for optimal proportion of filler in polymers for wear resistance. In K.C. Ludema (ed.), *Wear of Materials 1991*, American Society of Mechanical Engineers, New York, 1991, pp. 177–187.

[11] *ASTM D 4060*, Test method for abrasion resistance of organic coatings by Taber Abraser, American Society for Testing and Materials, W. Conshohocken, PA.

[12] *ASTM D 1242*, Test method for resistance of plastic materials to abrasion, procedure A, American Society for Testing and Materials, W. Conshohocken, PA.

[13] *ASTM D 1242*, Test method for resistance of plastic materials to abrasion, procedure B, American Society for Testing and Materials, W. Conshohocken, PA.

[14] *ASTM G 56*, Test method for the abrasiveness of ink-impregnated fabric printer ribbons, American Society for Testing and Materials, W. Conshohocken, PA.

[15] *ASTM G 132*, Standard test method for pin abrasion testing, American Society for Testing and Materials, W. Conshohocken, PA.

[16] Y. Yamaguchi, *Tribology of Plastic Materials*, Elsevier, Amsterdam, 1990, p. 125.

[17] *ASTM G 65*, Practice for conducting dry-sand rubber wheel abrasion tests, American Society for Testing and Materials, W. Conshohocken, PA.

[18] E. Rabinowicz, *Friction and Wear of Materials*, Wiley, New York, 1966, p. 168.

[19] G.M. Bartenev and V.V. Lavrentev, in K.C. Ludema and L.H. Lee (eds.), *Friction and Wear of Polymers*, Elsevier, Amsterdam, 1981, p. 239.

[20] B.J. Briscoe, S.K. Besines and S.S. Panesar, The scratch hardness and friction of a soft rigid material. In K.C. Ludema and R.G. Bayer (eds.), *Wear of Materials 1991*, American Society of Mechanical Engineers, New York, 1991, pp. 451–456.

[21] J.J. Bikerman, The nature of polymer friction. In L.H. Lee (ed.), *Polymer Science and Technology*, Vol. 5a, Plenum, New York, 1974, p. 168.

[22] S.B. Ratner, I.I. Farberova, O.V. Radyakenich and E.G. Lur, *Sov. Plast.*, (1964) 37.

[23] E. Hornbogen, The role of fracture toughness in the wear of metals, *Wear, 33* (1975) 251–259.

[24] P.J. Blau, *Friction Science and Technology*, Marcel Dekker, New York, 1996, p. 194.

[25] N. Viswanath and D.G. Bellow, Development of an equation for the wear of polymers, *Wear, 181–183* (1995) 42–49.

Instructions for Authors

Submission of papers

Three copies of papers should be sent to D. Dowson, c/o Editorial Office (UK) of *Wear*, Mayfield House, 256 Banbury Road, Oxford OX2 7DH, UK.
Alternatively, manuscripts may be sent to either Regional Editor:
K. Kato, Laboratory of Tribology, Department of Mechanical Engineering, Tohoku University, Sendai 980, Japan
or
K. C. Ludema, Department of Mechanical Engineering, University of Michigan, 2250 GG Brown Building, Ann Arbor, MI 48109-2125, USA.
Contributions are accepted on the understanding that authors have obtained the necessary authority for publication. Submission of an article is understood to imply that the article is original and unpublished and is not being considered for publication elsewhere. Upon acceptance of an article by the journal, the author(s) will be asked to transfer the copyright of the article to the publisher. This transfer will ensure the widest possible dissemination of information.

Manuscripts

Three copies should be submitted, in double-spaced typing on pages of uniform size with a wide margin at the left. Each table should be typed with its title on a separate page. Legends to illustrations should be typed widely spaced in sequence on a separate page or pages. Each paper should have a summary of 100–200 words. References should be numbered consecutively (numerals in square brackets) throughout the text and collected together in a reference list at the end of the paper. The reference list should be typed according to the style found in recent issues of *Wear*.

Authors in Japan please note: if you would like information about how to have the English of your paper checked, corrected and improved (before submission), please contact our Tokyo office who will inform you of the services provided by language correctors: Elsevier Science Japan, Editorial Service, 1-9-15 Higashi Azabu, Minato-ku, Tokyo 106, Japan; Tel: +81-3-5561-5032; Fax: +81-3-5561-5045; E-mail: KYF04037@niftyserve.or.jp

Submission of Electronic Text

The final text may be submitted on a 3.5 in or 5.25 in diskette (in addition to a hard copy with original figures). Double density (DD) or high density (HD) diskettes formatted for MS-DOS or Apple Macintosh compatibility are acceptable, but must be formatted to their capacity before the files are copied on to them. The files should be saved in the native format of the wordprocessing program used. Most popular wordprocessor file formats are acceptable although our preferred format is WordPerfect 5.1 (MS-DOS). It is essential that the name and version of the wordprocessing program, type of computer on which the text was prepared, and the format of the text files are clearly indicated.

Illustrations

Line drawings should be in a form suitable for reproduction, drawn in Indian ink on drawing paper (letter height, 3–5 mm). They should preferably all require the same degree of reduction, and should be submitted on paper of the same size as, or smaller than, the main text, to prevent damage in transit. Photographs should be black-and-white glossy prints and as rich in contrast as possible. Where magnifications are concerned, it is preferable to indicate the scale by means of a ruled line on the photograph. Material that can be included in the figure captions should not appear on line drawings.

Offprints

A total of 25 offprints of each paper will be supplied free of charge to the author(s). Additional offprints can be ordered at prices shown on the offprint order form which will accompany the proofs.

Further Information

All questions arising after the acceptance of manuscripts, especially those relating to proofs, should be directed to: Elsevier Editorial Services, Mayfield House, 256 Banbury Road, Oxford OX2 7DH, UK. (Tel: +44 1865 314900, Fax: +44 1865 314990, Telex: 837966)

Abstracting Services

This journal is cited by the following Abstracting Services: Current Contents — Engineering, Technology and Applied Sciences, Engineering Index, Physikalische Berichte, Chemical Abstracts, Metals Abstracts, Physics Abstracts, Applied Mechanics Reviews, Science Citation Index, Automatic Subject Citation Alert, PASCAL (Centre National de la Recherche Scientifique), American Petroleum Institute — Central Abstracting and Indexing Service.

Example of a Patent

US005315259A

United States Patent [19]

Jostlein

[11] **Patent Number:** **5,315,259**

[45] **Date of Patent:** **May 24, 1994**

[54] **OMNIDIRECTIONAL CAPACITIVE PROBE FOR GAUGE OF HAVING A SENSING TIP FORMED AS A SUBSTANTIALLY COMPLETE SPHERE**

[75] Inventor: **Hans Jostlein**, Naperville, Ill.

[73] Assignee: **Universities Research Association, Inc.**, Washington, D.C.

[21] Appl. No.: **889,060**

[22] Filed: **May 26, 1992**

[51] Int. Cl.⁵ G01R 27/26
[52] U.S. Cl. **324/690**; 324/662; 324/686
[58] Field of Search 324/658, 661, 662, 681, 324/686, 690

[56] **References Cited**

U.S. PATENT DOCUMENTS

2,842,738	7/1958	Warnick	324/690
2,935,681	5/1960	Anderson	324/690 X
3,452,273	6/1969	Foster	324/662
3,706,919	12/1972	Abbe	361/280
3,815,020	6/1974	Mayer	324/662
4,067,225	1/1978	Dorman et al.	73/1 DV
4,422,035	12/1983	Risko .	
4,482,860	11/1984	Risko .	
4,688,141	8/1987	Bernard et al.	324/662 X
4,695,789	9/1987	Lambertz et al.	324/690
4,766,389	8/1988	Rhoades et al.	
4,816,744	3/1989	Papurt et al.	324/661 X
4,908,574	3/1990	Rhoades et al.	324/675
4,918,376	4/1990	Poduje et al.	324/663
4,924,172	5/1990	Holmgren	324/664
5,189,377	2/1993	Rhoades et al.	324/662

FOREIGN PATENT DOCUMENTS

0169301	7/1989	Japan	324/690

OTHER PUBLICATIONS

Dean Campbell, Noncontact test and inspection, Jul. 1986, pp. 57–60.

Robert C. Abbe, A Brief Report on Non–Contact Gauging, Jun. 1975, pp. 16–20.

B. H. Knowles, Capacitance gauging: Survey of Recent Advances, Oct. 1988, pp. 14–22.

David J. McRae, Using Capacitive Sensing for Non-contact Dimensional Gauging, Oct. 1988, pp. 13–20.

Technical literature for an "OMNISIP" Probe as manufactured by American SIP Corporation, Nov., 1988, pp. 1–5.

Primary Examiner—Kenneth A. Wieder
Assistant Examiner—Glenn W. Brown
Attorney, Agent, or Firm—McAndrews, Held & Malloy, Ltd.

[57] **ABSTRACT**

A non-contact, omni-directional capacitive probe for use in dimensional gauging includes an electrically conductive spherical sensing tip that forms a capacitor with a workpiece, the capacitance of the capacitor being indicative of the distance between the spherical sensing tip and the workpiece.

6 Claims, 7 Drawing Sheets

FIG 3

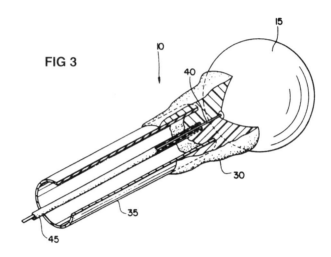

FIG 4

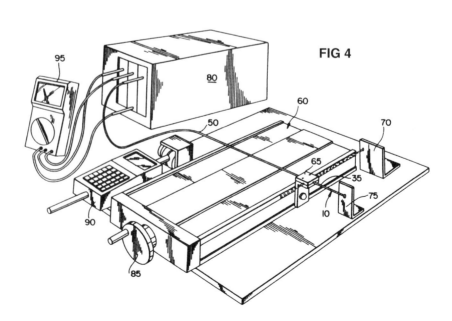

FIG 5

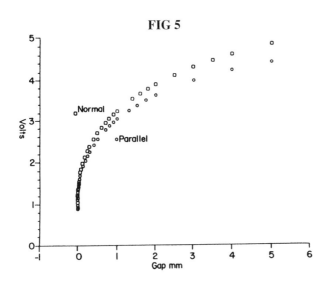

FIG 6

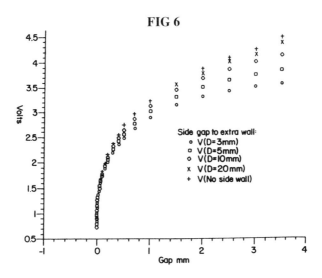

5,315,259

1

OMNIDIRECTIONAL CAPACITIVE PROBE FOR GAUGE OF HAVING A SENSING TIP FORMED AS A SUBSTANTIALLY COMPLETE SPHERE

This invention was made with Government support under Contract No. DE-AC02-76CH03000, awarded by the United States Department of Energy. The Government has certain rights in the invention.

BACKGROUND OF THE INVENTION

The present invention is directed to a capacitive probe for use in dimensional gauging. More specifically, the invention is directed to a non-contact, omni-directional probe for use with a coordinate measurement machine (CMM).

Computerized CMMs have become standard equipment in machine shops, QA laboratories, and precision engineering labs. The CMMs guide a sensing probe in a coordinate measurement system to obtain dimensional information from a workpiece.

Several different sensor probe types are available. In systems with mechanical contact type sensor probes, the sensor probe is guided into contact with the workpiece until a switch in the probe is opened or closed. The CMM computer acquires the dimensional information upon activation of the switch.

Such mechanical contact probes have several disadvantages. Mechanical contact probe typically require contact forces in excess of a tenth of a newton thus rendering them useless in, for example, gauging silicon structures which cannot be touched in a non-destructive manner.

Non-contact probes may be used in lieu of the mechanical contact probes described above. One type of non-contact probe is described in McRae, "Using Capacitive Sensing for Noncontact Dimensional Gauging", Sensors, pp. 13–20, October 1988. The article describes the use of a flat plate sensor that forms a parallel plate capacitor when placed adjacent a conductive target plate (e.g. workpiece). The capacitance of the parallel plate capacitor is inversely proportional to the distance between the sensor and the target plate. Thus,

$C = K/d$ where

C is the capacitance,

K is a constant, and

d is the distance between the sensor and target plate. Parallel plate capacitive probes have several disadvantages. They must be precisely aligned in parallel with the target plate so that the capacitive plate area remains relatively constant. Additionally, the unidirectional character of the parallel plate capacitance only allows positional measurement along a single coordinate axis that is normal to the flat probe plate.

It is therefore an object of the present invention to provide a non-contact, omni-directional capacitive probe for use in dimensional gauging.

SUMMARY OF THE INVENTION

The present invention is directed to a non-contact, omni-directional probe for dimensional gauging. The probe includes an electrically conductive spherical sensing tip that forms a capacitor with a target workpiece. The spherical sensing tip is mounted at one end of a hollow, conductive stalk and is insulated therefrom. A conductive wire is coaxially disposed through the center of the hollow stalk and is connected to the spherical sensing tip. The stalk and coaxial wire are each con-

2

nected to receive an A.C. voltage from a respective A.C. source. The A.C. voltages are generally in phase with one another and have generally equal amplitudes to reduce the effects of stray capacitance on the probe measurements.

BRIEF DESCRIPTION OF THE DRAWINGS

Other objects, features and advantages of the present invention will become apparent upon review of the description of the preferred embodiments taken in conjunction with the following drawings, on which:

FIG. 1 is a partial cross-sectional view of a capacitive probe constructed in accordance with one embodiment of the invention.

FIG. 2 is a cross-sectional view of a spherical sensing tip constructed in accordance with a further embodiment of the invention.

FIG. 2A is a partial cross-sectional view of a further embodiment of the spherical probe tip.

FIG. 3 is a further partial cross-sectional view of the capacitive probe of FIG. 1.

FIG. 4 is a perspective view of an apparatus for testing and calibrating the capacitive probe of FIG. 1.

FIG. 5 is a graph showing test data for two stalk orientations obtained using the apparatus of FIG. 4.

FIG. 5A is a graph showing the logarithmic characteristics of the probe at small gap distances.

FIG. 6 illustrates the increase in effective capacitive plate area as the probe of FIG. 1 is brought close to a target.

FIG. 7 is a graph showing the effect that an extra sidewall has on probe measurements.

FIG. 8 shows the probe of FIG. 1 incorporated into a CMM.

DESCRIPTION OF THE PREFERRED EMBODIMENT

Referring to FIG. 1, there is shown a capacitive probe 10. The probe 10 includes a spherical sensing tip 15 that may be constructed as a single unitary structure from, for example, steel, tungsten carbide, Inconel, or another hard conductive material which resists chipping and deformation. The diameter of the spherical tip 15 is dependent on the particular application of the probe. The radius of the spherical tip 15 should be much greater than the distance that is to be sensed between the tip and a target plate. For present purposes, however, the probe 10 and the corresponding test results will be described with respect to a 4.78 mm diameter spherical tip 15 made from steel.

The spherical sensing tip 15 can also be formed as shown in FIG. 2. As illustrated the spherical tip 15 has a central portion 20 formed, for example, from an insulating material such as a ceramic. An outer conductive shell 25 is disposed about the periphery of the central portion 20.

Referring once again to FIG. 1, an insulating neck 30 connects the spherical tip 15 to one end of a hollow stalk 35. The neck 30 may be formed, for example, from an epoxy material that insulates the tip 15 from the hollow stalk 35 and provides a sturdy mechanical connection therebetween. The stalk 35 may be formed, for example, from brass or stainless steel.

A wire 40 is coaxially disposed within the hollow stalk 35 and extends into electrical contact with the spherical tip 15. The wire 40 may include an insulating sheath 45 to ensure that the hollow stalk 35 and wire 40

5,315,259

5

could move the probe 10 to several sensing gap distances of 1 mm, 0.1 mm, and 0.01 mm. Based on these measurements, the stray capacitance due to, for example, sidewalls, can be extracted and corrected automatically by the CMM.

FIG. 8 shows an exemplary CMM 105 using the probe 10. The CMM 105 includes a control console 110, a calibration sphere 115, a workpiece table 120, and an arm 125 which holds and maneuvers the probe 10 under program control. The table 120 supports a workpiece 130, here shown as a silicon structure used in high energy physics applications. The arm 125 guides the probe 10 along the workpiece 130 to conduct the required measurements. Although the probe is shown with respect to one type of CMM, it will be recognized that the probe is suitable for use with numerous other CMMs, such as a Cordax 1800, or one of the CMM types shown in ANSI/ASME B89.1.12M-1985, incorporated herein by reference.

The probe 10 can be used to replace a contact switch probe normally used in such a CMM. To accomplish this, an interface box (not shown) can provide a short or open circuit signal that would otherwise be produced by a switch closure in the contact probe. This short or open circuit signal to the CMM would occur when the voltage output from the capacitive probe driver 80 (not shown in FIG. 8) indicates that the spherical tip 15 is a fixed distance away from workpiece, for example, when the gap therebetween is approximately 1 μm.

While the invention has been described hereinabove with respect to several embodiments, those of ordinary skill in the art will recognize that the embodiments may be modified and altered without departing from the central spirit and scope of the invention. Thus, the preferred embodiments described hereinabove are to be considered in all respects as illustrative and not restrictive, the scope of the invention being indicated by the appended claims rather than by the foregoing description. Therefore, it is the intention of the inventor to embrace herein all changes which come within the meaning and range of equivalency of the claims.

I claim:

1. An omnidirectional capacitive probe comprising:
an electrically conductive hollow stalk;
an electrically conductive spherical sensing tip disposed at one end of said hollow stalk, said spherical sensing tip formed as a substantially complete sphere;
a wire coaxially disposed within said hollow stalk and electrically connected to said spherical sensing tip;
a first voltage generating means for generating a first A.C. voltage between said spherical sensing tip and a workpiece; and
a second voltage generating means for generating a second A.C. voltage between said hollow stalk and said workpiece, said first and second voltages being generally equal in phase and magnitude thereby to reduce the effect of stray capacitance on measurements made with said capacitive probe;
said spherical sensing tip forming a capacitor with said workpiece when said spherical sensing tip is

6

located in close proximity with the workpiece, the capacitor formed by said spherical sensing tip and the workpiece having a capacitance that is indicative of the distance between said spherical sensing tip and the workpiece, the capacitance of the capacitor formed by said spherical sensing tip and said workpiece being generally independent of the orientation of the spherical sensing tip with respect to the workpiece at distances less than the radius of said spherical tip.

2. An omnidirectional capacitive probe comprising:
an electrically conductive hollow stalk;
an electrically conductive spherical sensing tip disposed at one end of said hollow stalk, said spherical sensing tip formed as a substantially complete sphere;
means for connecting said spherical sensing tip with said hollow stalk, said means for connecting electrically insulating said spherical sensing tip from said hollow stalk;
a wire coaxially disposed within said conductive hollow stalk and electrically connected to said electrically conductive spherical sensing tip;
first voltage generating means for generating a first A.C. voltage between said spherical sensing tip and a target workpiece; and
second voltage generating means for generating a second A.C. voltage between said hollow stalk and said target workpiece, said first and second voltages being generally equal in phase and magnitude thereby to reduce the effect of stray capacitance on measurements made with said capacitive probe;
said spherical sensing tip forming a capacitor with said target workpiece when said spherical sensing tip is located in close proximity with the workpiece, the capacitor formed by said spherical sensing tip and the workpiece having a capacitance that is indicative of the distance between said spherical sensing tip and the workpiece, the capacitance of the capacitor formed by said spherical sensing tip and said workpiece being generally independent of the orientation of the spherical sensing tip with the workpiece at distances less than the radius of said spherical tip.

3. An omnidirectional capacitive probe as claimed in claim 2 wherein said spherical sensing tip comprises:
a center portion formed from an insulating material; and
an electrically conductive outer shell disposed over said center portion.

4. An omnidirectional capacitive probe as claimed in claim 3 wherein said center portion is made from a ceramic material.

5. An omnidirectional capacitive probe as claimed in claim 2 wherein said spherical sensing tip is formed from a material selected from the group consisting of steel, tungsten carbide, and Inconel.

6. An omnidirectional capacitive probe as claimed in claim 2 wherein said hollow stalk is formed from brass.

* * * * *

Document
Review Checklist

Document Review Checklist					
Review Items	**Ranking**				
	1-Disagree		**Agree -5**		
	1	**2**	**3**	**4**	**5**
I. Technical Content: **(Does the document have substance?)**					
a) The message to the reader is clear.					
b) The experimental approach is logical.					
c) The author showed adequate research of previous work.					
d) There is adequate comparison of this work to the work of others.					
e) The conclusions are supported by the work.					
f) The value of the work is clearly stated.					
g) The work met the stated objective.					
h) The work is original.					
i) The document is free of plagiarism.					
j) The work is timely.					

Review Items	Ranking				
	1-Disagree			Agree-5	
	1	2	3	4	5
II. Style: **(Is it is written in an appropriate way for the application?)**					
a) The writing is objective.					
b) The sections are logical.					
c) The tone is neutral.					
d) The writing level is appropriate.					
e) The writing is free of jargon.					
f) The writing is free of commercialism (for papers)					
g) The use of English is satisfactory.					
h) The document is understandable.					
i) The document is concise.					
j) The document is interesting.					
k) The document is free of personal opinions.					
l) The document is free of figures of speech.					
m) The document does not over-explain.					
n) The document conforms to standard writing practices.					
o) The page layout and white space are acceptable.					
III. Report Mechanics: **(Does it have required elements?)**					
Introduction					
a) Sufficient background information is presented.					
b) The purpose of the work is clear.					
c) The objective of the work is clear.					
d) The purpose of the report is clear.					
e) The objective of the report is clear.					
f) The format of the report is stated.					
g) The work of others is adequately referenced.					
Body					
Procedure:					
a) The experimental steps are clearly outlined.					
b) There is adequate detail to all others to repeat the work					
c) The procedure is free of unnecessary trade names.					
d) Trademark attributions are made for proprietary materials.					
e) The procedure is free of unnecessary detail.					
f) Test standards are properly cited.					

Review Items	Ranking 1-Disagree Agree -5				
	1	**2**	**3**	**4**	**5**
Results:					
a) Results are clearly stated.					
b) Results are supported by illustrations where necessary.					
c) The results section is free of procedure details and discussion.					
d) Graphs are properly made.					
e) Graphs are necessary.					
f) Tables are clear.					
g) Tables are necessary.					
h) Sufficient results are presented.					
Discussion:					
a) The discussion is helpful.					
b) The discussion relates this work to findings of others.					
c) The discussion is too long.					
d) The discussion is too short.					
Conclusions:					
a) The conclusions logically follow from the results and discussion.					
b) The conclusions are clear and unambiguous.					
Best Practices:					
a) Sentence length is appropriate.					
b) Paragraph length and content are appropriate.					
c) The use of section headings is adequate.					
d) Reference attributions are proper.					
e) References are listed per journal guidelines.					
f) The acknowledgement section (if any) is properly written.					
g) The front matter is appropriate.					
h) The document is free of trade names.					
i) The use of illustrations is appropriate.					
j) The illustrations are drawn properly.					
k) There are adequate transitions between topics.					
l) The font is easy to read.					

Review Items	Ranking 1-Disagree Agree-5				
	1	2	3	4	5
Abstract:					
a) It states why the work was done.					
b) It states what was done.					
c) It states the outcome of the work.					
d) It is concise.					
Grammar/spelling :					
a) Does it meet accepted standards?					
b) Have you annotated the document to show misspelled words, punctuation?					
c) Did you suggest corrections for all grammatical errors?					

[1]Designed for formal paper, delete items that do not apply to informal documents or mark NA for not applicable.

Example of a Simple Project Planning Form

THE FOLLOWING is an example of a project planning form that can be used to track small projects and laboratory work. The tasks are numbered, identified, and assigned to an individual. Completion dates are indicated by placing the corresponding numbers on the timeline.

DATE: _1/12/01_

PROJECT PLANNING FORM

TITLE OF PROJECT: _Knife failure - glossy paper slitter_

OBJECTIVE: _Determine why the knife failed after_ _a short production run_

CUSTOMER: _G. Vanhelt_ PHONE: _716_ DEPT NO. _8211_

CHARGE NO: _22416H_

ESTIMATED COST: _4 hours_

TASK TIMELINE

MONTH: _January_

1. TASK: _Perform chemical analysis (Co, Cr, Ni, W) on three samples submitted_
   PERSON RESPONSIBLE: _RTO_

2. TASK: _Make metallurgical section of each of three knives (transverse)_
   PERSON RESPONSIBLE: _BDT_

3. TASK: _Measure hardness of three knives (six readings on each knife) HK_{100}_
   PERSON RESPONSIBLE: _TDT_

4. TASK: _Measure edge roughness with wedge profilometer_
   PERSON RESPONSIBLE: _TDT_

5. TASK: _Assemble laboratory data and write report_
   PERSON RESPONSIBLE: _KGB_

COPY TO GL: _Yes_   LOGGED IN : _Yes_

COMPLETION DATE: CONTRACTED: _1/30_   ACTUAL: _1/28_

ON TIME: YES _✓_   NO_____

Index